*Cancer Metastatis:
In Vitro and In Vivo
Experimental Approaches*

LABORATORY TECHNIQUES IN BIOCHEMISTRY AND MOLECULAR BIOLOGY

Edited by

P.C. van der Vliet — *Department for Physiological Chemistry, University of Utrecht, Utrecht, Netherlands*

and

S. Pillai — *MGH Cancer Center, Boston, Massachusetts, USA*

Volume 29

ELSEVIER
AMSTERDAM – LAUSANNE – NEW YORK – OXFORD – SHANNON – SINGAPORE – TOKYO

CANCER METASTASIS: IN VITRO AND IN VIVO EXPERIMENTAL APPROACHES

D. Rusciano
Friedrich Miescher Institute
Maulbeerstrasse 66, Basel
CH-4058 Switzerland

D.R. Welch
The Jake Gitlin Cancer Research Institute
The Pennsylvania State University of Medicine
Milton S. Hershey Medical Center
500 University Drive, Hershey
PA 17033-0850 USA

M.M. Burger
Friedrich Miescher Institute
Maulbeerstrasse 66, Basel
CH-4058 Switzerland

2000
ELSEVIER

AMSTERDAM – LAUSANNE – NEW YORK – OXFORD – SHANNON – SINGAPORE – TOKYO

ELSEVIER SCIENCE B.V.
Sara Burgerhartstraat 25
P.O. Box 211, 1000 AE Amsterdam, The Netherlands

© 2000 Elsevier Science B.V. All rights reserved.

This work is protected under copyright by Elsevier Science, and the following terms and conditions apply to its use:

Photocopying
Single photocopies of single chapters may be made for personal use as allowed by national copyright laws. Permission of the publisher and payment of a fee is required for all other photocopying, including multiple or systematic copying, copying for advertising or promotional purposes, resale, and all forms of document delivery. Special rates are available for educational institutions that wish to make photocopies for non-profit educational classroom use.

Permissions may be sought directly from Elsevier Science Rights & Permissions Department, PO Box 800, Oxford OX5 1DX, UK; phone: (+44) 1865 843830, fax: (+44) 1865 853333, e-mail: permissions@elsevier.co.uk. You may also contact Rights & Permissions directly through Elsevier's home page (http://www.elsevier.nl), selecting first 'Customer Support', then 'General Information', then 'Permissions Query Form'.

In the USA, users may clear permissions and make payments through the Copyright Clearance Center, Inc., 222 Rosewood Drive, Danvers, MA 01923, USA; phone: (978) 7508400, fax: (978) 7504744, and in the UK through the Copyright Licensing Agency Rapid Clearance Service (CLARCS), 90 Tottenham Court Road, London W1P 0LP, UK; phone: (+44) 171 436 5931; fax: (+44) 171 436 3986. Other countries may have a local reprographic rights agency for payments.

Derivative Works
Table of contents may be reproduced for internal circulation, but permission of Elsevier Science is required for external resale or distribution of such material.

Permission of the publisher is required for all other derivative works, including compilations and translations.

Electronic Storage or Usage
Permission of the publisher is required to store or use electronically any material contained in this work, including any chapter or part of a chapter. Contact the publisher at the address indicated.

Except as outlined above, no part of this work may be reproduced, stored in a retrieval system or transmitted in any form or by any means, electronic, mechanical, photocopying, recording or otherwise, without prior written permission of the publisher.
Address permissions requests to: Elsevier Science Rights & Permissions Department, at the mail, fax and e-mail addresses noted above.

Notice
No responsibility is assumed by the Publisher for any injury and/or damage to persons or property as a matter of products liability, negligence or otherwise, or from any use or operation of any methods, products, instructions or ideas contained in the material herein. Because of rapid advances in the medical sciences, in particular, independent verification of diagnoses and drug dosages should be made.

First edition 2000

Library of Congress Cataloging in Publication Data
A catalog record from the Library of Congress has been applied for.

British Library Cataloguing in Publication Data
A catalogue record from the British Library has been applied for.

ISBN: 0-444-82372-7 (library edition)
ISBN: 0-444-82358-1 (pocket edition)
ISBN: 0-7204-4200-1 (series)

∞ The paper used in this publication meets the requirements of ANSI/NISO Z39.48-1992 (Permanence of Paper).
Printed in The Netherlands.

Preface

The ability of cancer cells to give rise to secondary tumors in distant organs has been known for over a century. However, it is only in the last 30 years that a systematic approach to the study of the metastatic process has been developing, providing a deeper knowledge about many of the changes that contribute to malignancy and metastasis. Thus, it is only very recently that therapeutic approaches can be seen, such as an anti-metastatic breast cancer vaccine that shows some promise based on animal investigation, or anti-angiogenic therapy, which is beginning to enter clinical trials.

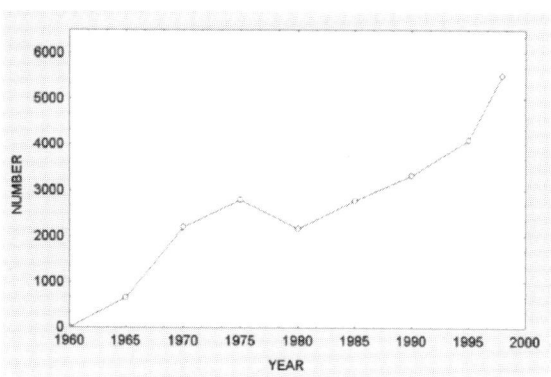

Medline search with the keywords 'cancer metastasis' shows an increase from the four papers published in 1960 to the 5508 published in 1998.

However, despite this great effort, cancer metastasis still remains the main cause of morbidity and death for cancer patients. This indicates that more studies are required to develop powerful strategies to fight what is known as the 'minimal residual disease', which will eventually develop into overt disease and finally kill its host. Therefore, a combination of new targets, new drugs and better experimental models to test their efficacy, will be required to improve our chances against this disease. It is therefore the aim of this book to become a practical guide for all those interested in metastasis research. Moreover, we believe that a clear understanding of the theory behind the practice is essential to achieve good results. Therefore, we have also tried to explain the rationale for all the protocols, indicating the limits of each one, which are mostly the natural limits set by any reductionist approach to a complex problem. Hence, this is not the usual book on cancer metastasis, reporting on the most recent advances in the field. We tried our best to mix some theoretical knowledge with considerable laboratory practice, to build a book that will not be outdated by the time it is published, but can hopefully stand on the laboratory shelves for quite a few years and provide information and help in such a burgeoning field.

Given the vast amount of different assays used in metastasis research, it was clearly impossible to write this book as a monograph. Therefore, different fields of expertise had to be combined, and I am very grateful to Dario Rusciano and Danny Welch, who respectively took care of the *in vitro* and *in vivo* parts of the book. There are also additional contributors—Spiridione Garbisa, Luigi Sartor and Alessandro Negro provided a valuable chapter on invasion assays; Adriana Albini, Marina Ziche, Domenico Ribatti and Roberto Benelli wrote about angiogenesis and metastasis. Lloyd Culp supplied the methods on gene tagging with β-galactosidase, and Tullio Giraldi contributed a description of procedures used to evaluate the influence of host stress on the metastatic behavior of tumor cells.

It is our hope that this book will not only provide a useful benchtop manual for those entering the field of metastasis research, but will also serve as a source of valuable information for the experienced scientist in this field.

Max M. Burger

CANCER METASTASIS: EXPERIMENTAL APPROACHES

Dario Rusciano
Danny R. Welch*
Max M. Burger
Friedrich Miescher Institute
Maulbeerstrasse 66
Ch-4058 basel, Switzerland
*The Jake Gittlen Cancer Research Institute,
The Pennsylvania State University College of Medicine,
The Milton S. Hershey Medical Center,
500 University Drive,
Hershey. PA 17033-0850

with contributions by:
Spiridione Garbisa
Luigi Sartor
Alessandro Negro
Institute of Histology and Embriology
Medical School, University of Padova, Italy

Adriana Albini
Roberto Benelli
Domenico Ribatti*
Marina Ziche**
Neoplastic Progression Unit
Advanced Biotechnology Center
Genoa, Italy
*Intitute of Human Anatomy, Histology and Embriology
University of Bari, Italy
**Dept. of Pharmacology
University of Florence, Italy

Tullio Giraldi
Pharmacology Institute
University of Trieste, Italy

Lloyd A. Culp
Department of Molecular Biology and Microbiology
Case Western Reserve University
School of Medicine
Cleveland, OH 44106-4960

Acknowledgements

The work of Dario Rusciano in Prof. Max M. Burger's laboratory was supported by the Friedrich Miescher Institute, and by grants from the EC, the Swiss Federal Office for Science and Education, and the Swiss Cancer League. A very special thank to Ms. Patrizia Lorenzoni, whose dedication to laboratory work during the long period of writing and arranging this book made this enterprise possible.

Dr. Danny R. Welch's work was supported by grants from the US Public Health Service (CA 62168), the US Army Medical Research Defense Command (DAMD 17-96-1-6152), the National Foundation for Cancer Research, and the Jake Gittlen Memorial Golf Tournament. Special thanks are extended to members of his laboratory, who assisted in preparing photographs of the animal procedures, especially John Harms.

Prof. Spiridione Garbisa's work was supported by grants from the Italian Association for Cancer research (AIRC), the Italian Ministry for Scientific Research and Technology at the University (MURST), and the University of Padua. He especially thanks Dr. Susan Biggin who helped with the revision of his manuscript.

Dr. Adriana Albini's work was supported by funds from the Italian Association for Cancer research (AIRC), the EC BioMed 2 (Angiogenesis and Cancer), the National Research Council (CNR), the Italian Ministry for Scientific Research and Technology at the University (MURST), the 'Compagnia di San Paolo', the AIDS Project by the Italian Superior Health Institute (ISS), and the Italian Health Ministry. A special thank goes to Prof. Leonardo Santi, Director of the Institute.

Contents

Preface v

Chapter 1.   Introduction 1

Chapter 2.   Homotypic and Heterotypic Cell Adhesion in Metastasis 9

2.1.   Release of malignant cells from the tumor mass: intercellular cohesion 10
2.2.   Malignant tumor cells in the blood stream: interactions with blood elements and platelet aggregation 22
2.3.   Adhesion to the target organ 29

Chapter 3.   Motility, Deformability and Metastasis 65

3.1.   Motility and metastasis 65
3.2.   The role of active and passive deformability in invasion and resistance to shear stress forces in the blood stream 90

Chapter 4.   ECM Degradation and Invasion 97

4.1.   Degradation 98
4.2.   Invasion 112

Chapter 5.   The Role of Growth Interactions in Cancer Metastasis 123

5.1.   Methods to evaluate growth interactions in vitro 129
5.2.   Growth interactions in vivo 154

Chapter 6. Selection of Metastatic Variants 161

6.1. Selection of organ-specific metastatic variants 163
6.2. Selection of metastatic variants with enhanced or decreased
metastatic abilities 168

Chapter 7. Genetic Tagging as a Means of Studying Tumor Progression or Metastasis-related Genes 185

7.1. Clonal dominance in tumor progression 185
7.2. Visualization of cancer metastasis 189
7.3. Genes controlling the metastatic phenotype: use of gene tags to
identify metastasis-related genes 199

Chapter 8. In Vivo Cancer Metastasis Assays 207

8.1. Why study metastasis in vivo? 208
8.2. What defines an appropriate model of metastasis? 209
8.3. Cell lines 210
8.4. Considerations regarding animals 214
8.5. Site of injection 219
8.6. Materials needed 222
8.7. Spontaneous metastasis assay 222
8.8. Experimental metastasis assay 228
8.9. Enumeration of metastases 231
8.10. Statistical considerations 236
8.11. The influence of stress 236
8.12. Concluding remarks 242

Chapter 9. Angiogenesis and Metastasis 243

9.1. The corneal assay for angiogenesis 244
9.2. The chick embryo chorioallantoic membrane assay 252
9.3. Subcutaneous implant assay 263

References 271

Subject Index 349

CHAPTER 1

Introduction

Metastasis (from the Greek verb μετίσταμαι, composed of μετα and ιστάμαι), means change of position, migration from one original place to another. Therefore, in biology and medicine it is possible to find migratory movements of cells that are physiologic, such as during embryogenesis, or when leukocytes migrate to the thymus and lymphnodes for maturation, or when they reach a distant inflammatory site; however, pathologic migrations may also occur, for instance when pathogenic microorganisms migrate from the initial invasion site to their target organ(s), or when tumor cells move from their primitive site to other distant organs. The latter process is referred to as 'cancer metastasis' (first used in this meaning by the French physician Joseph Claude Recamier in his 1829 treatise *Recherches du Cancer*).

Cancer metastasis represents the major cause of morbidity and death for cancer patients. In fact, whereas the primary tumor is in most cases susceptible to eradication by combined surgical and radiochemical treatments, its metastases, when distributed throughout the body, are more difficult to treat by any therapeutic mean, and finally cause the death of the patient. Moreover, metastases tend to diversify from the original tumor, and from each other, complicating even further a possible therapy, because of their heterogeneity in responding to any given medical treatment. Besides this, the different organ environments in which metastases may develop can also alter the efficiency with which anticancer agents are delivered, thus directly modifying the metastatic tumor cells' response to therapy.

Despite the improvements in early tumor detection, still many solid tumors in internal organs reach the size of 1 cm^3 ($\sim$1 g) before diagnosis. Such a tumor already contains 10^9 cells (resulting from about 30 duplication events starting from one cell; it will take only 10 doublings more to reach the size of 1 kg, which is the maximum estimated tumor size compatible with life), and studies in experimental rodent mammary cancer have shown that it can release millions of cells per day into the circulation (Weiss, 1986). Although it has been determined that $\sim$0.01% of tumor cells that enter the circulation have the potential to form secondary tumors (Fidler, 1970; Butler and Gullino, 1975; Mayhew and Glaves, 1984), this means that still hundreds of viable tumor cells each day have the possibility to lodge in distant organs, where they could undergo new growth and diversification. Clinically, it has been observed that, while large numbers of cancer cells can be shed into the circulation, only very few of these cells will result in clinically apparent metastases (Tarin et al., 1984a, b), thus indicating a strong selectivity of the metastatic process.

The process of cancer metastasis consists of a cascade of events, each of which can be rate-limiting, since failure at any of them blocks the process. The final result at the end of the cascade depends on both the intrinsic properties of the tumor cells and the host response. Since the balance between these two aspects can vary among different patients, the metastatic behavior of individual tumors is often unpredictable. Theoretically, the steps required to progress from a primary tumor to a distant metastasis are similar in all tumors (Fig. 1.1):

- First, transformation events have to occur, changing a normal cell into a tumor cell, which has to grow progressively to form a small tumor mass, initially supported by simple diffusion of nutrients.
- In order for the tumor mass to grow beyond the size of 1 or 2 mm, it has to be able to induce its own vascularization by secretion of angiogenic factors (Folkman, 1986).

Ch. 1

INTRODUCTION

3

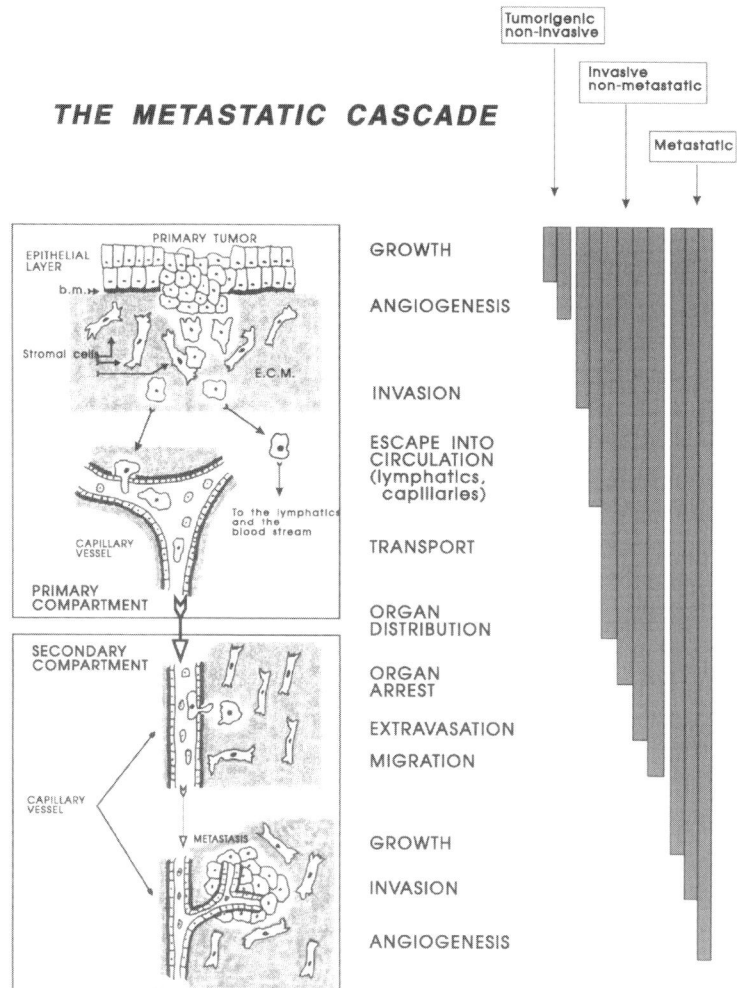

Fig. 1.1. Scheme of the metastatic process, seen as a sequential cascade of events, as indicated on the left-hand side of the picture. The figure stresses the two most relevant facts in cancer metastasis. First, only those cells that can go through the whole metastatic cascade succeed in forming clinically relevant metastatic colonies. Second, metastatic cells have to deal with different organ microenvironments in the primary, and the secondary organ compartment. The nature of these different interactions will strongly influence the metastatic ability of malignant cells. (Modified from Rusciano and Burger, 1993.)

- Local invasion through the basement membrane and into the surrounding stroma is enabled by increased production of proteases by tumor cells themselves (Duffy, 1996), or by their ability to induce such a response in host stromal cells (Basset et al., 1990; Ahmad et al., 1997). Thin-walled vessels, such as newly forming capillaries or lymphatic channels, offer very little resistance to penetration by tumor cells and provide the most common pathways for tumor cell entry into the circulation.
- Detachment of tumor cells from the tumor mass can be facilitated by a partial loss of intercellular adhesion molecules, such as cadherins (Bracke et al., 1996). Tumor cells, either single or as aggregates, enter the circulatory system, where many of them could be rapidly destroyed by unfavorable conditions, including the effects of hemodynamic forces (Weiss, 1991) and the immune system in immunocompetent animals (Hanna et al., 1982; Key, 1983). However, most recent observations by in vivo videomicroscopy have shown that despite the shear forces and the stretching through blood capillaries, over 97% of tumor cells injected in the blood stream maintain membrane integrity for at least 2 h after injection, and that more than 89% of the cells under observation had extravasated after 24 h, thus suggesting that the strongest selection of intravasated tumor cells does not take place in the blood stream, but afterwards (Morris et al., 1997).
- Tumor cells that survive the circulation may arrest in the capillary beds of distant organs by either mechanical trapping, or adhesion to capillary endothelial cells. After their arrest, malignant cells stimulate endothelial cell retraction and exposure of the subendothelial basement membrane (Kramer and Nicolson, 1979), to which they can adhere strongly (Crissman et al., 1985; Lapis et al., 1988). Since tumor cells often adhere better to the subendothelial matrix than to the surface of endothelial cells, there is usually a net movement of malignant cells to the subendothelial matrix. Some malignant cells, however, arrest

in the microcirculation and grow expansively until they rupture the vessel wall (Kawaguchi et al., 1983).
— Once arrested, tumor cells have to invade into the surrounding parenchyma, a step probably occurring through mechanisms similar to those working at the primary invasion site.
— Finally, and most importantly, these cells must now be able to proliferate within the new organ parenchyma, if they are to give rise to a clinically relevant metastatic colony. To expand further, the micrometastases must develop a vascular network and evade destruction by host defenses. Malignant cells can then invade into the blood vessels, and start the metastatic process again.

Figure 1.1 schematically depicts the steps required for metastasis development. On the right-hand side of the figure, the fact is stressed that only those malignant cells capable of successfully accomplishing all the required steps will finally produce metastatic colonies. If any of the intermediate steps do not take place, no metastases can develop from those cells. Another critical fact emphasized in the figure is that malignant tumor cells interact with different microenvironments at the primary original site, compared with the secondary metastatic location(s). The outcome of these interactions, depending on both tumor cell properties and the different organ milieus, will dictate whether a metastatic tumor colony will, or will not, develop at a certain specific site. The nonrandomness of the metastatic process was already recognized more than 100 years ago by Paget (1889) in a classical study on the organ distribution pattern of breast cancer metastasis. His observations led him to formulate the hypothesis—now generally known as the 'seed and soil' theory—that a proper match should exist between tumor cells (the seed) and their target organs (the soil). Metastases result only when seed and soil are compatible. In 1928 Ewing challenged the 'seed and soil' theory of Paget, and hypothesized that metastatic dissemination occurs by purely mechanical factors, and thus would be completely accounted for by the vascular connections of the primary tumor: intravasating tumor cell emboli are much more likely to be mechanically trapped in the circulatory network of the first connected organ, which will then sustain the highest burden of metastatic colonization. Other organs receive less tumor cells, and develop fewer metastatic colonies. Of course, there is no question of whether, or not, the circulatory anatomy influences the dissemination of malignant cells (Willis, 1972; Sugarbaker, 1979; Weiss, 1985). However, as Ewing proposed, it cannot fully explain the dissemination pattern of numerous tumors. According to Sugarbaker (1979), locoregional spread of tumors to lymphnodes depends mainly on mechanical events. However, distant organ colonization by metastatic cells relies mainly on specific interactions between tumor cells and the organ environment.

Many experimental data gathered so far give strong support to the 'seed and soil' hypothesis (reviewed in Nicolson, 1988b; Rusciano and Burger, 1992). Murine melanoma cell lines tested for metastatic ability in syngeneic animals have clearly shown that, upon the same injection route, cells from the same line colonize different organs with different efficiency, while cells from different lines may preferentially colonize different regions of the same organ. After intravenous injection, B16 melanoma cells gave metastatic colonies in the lungs, and in fragments of pulmonary or ovarian tissue surgically implanted intramuscularly. In contrast, metastatic lesions did not develop in renal tissue implanted as a control, or at the site of surgical trauma (Hart and Fidler, 1980). On the other hand, two different mouse melanoma cell lines differed in their metastatic pattern to the brain after direct inoculation via the carotid artery: K-1735 melanoma produced metastasis only in the brain parenchyma, while B16 melanoma produced only meningeal growths (Schackert and Fidler, 1988a, b). Similarly, different human melanomas (Schackert et al., 1989) or carcinomas (Schackert et al., 1990) injected into the internal carotid artery of nude mice, despite showing the same distribution of initial arrest in the brain microcirculation, eventually produced unique patterns of brain metastasis. Therefore, the lodgement of viable tumor cells in a given organ is a necessary, but by itself an insufficient condition for metastases to develop. Thus, the cells of the M5076 reticulum cell sarcoma, for instance, are rapidly arrested in the lungs, and retained there for 3–4 days. They then slowly detach, recirculate and arrest in the liver, where they find the right conditions to develop into tumor nodules (Hart et al., 1981). Ethical considerations rule out the experimental analysis of cancer metastasis in patients as studied in laboratory animals. However, the introduction of peritoneovenous shunts for palliation of malignant ascites provided a way of testing the metastatic behavior of some cancer cells in humans. The draining of malignant ascites into the venous circulation introduced many viable tumor cells into the jugular veins, and resulted in good palliation with minimal complications for 29 patients with different neoplasms. Autopsy findings in 15 patients showed that, despite continuous entry of millions of tumor cells into the circulation, metastases in the lungs (the first downstream capillary network) were rare (Tarin et al., 1984a–c), thus providing a compelling verification of the 'seed and soil' theory also in humans.

Theoretically, each step during the metastatic cascade might contribute to the organ-specificity of cancer metastasis. Paracrine induction of protease production (relevant to the invasive step) has been reported for both tumor (Borchers et al., 1997), and host cells (Swallow et al., 1996; Ahmad et al., 1997; Leber and Balkwill, 1998). The invasive phenotype is also susceptible to paracrine modulation, as in the case of IL-4 mediated inhibition of hepatocyte growth/scatter factors (HGF/SF)-induced invasion and migration of colon cancer cells (Uchiyama et al., 1996). It is likely that these paracrine interactions may occur with different efficiency in the different districts of the body, with the result that tumor cell invasion could be more or less facilitated in an organ-specific way. Interactions with components of the immune system, whose activity can have different relevance at different organ sites, may also influence the organ distribution of metastases. Two different murine T-cell hybridoma cell lines showed different organ specificity (kidney or spleen and liver) upon intravenous inoculation, depending on their different response to macrophages within the splenic microenvironment (Schmidt et al., 1994). Similarly, liver colonization ability of B16 murine melanoma cells may also depend on the ability of their interacting with Kupffer cells (Wiltrout et al., 1985; Calorini et al., 1992). Specific adhesion of tumor cells to target organ components, such as endothelial cells, sub-endothelial and extracellular matrix, or organ parenchymal cells is a common mechanism for organ-specific metastasis, together with specific growth response to organ-derived paracrine factors (Rusciano and Burger, 1992). Finally, specific interactions between tumor cells and the local host microenvironment may also influence the production of angiogenic cytokines, thus resulting in a different ability of tumor cells to promote angiogenesis, and therefore their own growth beyond a certain limit, in different districts of the body (Singh and Fidler, 1996). Each of these specific interactions will be dealt with in the following chapters of this book, where methods are illustrated that enable the reader to perform experiments aimed at the evaluation of the role of each metastatic step in the experimental model under consideration. Moreover, since cancer metastasis as such can only be evaluated as a whole in the animal, a central chapter of the book describes in vivo experimental model systems in which the efficiency of the metastatic cascade can be analyzed in its entirety.

CHAPTER 2

Homotypic and heterotypic cell adhesion in metastasis

Adhesive interactions play a very critical role in the process of metastatic tumor dissemination, and the abnormal adhesiveness that is generally displayed by tumor cells appears to contribute to their metastatic behavior. Both positive and negative regulation of cell adhesion are required in the metastatic process, since metastatic cells must break away from the primary tumor, travel in the circulation and then adhere to cellular and extracellular matrix elements at specific secondary sites. Along this line, several authors have originally found that tumor cells are more easily separated from solid tumors than normal cells are from corresponding tissues (Coman, 1953; McCutcheon et al., 1948; Tjernberg and Zajichek, 1965).

Once they are released into the circulation, adhesion of malignant cells to one another, to circulating normal elements, such as lymphocytes and platelets, to specific microvessel endothelial cells and their underlying subendothelial matrix basement membrane (BM) are all important events for blood borne metastasis (Nicolson, 1988b; Weiss et al., 1988, Rusciano and Burger, 1993).

A variety of data have implicated cell surface glycoconjugates and endogenous lectins that can bind to oligosaccharide sequences as key structural determinants of the tumor cells that participate in cell interactions during the metastatic process. The pioneering work of Tao and Burger clearly demonstrated that alterations of cell surface carbohydrates in variants of murine B16 melanoma cells, selected for resistance to toxic doses of lectins, had a strong influence on their metastatic (but not tumorigenic) properties (Tao and Burger, 1977, 1982; Tao et al., 1983; Finne et al., 1982). Similarly, modification of cell surface oligosaccharides of B16 melanoma cells by treatment with the drugs tunicamycin (Irimura et al., 1981) and swainsonin (Humphries et al., 1986a) also altered both their adhesive and metastatic properties (see Humphries and Olden, 1989 for a comprehensive review).

What follows is a selection of methods to measure the different and various interactions that metastatic cancer cells may entertain with one another or with normal host cells.

2.1. Release of malignant cells from the tumor mass: intercellular cohesion

The detachment of malignant cells from the primary tumor is an essential step for the initiation of the metastatic cascade. Most normal adult cells are restricted by compartment boundaries, and so changes on the cell surface leading to a weakening of the cellular constraints might favor the release of such 'mutant' cells from the primary tumor mass (Steinberg and Foty, 1997). Indeed, certain genes coding for cell adhesion molecules might well be considered as 'tumor or even metastasis suppressor genes' in that their loss or functional mutation can strongly contribute to the acquisition of the malignant phenotype (Hedrick et al., 1993). For instance, the *fat* gene, a cadherin homologue in *Drosophila*, is required for correct morphogenesis, and functions as a tumor suppressor gene (Mahoney et al., 1991; Bryant et al., 1993); the DCC (deleted in colorectal carcinoma) gene, a member of the immunoglobulin superfamily of adhesion receptors, is located on a region of chromosome 18q that is often lost in malignant colorectal carcinomas and in intestinal type gastric cancer (Fearon et al., 1990; Uchino et al., 1992); the recently isolated von Hippel-Lindau disease tumor suppressor gene may be involved in signal transduction or cell adhesion (Latif et al., 1993), beside having a role as a differentiation/morphogenetic factor (Lieubeau-Teillet et al., 1998). More specifically, the perturbation of critical components of adherens junctions, such as E-Cadherin or catenins, has been shown to prompt invasive behavior in several cell lines in vitro (Birchmeier and Behrens, 1994). Also, in cases where E-cadherin was involved, re-introduction of a wild type copy of the gene could revert the invasive phenotype (Vleminckx et al., 1991, Frixen et al., 1991). Cadherins expression has been shown to influence intercellular cohesion, in direct correlation with invasive behavior (Foty and Steinberg, 1997; Foty et al., 1998). Most recently, strong in vivo evidence has emerged that alterations in E-Cadherin also play a major role in the development and progression of human cancers (Birchmeier and Behrens, 1994; Becker et al., 1995; see also Christofori and Semb, 1999, for a recent comprehensive review).

A different type of cellular constraint is provided by gap junctional communication. Gap junctions play an essential role in the integrated regulation of growth, differentiation and function of tissues and organs, so that disruption of gap junctional communication can cause irreversible damage to the integrity of the tissue, and finally contribute to tumor promotion and malignant progression by favoring local cell isolation. Indeed, experimental evidence shows that the loss of intercellular junctional communication affects the metastatic potential of cell lines by a mechanism that appears to be related to growth control exerted by normal cells on tumor cells via the gap junction (Nicolson et al., 1988; Hamada et al., 1988; Mehta et al., 1986; Ren et al., 1990). Most recently, quantitative and qualitative changes in gap junction proteins (connexins) expression have been found to be associated with tumor progression during multistage skin carcinogenesis in the mouse model system (Kamibayashi et al., 1995), and with tumorigenesis in a rat bladder tumor cell line (Krutovskikh et al., 1998).

2.1.1. Intercellular cohesion assay

Intercellular cohesion is a difficult parameter to quantitate, and reliable results depend very much on a careful set up of the experimental conditions. The intercellular cohesive ability of cells can be estimated either by: (1) Measuring the shearing force required to disrupt cell aggregates; or by (2) Evaluating the propensity of single cells in suspension to form aggregates under controlled conditions.

Shearing of cell aggregates. This method was described by Tullberg and Burger (1985) to measure subtle changes in the intercellular cohesive ability of B16 melanoma cells selected for their ability to penetrate filter pores 10 times smaller than their cell diameter.

The method is based on particle counting by Coulter counter, under conditions that slowly disrupt cell aggregates. Shearing forces applied to the cell suspension by means of repeated pipetting will produce only a certain degree of disaggregation, depending on intercellular cohesion forces, giving thus a plateau of cell counting at the Coulter. From this time on, trypsin is added to the cell suspension to produce full disaggregation and give the total cell count. The ratio between the plateau and the final counting gives a value that is proportional to the intercellular cohesion strength: the lower the value, the higher the cohesion strength.

Materials

Coulter counter (model Z_B, Coulter Electronics Inc., Hialea, FL);
pipet device (Pipet Aid) with controlled aspiration power;
graded pipettes (2.5 ml pipette, orifice ø 0.8 mm; 25 ml pipette, 42 ml total volume, orifice ø 1.2 mm);
pipet aid and pipettes were fixed vertically on a stand during the shearing procedure.

Procedure

Cells from a subconfluent dish are rinsed twice in PBS without Calcium and Magnesium (PBS-CMF), and detached by a short incubation (around 5 min at 37°C) with PBS-CMF (2 ml/10 cm^2). Cells are then resuspended with a hand-drawn 2.5 ml pipette, by aspirating and releasing them twice on the dish, once in the middle and once along the border, so that most of the cells are recovered as aggregates. After transfer into 38 ml PBS-CMF, cell counts are determined immediately on the Coulter counter with 100 mm aperture. The whole suspension is then sheared and counted at 2 min intervals (1× aspiration and release) with the 25 ml pipette at a power which fills the pipette to 25 ml in exactly 9 sec. When a plateau in counting is reached (meaning that the applied shear forces cannot break the aggregates to a smaller size), trypsin is added to the remaining volume at a final concentration of 0.05%, and the shearing is continued as before until no further increase is observed in cell number. The ratio between the maximal value (at plateau) before trypsin treatment and the final value is taken as a measure of cohesion between cells.

This method has been set up by using the B16 mouse melanoma cell line. A typical experiment is illustrated in Fig. 2.1(A, B). The application of this method to other cell lines might require little adjustments of some parameters, such as the time of incubation in PBS-CMF to detach the cells from the substrate as aggregates, and the shearing force to apply to reach a plateau value before trypsin addition.

Aggregation of cells in suspension. The method was originally set up by Orr and Roseman (1969), and next modified by Tao et al. (1983) to measure differences in homotypic adhesiveness due to cell surface carbohydrates changes. A similar assay was also set up by others in order to evaluate how different parameters could influence the aggregation ability of BHK cells in culture (Urushihara et al., 1976; Takeichi, 1977). These methods measure the propensity of a

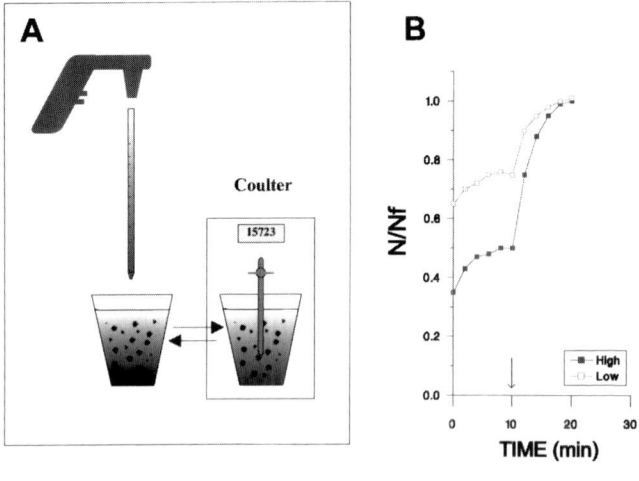
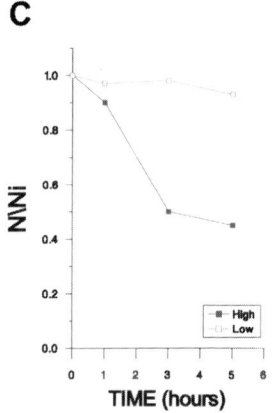

Fig. 2.1. (**A**) The shearing assay assembly as described by Tullberg and Burger (1985). (**B**) Hypothetical results from the shearing assay for a high and a low cohesive cell line. The ordinates report the ratio between actual and final particle count in function of time. The arrow indicates addition of trypsin to the suspension. (**C**) Hypothetical results from the aggregation assay for a high and a low cohesive cell line. The ordinates report the ratio between the actual and the initial particle count in function of time.

single cell suspension to form aggregates, and can be regarded in a way as complementary to the 'shearing assay' previously described.

Materials

Coulter counter (model Z_B, Coulter Electronics Inc., Hialea, FL);

gyratory shaker;

Cell strainers, 70 μm pores (Falcon, cat. # 2350);

BSA-coated plates: the treatment serves to prevent cell attachment to plastic. Plastic wells of a multiwell plate are coated with 0.5% BSA in PBS, 1 h at 37°C (Takeichi and Okada, 1972).

Procedure

A single cell suspension has to be obtained, either by trypsin or by EDTA treatment of a cell monolayer. In case trypsin treatment is used, cells have to be allowed to recover for 1–3 h at 37°C by keeping them in suspension in culture medium (with or without divalent cations). This can be accomplished through the use of a 50-ml sterile tube kept on gentle rotation, or plating in a large dish (5 cm or more) coated with BSA to prevent adhesion, and kept on the gyratory shaker at 90 rpm will serve the scope. After the recovery time cells are collected and aggregates, if any have formed, can be removed by filtering through a cell strainer. Cells are then washed once in PBS-CMF/EDTA, twice in PBS-CMF and finally resuspended in culture medium without divalent cations. A defined number of cells (usually 10^6 in 3 ml medium for a 6 multiwell plate) is then inoculated in BSA-coated plates. After adding an appropriate amount of divalent cations (usually 1 mM $CaCl_2$ and/or 1 mM $MgCl_2$), dishes are incubated on a gyratory shaker at 80–90 rpm at 37°C. At defined times cells are fixed by adding 2 ml of a 5% glutaraldehyde solution in PBS-CMF (2% final concentration), and the total particle number is counted with the Coulter counter with 100 μm aperture. The ratio between the counting at a certain time and the initial counting gives a measure of the aggregation and thus of the intercellular cohesion ability: the lower the ratio, the higher the aggregating forces. Figure 2.1(C) shows theoretical curves that can be expected from low and high cohesive cell populations.

Also in this case, some parameters might need specific adjustments to the particular cell line that is used. Particularly, the method to choose to detach cells has to be considered very carefully, since the aggregation kinetics depends quite strongly on this. Aggregation proceeds pretty fast for EDTA-dissociated cells (usually a plateau value is reached in 90 min), while it is somewhat slower for trypsin-dissociated cells (up to 5 h).

This method is well suitable to study the specific requirements of different cations for cell aggregation. Moreover, different amounts of fetal calf serum or other macromolecules such as extracellular matrix components, or lectins could also be added to the cell suspension to study their influence on cell aggregation.

An alternative assay to estimate intercellular cohesion is the monolayer adhesion assay, originally described by Walther et al. (1973) and modified by Tao et al. (1983). Details on this assay can be found in Section 2.3 below.

2.1.2. Evaluation of gap junctional communication

Gap junctions are impermeable to extracellular substances but, by virtue of the small, regularly-sized channels that pass through them, allow both metabolic and physiological coupling of the cell interiors by diffusion of low molecular weight components. Gap junctions between mammalian cells permit the passage of molecules as large as 2 nm in diameter, meaning that molecules with a molecular weight below 1200 pass freely, while those bigger than 2000 do not pass; the passage of intermediate-sized molecules is variable and limited. Therefore ions, amino acids and nucleoside phosphates can be exchanged between cells.

Methods designed to evaluate gap junctional communication between cells take advantage of this size-selective property of the junction, and use small fluorescent molecules of various sizes to load some cells in a culture dish, and observe whether or not they pass into neighboring cells by using a fluorescence microscope.

Different techniques have been established to load suitable molecules in cells, some of which will be subsequently illustrated.

2.1.2.1. Microinjection of fluorescent dyes

Cell microinjection is a very delicate technique, requiring considerable expertise, that can hardly be conveyed by a mere description of the technique itself. Since this is not within the scope of this book, the interested reader is thus referred to more comprehensive reviews, describing in details various equipment and protocols (Graessman et al., 1980; Wadsworth, 1999). Here, we will only give a general description of the application relative to the study of gap junctional communication between cells.

Materials

    Microneedle puller;
    micromanipulator and needle holder;
    inverted microscope, preferentially equipped with both short ($\times 63$ or $\times 100$) and long ($\times 40$) working distance phase objective lens;
    glass capillaries;
    a source of positive pressure (e.g. a syringe) and tubing for airtight coupling of the syringe to needle or needle holder.

Procedure

Cells to be injected have to be grown in the medium of choice to a density at which extensive cell-to-cell contact can be observed. In order to allow observation with high magnification, short working distance objectives, cells can be plated on square glass coverslips (23 mm) which have been glued to cover a hole (15 mm ø) drilled at the bottom of a 60-mm dish. Care has to be taken to drill the hole from the external side of the dish, in order to create smooth edges to which the coverslip is then glued. Dishes so prepared can be sterilized by overnight exposure under the u.v. light of a laminar flow hood. If dye transfer among two different cell populations has to be studied, they have to be grown in co-culture until the right density is achieved, and the two cell types must be clearly distinguishable under the microscope, preferentially by morphologic criteria.

Before the injection, cells should be transferred to a buffered medium, in order to avoid drastic changes of pH which might influence results.

Two different fluorescent dyes have been successfully used to microinject cells and visualize dye transfer to neighboring cells coupled by gap junctions.

Lucifer Yellow CH is a highly water-soluble compound with a molecular weight of 457.3 Daltons, excitation and emission wavelengths of 430 and 540 nm, respectively. Upon injection, Lucifer Yellow CH spreads through the injected cell within seconds. It is visible in the cell at a concentration as low as 10^{-6} M and, unlike fluorescein, it does not leak out the cell (Weinstein et al., 1977). Lucifer dyes are stable in the dark, but fade under intense illumination. However, even though 75% of the original intensity is lost in the first 40 min under illumination, further decay is much slower, making the labeled cells still clearly visible after 7 days of illumination (Weinstein et al., 1977). Finally, it resists fixation, dehydration and embedding, and is thus suitable also for more accurate microscopical observations.

Usually, a 4–10% solution of Lucifer Yellow CH in lithium chloride is iontophoretically injected through glass capillaries by passing current (1–3 nA) pulses (50–500 nsec) at 10 or 1 Hz, respectively (Miller and Goodenough, 1985; Schuetze and Goodenough, 1982). Transfer of the dye into the surrounding cells can be observed under a fluorescence microscope about 10 min after the injection. If required, cells can be fixed after dye injection by incubation for at least 20 min in 4% paraformaldehyde in PBS (Ren et al., 1990).

Carboxyfluorescein has been introduced as a fluorescent dye marker by Weinstein et al., to measure the liposome-mediated transfer into the cell. It has a molecular weight of 376.3 Daltons, excitation and emission wavelengths of 490 and 520 nm, respectively.

For dye injection, tips are backfilled with a solution of 5 mg/ml carboxyfluorescein that has been dissolved at pH 10 and then adjusted to pH 7.4 in 0.2 M KCl (Goodall and Johnson, 1984), while the shaft is filled with 0.2 M KCl. Dye and current injection are performed using 500 msec hyperpolarizing current pulses, with 500 msec intervals, at a magnitude of 2 nA (Goodall and Maro, 1986).

A clear disadvantage of microinjection is that it is capable of loading only a relatively small number of cells. When a statistically relevant number of independent observations has to be done by evaluating junctional coupling in many cells, then other techniques are available that allow loading of many more cells.

2.1.2.2. Scrape and scratch loading techniques

Scrape loading. By scrape loading, adherent cells are scraped off their substratum with a cell scraper in the presence of the macromolecule(s) (up to 500 kD mol weight) to be loaded (McNeil et al., 1984). Thus loaded cells are obtained in suspension, and have to be replated and allowed to spread before microscopic observation.

Replating in co-culture with a different cell type allows evaluation of heterologous gap junction formation.

Materials

    Inverted fluorescence microscope;
    tissue culture dishes;
    glass coverslips;
    cell scrapers (Costar, N. 3010);
    Lucifer Yellow CH and Rhodamine dextran (Molecular Probes, Inc., OR).

Procedure

Cells are grown to confluency in 35-mm dishes, rinsed three times in PBS-CMF (at 37°C), and then layered with a solution (100 to 500 µl) of 0.05% of Lucifer Yellow CH and Rhodamine dextran (mw 10,000) in PBS at room temperature. The dish is swirled to distribute the solution evenly, and then cells in the culture are scraped off the dish with a cell scraper. Cells are immediately resuspended in complete culture medium, rinsed at least twice by centrifugation to remove unincorporated macromolecules, and replated on a new dish containing a culture of cells which are the potential recipients of the dye via gap junction formation. After 1 h of incubation at 37°C, the dish can be examined under an epifluorescence phase microscope at the appropriate wavelengths.

The concurrent introduction of both Lucifer Yellow and Rhodamine dextran (which emits red fluorescence with a spectrum distinct from that of Lucifer Yellow) into cells allows the identification of the primary loaded cells, and thus serves to control that Lucifer Yellow transfer into neighboring cells happens via gap junctions.

Scratch loading. Scratch loading is a variation of the scrape loading technique (Swanson and McNeil, 1987). An artificial wound is made on a cell monolayer with a sharp instrument in the presence of a solution of the molecule(s) to be loaded (up to 150 kD mol weight). Cells lining the edge of the wound are loaded.

Materials
Same as above.

Procedure
Cells are grown to confluency either in a dish or on a glass coverslip. Using glass coverslips allows to reduce the amount of dye solution required per experiment. Cells are then rinsed in PBS and overlaid (if on a glass coverslip) with 100 µl of the Lucifer Yellow/Rhodamine dextran solution. This is repipetted several times onto the coverslip to ensure mixing at the liquid interface with the cells. One or more artificial wounds on the monolayer are then produced by using a sharp instrument (Pasteur pipette, jeweller's

forceps, pipette tip). Cells are left 1–5 min in the dye solution at room temperature, then coverslips are rinsed several times in PBS and returned to culture medium at 37°C. Then, 5–30 min later the cells are rinsed once more in culture medium, and finally observed with a phase contrast microscope equipped for epifluorescence. The number of cells coupled by gap junction can be determined by estimating the number of cells containing Lucifer Yellow in the cell monolayer adjacent to scrape-loaded cells lining the scraped edge (Nicolson et al., 1988; El-Fouly et al., 1987).

The mechanical perturbation of the plasma membrane consequent to scraping or scratching might result in transient opening of holes (~48 nm ø) at those sites of tightest cell adherence to the substrate. Exogenous macromolecules would therefore enter the cytoplasm by diffusion through such holes (McNeil et al., 1984). Therefore, the strength with which the cells adhere to the substratum could influence the opening of the holes, and thus the loading efficiency.

Permeabilization with carboxyfluorescein-diacetate. 6-carboxyfluorescein-diacetate (CFDA) is a nonfluorescent, nonpolar reagent which enters the cell freely, where it becomes entrapped following enzymatic conversion to the hydrophilic fluorophore 6-carboxyfluorescein (CF, mw 370). Thus, labeled cells can then be replated onto unlabeled test cells to measure dye transfer via gap junction. This method has been successfully used to measure gap junction communication in early embryo cells (Goodall and Johnson, 1982; Kidder et al., 1987), and also to label tumor cells in culture (Price et al., 1995). We tried to extrapolate it for detection of gap junction formation among cultured cells.

Materials

Inverted fluorescence microscope;
Issue culture dishes;
CFDA (Molecular Probes Inc., OR) 10 mg/ml stock solution in acetone.

Procedure
Cells grown to subconfluence are detached by EDTA or enzymatic treatment, appropriately rinsed by centrifugation, and resuspended in Hank's Balanced Salt Solution (HBSS) containing CFDA at 40–100 μg/ml. After 30 min of incubation at 37°C with gentle agitation, they are rinsed twice in HBSS and once in complete culture medium, and finally plated in complete culture medium in co-culture with unlabeled test cells for the evaluation of gap junction formation. Examination of the cells under the fluorescence microscope to evaluate transfer of the fluorescent dye can be started after 1 h of incubation at 37°C.

2.2. Malignant tumor cells in the blood stream: interaction with blood elements and platelet aggregation

Blood-borne tumor cells undergo various homotypic and heterotypic interactions, the effect of which will also influence their metastatic behavior. Some of these interactions may be detrimental to circulating tumor cells, such as tumor cell recognition by natural killer (NK) cells (Arisawa et al., 1990; Whiteside and Herberman, 1989), or by tumor infiltrating lymphocytes (TIL: Foon, 1989; Liebrich et al., 1991). Others, to a certain extent, may provide a protective effect and/or contribute to metastatic spreading, such as interactions with platelets or, in certain cases, with leucocytes. For instance, de novo expression of the cell adhesion molecule ICAM-1 by melanomas might lead to heterotypic adhesion between melanoma cells and leukocytes bearing the relative receptor (LFA-1). Such interaction might thus enhance tumor cell adhesion to migratory and invasive leukocytes, contributing thereby to further dissemination of malignant tumor cells (Johnson et al., 1989; Johnson, 1991). In this regard, Glinsky (1993) has suggested the intriguing hypothesis that site specificity of cancer metastasis might be, at least partly, consequence of the formation of 'multicellular metastatic units' (MSU) consisting of tumor cells, platelets and leukocytes. A subset of leukocytes within the 'MSU' would be responsible for site-specific endothelium recognition, adhesion and stable attachment, thus serving as 'carrier cells' targeting the metastatic 'spheroids' to specific sites of secondary tumor foci formation.

Several lines of evidence have provided strong support for the concept that tumor cell-platelet interaction significantly contributes to hematogenous metastasis. The first indication came with experiments showing that thrombocytopenia induced in mice by either neuraminidase or antiplatelet antibodies resulted in a decrease of experimental metastasis following the injection of TA_3 ascite tumor cells (Gasic and Gasic, 1962; Gasic et al., 1968). Following these initial observations, numerous clinical and experimental studies have provided convincing supportive data that platelets contribute to cancer metastasis: thrombocytosis occurs in 30–60% of patients with cancer, and particularly in those with advanced and disseminated tumors (Grignani et al., 1986b; Constantini et al., 1990); the ability of tumor cells to induce platelet aggregation has been correlated with their metastatic potential in mouse, rat and human models (Mahalingam et al., 1988; Zoucas et al., 1986; Mehta et al., 1987; see also Honn et al., 1992a for a comprehensive review); antiplatelets agents have strong antimetastatic effects (Honn et al., 1981; Al-Mondhiry, 1984; Honn et al., 1983).

Mechanistically, platelets may contribute to metastasis by stabilizing tumor cell arrest in the vasculature, shielding tumor cells from physical damage, providing additional adhesion mechanisms to endothelial cells and subendothelial matrix, and serving as a potential source of growth factors (Honn et al., 1992b). Interestingly enough, if tumor cell interaction with host platelets occurs while tumor cells are circulating (Lapis et al., 1988), it might influence the organ-specific colonization ability of blood-borne tumor cells. In fact the resulting embolus will be more easily arrested in the vasculature of the first organ downstream from the primary tumor site. If this organ represents a favorable milieu for tumor growth, then interaction with platelets will enhance tumor metastasis at that site; if not, then it may prevent tumor cells from reaching their preferred organ, and thus cause a reduction of the metastatic potential (Gasic, 1984). It seems however that in most cases platelets are involved only after tumor cells have arrested, and then platelet activation may stabilize the initial tumor cell arrest in the microvasculature (Crissman et al., 1985; Crissman et al., 1988).

Multiple factors can trigger tumor cell induced platelet aggregation (TCIPA) and their subsequent activation, and different tumor cells may use different mechanisms. Even though platelet aggregation could be theoretically induced without direct contact between platelets and tumor cells, via soluble factors produced by the tumor cell, the vast majority of experimental data suggest that aggregation occurs with the initial formation of focal tumor cell-platelet aggregates, and subsequent platelet recruitment to form further platelet-platelet aggregates. Sialylation seems to be a general requirement for TCIPA, and sialoglycoconjugates present on both tumor cells and platelets have been involved in tumor cell-platelet interactions (reviewed in Honn et al., 1992a, b). More specifically, two categories of molecules are involved in TCIPA: soluble mediators and adhesion molecules. These latter are likely responsible for the initial tumor cell-platelet contact, and may further stabilize the established interaction. P-selectin and $\alpha IIb\beta 3$ on the platelet surface may bind Le^x carbohydrate determinants and fibrin on the surface of tumor cells, thus triggering platelet activation. Tumor cells also possess an unusual ability to generate soluble platelet agonists. Thrombin can be generated by tumor cells through several different mechanisms (see Honn et al., 1992a, b), and is a potent inducer of platelet activation and a potent stimulator of metastasis (Nierozdik et al., 1991; Wojtukievicz et al., 1993). Alternatively, ADP- or thromboxane-induced platelet aggregation can be relevant in those cases in which thrombin inhibitors do not appear to affect platelet aggregation and metastatic potential (Bastida et al., 1982; Ugen et al., 1988; Pacchiarini et al., 1991).

2.2.1. Tumor cells/platelets interaction

Interactions between tumor cells and platelets can be analyzed by different methods. The ability of tumor cells to induce platelet aggregation ('tumor cells induced platelet aggregation' : TCIPA) is evaluated by the turbidometric assay (Menter et al., 1987; Watanabe et al., 1988; Sugimoto et al., 1991; Tang et al., 1993; Belloc et al., 1995). Besides, the ability of tumor cells to bind platelets can be evaluated by the classic adhesion assay (see below).

2.2.1.1. Tumor cell induced platelet aggregation (TCIPA)

Materials
    Platelet rich and platelet poor plasma (PRP and PPP);
    apparatus to measure platelet aggregation (dual channel aggregometer, model DP-247E, Sienco, Morrison, CO; Turbidometer, NKK HEMA TRACER I, Niko Bioscientific Co., Tokyo, Japan).

Procedure
PRP and PPP preparation. If rats or mice are used, they are anaesthetized by an i.p. injection of sodium pentobarbital (50 mg/Kg), and blood is drawn into heparin (5 units/ml final) or citrate/dextrose (final concentration 0.38% and 0.48% (wt/vol), respectively) anticoagulant from the dorsal aorta or by heart puncture, with a 22-gauge needle (the choice of heparin as anticoagulant appears to be more appropriate, because chelation of divalent cations by trisodium citrate or EDTA might affect platelet-tumor cell interaction). Blood is then diluted with an equal volume of 0.9% NaCl, and centrifuged for 7 min at $400 \times g$ on preformed gradients of 70% Percoll containing 0.9% NaCl. The yellowish supernatant is the PPP, while PRP forms a white band between PPP and the percoll layer. Platelets in PRP are counted with a Coulter counter and diluted with PPP to a concentration of 10^9/ml. PPP is a source of prothrombin, and should thus be included in the assay.

Preparation of tumor cells. Cultured tumor cells are detached by EDTA treatment, resuspended and rinsed in HBSS buffered with 10 mM HEPES at pH 7.3, and finally adjusted with the same buffer at a concentration of 10^7/ml.

In certain cases the aggregating activity can also be secreted by tumor cells in the culture medium (Mogi et al., 1991). Therefore, testing of conditioned medium is also advisable. Conditioned medium can be collected between 15 min (Mogi et al., 1991) and 72 h (Grignani et al., 1986a) from confluent plates, centrifuged 15 min at 500 g to remove free cells and cell debris, and tested as such, or after further ultracentrifugation at 100,000 g for 60 min at 4°C to remove membranous fractions.

Aggregation assay. 200 μl of PRP are placed and incubated in a siliconized glass cuvette (to prevent adhesion of platelets to the glass surface of the cuvette, which would activate them) at 37°C under constant stirring at 1000 rpm in the aggregometer for at least 5 min, to produce a stable baseline. Tumor cells (or their supernatant) are added in 10 μl, and change in light transmittance is monitored for at least 15 min. Platelet aggregation for the calibration of the aggregometer can be induced by addition of 0.2–0.5 U/ml of thrombin.

If platelet aggregation by tumor cells can be inhibited by the presence of 10–100 IU/ml of hirudin, this would suggest that generation of thrombin by tumor cells is responsible for the aggregation. If thrombin inhibition does not prevent platelet aggregation, the assay can be carried out in the presence of apyrase (0.5 U/ml) in order to degrade ADP which might also be inducing platelet aggregation. Tumor cells should be incubated with apyrase 20 min on ice before being added to the platelet suspension, also in the presence of the same concentration of apyrase.

If neither hirudin nor apyrase can prevent TCIPA, then other mechanisms should be considered for the platelet aggregating activity of tumor cells, which might be disrupted by enzymatic treatments, such as trypsin treatment (200 U/ml, 30–60 min at 37°C;

enzymatic reaction is then stopped by addition of soybean trypsin inhibitor at 10 μg/ml), neuraminidase treatment (0.2 U/ml with 0.2 mM PMSF, 1 h at 37°C); phospholypase A_2 (56 U/ml, 30 min at 37°C). Enzymes can be prepared in HBSS buffered at pH 7.3 with 10 mM HEPES. After enzymatic treatment cells are rinsed three times with HBSS pH 7.3, resuspended in the same buffer at the desired concentration, and finally used in the aggregation assay as described above (Grignani et al., 1986a; Watanabe et al., 1988).

Upon activation, platelets release several factors, which are normally involved in hemostasis and coagulation, and are also implicated in TCIPA (Honn et al., 1992b).These factors can be recovered from activated platelets and tested on tumor cells.

Platelets are activated by addition of thrombin (0.2–0.5 U/ml, 15 min at 37°C), and the enzymatic reaction is quenched by addition of either 25 μM leupeptin, or 10 U/ml hirudin. Aggregated platelets are pelleted by centrifugation at 2500 g for 10 min, and the supernatant used as source of platelet-secreted factors. The pellet containing degranulated platelets can be resuspended in HBSS buffer and used as control.

2.2.1.2. In vitro platelet-tumor cell adhesion assay

If the ability of tumor cells to induce platelet aggregation has a prominent role in metastasis formation, it could be expected that its inhibition by specific drugs would reduce the amount of metastases produced by tumor cells with a pronounced TCIPA activity.

However, while an adequate platelet number appears to be a necessary requirement for metastasis formation, since induction of thrombocytopenia is associated with a decrease in metastases number (Gasic et al., 1968, 1973; Karpatkin et al., 1988), there is considerable controversy regarding the effects of antiplatelet aggregating agents on metastases formation (Karpatkin et al., 1988, 1991). Therefore, a different approach was taken in delineating the mechanism of platelet requirement in tumor metastasis, directing attention to adhesion of tumor cells to activated platelets and platelet adhesive proteins (Karpatkin et al., 1988; Nierodzik et al., 1991,

1992). It was found that competing platelet-tumor cell adhesion could indeed decrease the amount of experimental lung metastasis, whereas enhancing platelet tumor cell adhesion by means of thrombin treatment of either platelets or tumor cells had stimulating effects on metastasis formation (Nierodzik et al., 1991, 1992).

Materials
Platelets prepared from PRP;
Stractan (arabinogalactan, Sigma);
96-well flat-bottomed microtiter plates;
tissue culture material.

Procedure
Preparation of platelets. PRP (prepared as described above) is placed on a discontinuous gradient of 10 and 20% Stractan dissolved in 13.6 mM sodium citrate, 117 mM NaCl, 11 mM glucose, 10 mM sodium phosphate buffer, pH 7.4, 290 mosmol/l and centrifuged at 900 g for 25 min. Plasma-free platelets were recovered at the interface, resuspended in the same buffer, counted and diluted at 10^9/ml.

Adhesion assay. 100 µl of the platelet suspension are plated into flat-bottomed plastic microtiter wells and allowed to adhere for 24 h at 4°C. Nonadherent platelets are removed by washing with PBS/1% BSA. 100 µl of this solution are added to each well and incubated for 1 h at 37°C to block 'free adherent sites' on plastic.

If required, adherent platelets can now be treated with the desired reagent(s), and then rinsed again in PBS/BSA.

Tumor cells are detached by EDTA treatment, rinsed with PBS and diluted in PBS at approximately 10^6/ml. 100 µl of the cell suspension are plated onto platelets and incubated for 1 h at 37°C. After washing three times with PBS, adherent tumor cells are removed by trypsin/EDTA treatment, and counted with an hemocytometer under phase microscopy. Alternatively, tumor cells can be radioactively labeled, and quantitation done by counting the radioactivity associated with adherent tumor cells (see next section).

2.3. Adhesion to the target organ

Circulating tumor cells, either as single cells, or most likely as homotypic and/or heterotypic aggregates, which have escaped killing by the host immune system and lysis by mechanical shear forces associated with passage in the blood stream, need now to arrest in the microvasculature and extravasate into the organ parenchyma. In fact, the survival time of tumor cells entering the circulation is very short, usually below 60 min, and therefore those cells that can rapidly arrest and get out the blood stream might have a selective advantage in giving rise to metastatic colonies.

Specific adhesion in the target organ has been proposed as a critical determinant of organ specific metastasis, and many experimental data indicate that malignant tumor cells preferentially adhere to organ-specific adhesion molecules (see review in Zetter, 1990). Tumor cells, for instance, adhered more efficiently to disaggregated cells or to histologic sections prepared from their preferred site of metastasis than from other organs (Nicolson and Winkelhake, 1975; Netland and Zetter, 1984). These type of assays, however, do not accurately mimic the physiological situation in vivo, where the first contact of circulating tumor cells happens with the luminal surface of the vascular endothelium, and, after endothelial retraction, with the subendothelial BM. In an accurate morphological study of the interaction of intravascular tumor cells with endothelial cells and subendothelial matrix in an experimental murine model system (Crissman et al., 1988), the following sequence of events has been described. After tail vein injection, initial arrest of tumor cells in the lung microvasculature is characterized by an intimate tumor endothelial cell contact. Arrested tumor cells are immediately covered by a platelet thrombus. Four hours after injection, endothelial cell retraction starts, with tumor cells extending towards and making contact with the subendothelial matrix. Tumor associated thrombus dissolves between 8 and 24 h after tumor cells inoculation. Dissolution of the BM and penetration through the subendothelial matrix follow after 24–48 h.

We will describe next the different adhesion assays which include: (1) adhesion to whole organ derivatives; (2) to vascular and microvascular endothelial cells; and (3) to the subendothelial extracellular matrix and its constituents.

2.3.1. Heterotypic aggregation with dissociated organ cells

The assay was originally described by Nicolson and Winkelhake (1975) in order to test the specific adhesion ability of B16 murine melanoma cells selected in vivo for lung-specific colonization (Fidler, 1973a). The test measures the ability of a tumor-cell suspension to bind to and aggregate a similar suspension of organ-derived host cells. The obtained results showed that such heterotypic aggregation occurred in a specific way, and it was related to the preferred organ site for implantation.

Tumor cells and organ-derived cells (epithelial and mesenchymal cells, plus endothelial cells derived from the organ vasculature) are artificially brought together by this assay, in order to measure their ability to form heterotypic aggregates. It has, however, to be kept in mind that the first adhesive interaction that tumor cells have at a given organ site is with vascular endothelial cells and that only after they have invaded into the organ parenchyma they will contact epithelial and mesenchymal organ cells. Therefore, adhesion to these host cells should not influence their homing as it might affect adhesion to microvascular endothelial cells. It is possible, however, that it might influence their further response to the organ microenvironment in terms of migration and/or growth.

Materials

Gyratory shaker;

surgical instruments for animal dissection;
sterilized stainless steel filter with 63 μm pores;
cell strainers (Falcon).

Procedure

Preparation of single organ-cell suspension. Organs (lung, liver, kidney, brain and spleen) are removed from adult mice (50–60 days old), thoroughly rinsed in cold PBS, and pooled according to organ type. After cutting in small pieces (2–4 mm: keep the organs wet with some cold serum free-DMEM), they are further rinsed at 4°C in serum free-DMEM in order to get rid as much as possible of contaminating blood elements. Single cells are freed from the tissue by gently teasing the pieces over the stainless steel filter with a teflon policeman. Cell suspensions are rinsed twice with PBS by centrifugation (800 g, 10 min). Single, viable organ cells were isolated from cell debris, erythrocytes, platelets and cell aggregates by the two-step isopycnic-rate zonal gradient technique of Perper et al. (1968). After the isopycnic centrifugation in the above Ficoll-Hypaque gradient, viable single cells are obtained, with some membrane contamination. A further centrifugation in a discontinuous sucrose density gradient (5, 20 and 30% sucrose in PBS) yields single organ cells in the 20% layer; erytrocytes, platelets and membrane debris in the 5% layer; and some aggregates in the 30% layer. Sucrose is finally removed from the single organ cell suspension by centrifugation in PBS, leaving less than 5% erythrocytes and 7% phagocytic cells contaminating the suspension. If further removal of aggregates is required, the suspension can be passed through a cell strainer. Cells are finally diluted in DMEM-2%FCS at 3 to 5 $\times$ 10^6/ml.

Aggregation assay. Tumor cells are detached from the culture dish by EDTA treatment (if trypsin has to be used, then a time for recovery in serum containing medium should be allowed before the aggregation assay) and diluted at 10^6/ml in DMEM-2%FCS.

Then, 0.2 ml of each cell suspension (tumor and organ-derived) are placed together in each well of a 24 well multiwell plate in quadruplicates, and shaken at 100–150 rpm for various times (up to 30 min) at 37°C. Controls include tumor and organ-derived cells alone to check for self-aggregation and erythrocytes (10^7/ml) to rule out an effect mediated by red blood cells.

Aggregation can be scored visually, and expressed as relative values ranging from 0 to 4+ (Nicolson, 1973), or quantitated by particle counting at the Coulter counter.

2.3.2. Adhesion to cryostat sections

The application of this method, originally developed to study the mechanism of lymphocyte homing into lymph nodes (Stamper and Woodruff, 1976, modified by Butcher et al., 1979), to the study of organ-specific metastasis was described by Netland and Zetter (1984). Their report showed that two metastatic murine cell lines (the B16-F10 melanoma, specific for lung colonization, and the M5076 reticulum cell sarcoma, specific for liver colonization) demonstrated an organ specificity in their binding in vitro that reflected the organ specificity of their metastatic distribution in the organ colonization assay in vivo (see next chapters). In a later paper (1985), the same authors found that adhesion to cryostat sections of host organs could also be used to select in vitro metastatic variants with increased adhesion to a specific syngeneic tissue, which also correlated with an increased metastatic colonization of that tissue. More recent papers described adhesion to cryostat sections to study lymph node metastasis by solid tumors (Whalen and Sharif, 1992; Nip et al., 1992), and the role of integrin molecules in binding of metastatic tumor cells to organ sections (Vink et al., 1993; Ruiz et al., 1993).

Materials

Mini-gyratory shaker (Kuehner, Basel, Switzerland);
surgical instruments for animal dissection;
cryostat with a freezing head with variable temperature

control;
OCT Tissue Tek from Miles Laboratories;
wax pen (Marktex, Tech Pen, Scientific Products, McGraw Park, Ill.), or PapPen (SCI, Science Services, Munich, Germany).

Procedure

Preparation of cryostat sections. Fresh, unfixed tissues are cut into small blocks, embedded in Tissue-Tek OCT, and frozen in isopentane at $-120/130°C$. Tissue blocks are stable for several years at $-70°C$. Cryostat sections (5–10 μm) are prepared on clean glass slides (or coverslips, if the adhesion assay is carried out with radioactively labeled cells), and used in the next 2 h, since prolonged storage of sections may result in high background adherence. Sections are outlined by the wax pen, dried 15 min at room temperature and immediately placed for 10 min in PBS (or HBSS) containing 1% BSA. If the adhesion assay has to be conducted under sterile conditions, it is possible to sterilize cryostat sections by placing them 5 inches under a germicidal lamp (2×15-W General Electric Co., Cleveland, OH) for 5 min at 5–7°C (Netland and Zetter, 1985).

Adhesion assay. Slides are dried around the outlined area, and a suspension of test cells in 100–200 μl of Hepes-buffered culture medium, containing 10^5 to 10^6 cells is loaded onto the tissue section. Binding is allowed to occur for 30–60 min at 8–15°C on a mini-skaker at 50 rpm. After incubation, the medium is removed by gently tapping the edge of the slide against an absorbent towel, and slides are then placed vertically (so to allow unbound cells to detach) into PBS containing 0.5% glutaraldehyde and 2% formaldehyde to fix adherent cells to the section. After 60 min of fixation at room temperature (if at 4°C, fixation can be between 4 h and overnight), loosely adherent cells can be removed with a gentle stream of PBS, and the slide is ready to be analyzed.

For appropriate quantitation by microscope analysis, cells have to be counterstained (0.5% thionine acetate or 0.5% toluidine blue in 20% ethanol for few minutes, or hematoxylin/eosin), placed in mounting medium (e.g. Mowiol) and covered with a coverslip. Adherent cells in several random chosen fields are then counted under the microscope, with the aid of a microscope grid placed in the eyepiece. Alternatively, radioactive-labeled cells (^{125}I or ^{51}Cr) can be employed in the assay, and quantitation done by counting the radioactivity associated with the tissue sections in a gamma counter (it is preferable in this case to have the tissue sections on coverslips that fit directly into scintillation vials). However, since in this case all the adherent cells are counted, an appropriate comparison among different tissue sections requires that data are normalized to the area of the section.

Even though this adhesion assay gives reproducible results on a fairly wide range of pH (between 6.7 and 7.9), bicarbonate-buffered solutions should be avoided if the assay is to be carried out in open air, because of bicarbonate poor buffering quality under such conditions (Butcher et al., 1979).

The presence of divalent cations in cell dilution medium is important for adequate adhesion.

As far as temperature is concerned, adherence is maximal in the cold (7 to 9°C), and is markedly reduced at 25°C (Stamper and Woodruff 1976, 1977 and Butcher et al., 1979).

2.3.3. Adhesion to endothelial cells

The arrest of tumor cells in the capillary bed of secondary organs and their subsequent extravasation occur through interactions with the local microvascular endothelium and the subendothelial matrix. The specificity of these interactions, depending on the heterogeneity of both microvascular endothelial cells (EC) and tumor cells, may favor in a selective way the initial adhesive events in preferred metastatic sites, and may consequently facilitate metastatic dissemination to those organs (Blood and Zetter, 1990; Pauli et al., 1990; Rusciano and Burger, 1992), similarly to what happens for the extravasation of lymphocytes from high endothelial venules of lymphoid tissues. Lymphocytes 'homing' represents in fact the paradigm for organ-specific cell adhesion, and it has been shown to follow specific interactions between surface 'homing' receptors on lymphocytes with vascular 'addressins' expressed on the high endothelial venule surface (Michie et al., 1993; Imhof and Dunon, 1995). In a similar way, tumor cells express various combinations of cell surface molecules that may serve as ligands for EC surface receptors, which are typically induced upon stimulation by mediators of inflammation (Rice and Bevilacqua, 1989; Lauri et al., 1991; Bevilacqua and Nelson, 1993; Walz et al., 1990; Aruffo et al., 1992; Majuri et al., 1992). A local inflammatory response might thus facilitate circulating tumor cells adhesion and arrest. The relevance of this type of interaction in directing tumor metastasis has been recently demonstrated in vivo with strains of transgenic mice designed to constitutively express cell surface E-selectin in all tissues, or in the liver alone (Biancone et al., 1996). Metastatic tumor cells which do not express the ligand colonized mostly the lung, while after induction of ligand expression they were redirected to colonize the liver with tremendous efficiency (Biancone et al., 1996).

Biochemical heterogeneity of EC has been revealed, that depends on both the tissue of origin and the size of the vessel (reviewed in Zetter, 1988). Heterogeneity is seen in the differential expression of plasma membrane glycoproteins (Belloni and Nicolson, 1988), cytoskeletal proteins (Dodge et al., 1991) and surface receptors (Kuzu et al., 1992) in microvascular endothelium of different organs. This diversity appears to be an expression of heterogeneous microenvironments within tissues (Pauli et al., 1990), as well as dynamic alterations due to environmental agents, host and cancer cells, that may lead EC to synthesize cell surface adhesion molecules and chemoattractants, with the effect of enhancing adhesion and motility of cancer cells (Lafrenie et al., 1992a). Such heterogeneity of endothelium has underscored the importance of using organ-specific capillary endothelium in studying the role of organ-specific tumor cell adhesion in metastasis.

In the following sections, we will describe methods to prepare vascular and organ-specific microvascular EC, methods for phenotypic modulation of EC in vitro (organ-specific modulation by subendothelial matrix, and activation by cytokines), and different types of adhesion assay of tumor cells on cultured EC.

2.3.3.1. Vascular endothelial cells
The main source of large vessel EC are usually animal aortae (from calves or pigs) or human umbilical veins. EC lining the internal wall are loosened by collagenase treatment, and released by flushing the vessel with medium (Booyse et al., 1975; Balconi and Dejana, 1986).

Materials
Sterile surgical instruments, including clamps to close vessels; collagenase and dispase;
sterile tissue culture medium (RPMI-1640, 15 mM HEPES, 100 U/ml penicillin, 50 μg/ml streptomycin, 200 U/ml neomycin, 2 mM glutamine, and FCS between 20 and 35%, pH 7.35) and buffers (PBS, HBSS).

Procedure
Untraumatized, freshly excised vessels are placed in cold HBSS and processed in the next few hours. Small arteries branching out from aorta have to be closed by clamping or suturing. Umbilical cord ends that have been in contact with clamps have to be cut away to ensure sterility. The vessel is flushed with PBS (100–200 ml; use blunt needles to cannulate umbilical vein) until no traces of blood are visible, then its lower end is clamped and the vessel is filled with collagenase (0.5–1.0 mg/ml) in PBS prewarmed at 37°C. The upper end is then clamped, and the vessel segment is incubated for 15 min at 37°C or 40 min at room temperature. After incubation, the vessel content is poured out and the vessel is refilled a few times with 50% of its volume capacity with complete culture medium, clamped again at both ends, and agitated back and forth several

times along its longitudinal axis in order to efficiently detach EC from their matrix. All the effluents are collected together and plated in tissue culture dishes or flasks at 37°C in complete tissue culture medium at a cell density between 2 and $4 \times 10^4/cm^2$. Cell counting, however, is not easy, because EC come off in clumps and not as single cells, and the success of the culture appears to be related to the presence of such clusters. Reaching of confluence (around 10^5 cells/cm^2) can be expected in about 1 week.

A different strategy to prepare EC has been also described, which is suitable to derive cells from tissues, like the brain, that are more difficult to perfuse, and, at the same time, also allows recovery of EC from large vessels (Gordon et al., 1991). The aorta is cut in little rings, while brain microvessel EC are prepared from a brain homogenate. The method is then based on a stepwise enzymatic dissociation of EC (dispase first, collagenase/dispase next), and isolation of dissociated EC from contaminating cells by a Percoll gradient (1650 g, 10 min, 4°C: EC are recovered in the middle third of the gradient, at a density range between 1.04 and 1.07 g/ml). Further enrichment in EC is achieved by plating the cells for 4 h on a poly-D-lysine coating, which favors attachment of fibroblasts and other contaminating cells, and next the unattached cells are plated on a fibronectin coating, which favors adhesion of EC.

Subculturing of EC is possible for a limited number of times (unless additional growth factors are supplemented into tissue culture medium), by using trypsin/EDTA (Gimbrone et al., 1974b; Maciag et al., 1981) or collagenase/EDTA (Jaffe et al., 1973; Gordon et al., 1983). Peptide growth factors such as FGF, EGF, ECGF and PDGF are all potent EC mitogens (Berliner, 1981; Maciag et al., 1981, 1982; Gordon et al., 1983; Knauer and Cunningham, 1983), and clonal growth of bovine vascular EC has been shown to be possible in the presence of exogenous FGF added to the culture medium (Gospodarowicz et al., 1976). The utility of PDGF in the culture medium is, however, questionable, since it will also strongly promote growth of contaminant cells, like fibroblast, which can shortly overcome growth of EC (Gordon et al., 1991).

Bovine EC have been studied more extensively because of the ease with which they can be serially subcultivated. Human EC, on the other hand, are more exigent in their growth requirement, and little progress has been made in long-term serial subcultivation of these cells. The use of endothelial cell growth supplement (ECGS) in association with heparin, hydrocortisone, EGF and b-FGF in MCDB 131 medium (a mixture sold by Promo Cell, Dr Hinz and Huettner GdbR, Handschuhsheimer Landstrasse 12, D-69120 Heidelberg, Germany, Fax (+) 49 6221 484943) has allowed the lowering of FCS requirement to 2%. Such mixture can also efficiently support growth of other vascular and microvascular human and animal EC. The same company is also able to provide human umbilical vein EC (HUVEC) already in culture, which are usable within four passages from delivery. HUVEC can also be obtained from Clonetics Corporation. Another 'EC culture medium', containing ECGF and suitable for cultivation of HUVEC is sold by Cell Systems Inc., Kirkland, WA.

2.3.3.2. Microvascular endothelial cells
Microvascular EC have been isolated and cultivated from a variety of animal tissues, including rat (Bowman et al., 1980; Diglio et al., 1982; Irving et al., 1984; Ager, 1987), mouse (DeBault et al., 1981; Auerbach et al., 1982, 1987), rabbit (Ryan et al., 1982; Davison et al., 1980), cow (Folkman et al., 1979; Gitlin and D'Amore, 1983) and human (Sherer et al., 1980; Kern et al., 1983; Striker et al., 1984). Although the methods used to prepare microvascular EC were similar, their growth requirements appeared to be quite different and peculiar, rendering difficult the establishment of organ-specific cultures suitable for biochemical studies. More recently, microvascular EC organ-specific cultures have been produced from mouse brain, lung and liver (Belloni et al., 1992), and culture conditions established that support their growth and maintainance of differentiated properties. DMEM/F12 containing 5% FCS, 2% platelet-poor plasma-derived human serum (PPHS) and ECGF (100 mg/ml) was a basic requirement for all EC. Addition of

heparin (10 U/ml) was helpful for the maintenance of a typical EC morphology and achievement of high cell densities.

We will describe next preparation of microvascular EC from liver (Vidal-Vanaclocha et al., 1993; Rusciano et al., 1993) and lung, which represent two frequent targets of metastatic cancer cells.

Materials

Sterile surgical instruments;

peristaltic pump (with setting between 0 and 6 ml/min) with sterile tubing (rinsed first with 250 ml of 70% etanol, and next with 100 ml of sterile GBSS);

water bath at 37°C;

butterfly needles (25G × 0.75″) Terumo Medical Corporation, Elkton, MD 21921, or syringe needles 21G × 0.5″, made blunt;

sterile nylon mesh;

Gey's balanced salt solution (GBSS): NaCl, 7.0 g; KCl, 0.37 g; $CaCl_2$, 0.17 g; $MgSO_4$ $7H_2O$, 0.07 g; $MgCl_2$ $6H_2O$, 0.21 g; Na_2HPO_4 $2H_2O$, 0.15 g; KH_2PO_4, 0.03 g; glucose 1.0 g; hepes, 2.4 g; phenol red, 0.01 g; pH 7.4. (NaCl is omitted in the preparation of metrizamide stock solution);

stock solutions (in GBSS): pronase E 1%; collagenase type I 0.5%; Dnase I 0.01%; metrizamide 30% (in GBSS without NaCl); sodium pentobartital stock solution in 0.9% NaCl (50 mg/ml);

growth factor supplement for EC culture: ENDO GRO™ (Vec Tec Inc., Albany, NY); Type I collagen-coated plates.

Procedure

Liver perfusion. Mice are anaesthetized by intraperitoneal injection of pentobarbital (0.07 mg/g bodyweight: dilute pentobarbital stock solution 7× with 0.9% NaCl, and inject 0.1 ml/10 g bodyweight). Abdomen is opened, portal vein exposed and loosely tied off with sterile surgical thread. Portal vein is then cannulated with the blunt needle, which is tied securely in place with a knot. Inferior vena cava is cut, and perfusion is started at a flow rate of 4.0 ml/min with sterile GBSS until the liver is clear of blood (about 3 min). Tissue digestion is then carried out by switching the perfusion (now at 2.0 ml/min) to 0.01% pronase E in GBSS for 5 min, then to pronase 0.01%, collagenase 0.05%, Dnase 0.0003% in GBSS at a flow rate of 1.0 ml/min for 15 min. A homogeneous perfusion of all liver lobes is indicated by a similar change in color (they become pale because of loss of blood). The liver can now be removed from the intestinal cavity and placed in a sterile petri dish.

Alternatively, perfusion can be carried out by cannulation of the inferior vena cava through the right atrium, and cutting of the portal vein to allow the flow through.

Lung perfusion. Mice are anaesthetized as above, abdomen opened, and the heart exposed. The right ventricle is cut and punctured in the direction of the right atrium with a blunt needle connected to the peristaltic pump. Perfusion is carried out as before, and the abdominal aorta is cut to allow the flow through. After blood clearing and tissue digestion, lungs are removed and placed in a sterile petri dish.

Care has to be taken when switching the inlet tube through different solutions not to let air bubbles in. Best way is to switch off the pump, reset the flow and move the tube to the next solution. When more than one mouse is used, the perfusion apparatus has to be rinsed between mice with at least 100 ml of sterile GBSS.

Preparation of endothelial cells. Organs (lung or liver) are cut under sterile conditions into small pieces using fine scissors or a blade. Tissue pieces are transferred into an Erlenmeyer flask containing 20 ml of 0.02% pronase E, 0.05% collagenase and 0.0003% Dnase in GBSS, and stirred at 37°C for 10 min. The cell suspension is then filtered through a nylon gauze, and centrifuged twice in GBSS at 300 g for 10 min. The pellet is resuspended in GBSS, and enough metrizamide (from the 30% stock solution) is added to the cell suspension to obtain a final concentration of 17.5% (wt/vol). Cells are gently dispersed in this solution, which is then layered with 1.0 ml PBS and spun at 1400 g for 15 min. EC are recovered at the interface between PBS and metrizamide, washed twice with RPMI, 10% FCS, and dispensed in multiwell plates, previously coated with collagen type I, in the same medium supplemented with ENDO GRO^{TM}.

When EC form a confluent monolayer, they are ready to be used as a substrate for the adhesion assay.

2.3.3.3. Characterization of endothelial cells

Because of their phenotypic heterogeneity, it is difficult to find general criteria do define unequivocally the endothelial nature of cells prepared from large or small vessels. Cobblestone morphology on nude or type I collagen-coated plastic is often observed in confluent EC cultures (Belloni et al., 1992). Capillary EC, however, are more plastic than large-vessel endothelium, and tend to differentiate into tubular structures when plated on type IV collagen (Madri et al., 1983; Li et al., 1991) or on a reconstituted extracellular matrix like 'Matrigel^{TM}'; or they show a fibroblast-like morphology in the presence of fibroblast growth factor (McCarthy et al., 1991). A nonthrombogenic surface appears also to be a good distinctive way of discriminating between a pure and grossly contaminated EC culture (Belloni et al., 1992). Expression of Factor VIII-related antigen (von Willibrand factor: vWF), angiotensin-converting enzyme and uptake of acetylated low density lipoproteins (LDL) are all widely variable among EC derived from different sources: hepatic sinusoidal EC take up LDL but are almost negative for vWF expression, while lung microvascular EC show a higher positivity for vWF expression, and have a much lower reduced ability for acetylated-LDL uptake (Belloni et al., 1992). Angiotensin-converting enzyme activity was initially considered to be a characteristic tract of lung EC (Caldwell et al., 1976), while later papers have shown variable activities among EC derived from different organs (Auerbach et al., 1982; Gumkowski et al., 1987).

Nonthrombogenic cell surface. This is evaluated with a platelet adhesion assay (Belloni et al., 1992). Platelets, isolated from fresh blood according to Tollefsen et al. (1974), are washed and resuspended at 5×10^8/ml in PBS containing 5 mM glucose and 5 mg/ml BSA. Confluent cultures of EC (bovine aortic EC as a positive control of a nonthrombogenic surface can be used) and fibroblasts (negative control) in 35 mm dishes are washed and incubated for 1 h at 37°C with DMEM containing 25 mM hepes, 5 mg/ml BSA to decrease the amount of serum proteins bound to the cell surface, which might interfere with the test. Platelets (10^8) are added to each dish, and incubated for 30 min at 37°C in a humidified incubator. Nonadherent platelets are removed by extensive washing with DMEM/BSA, and platelet adhesion is evaluated by phase contrast microscopy. Very low platelet adhesion is expected on a nonthrombogenic surface.

Uptake of acetylated LDL. Visualization of Ac-LDL uptake is attained by using the fluorescent probe 1'-dioctadecyl-3,3,3',3'-tetramethyl- indocarbocyanide perchlorate conjugated to Ac-LDL (DiI-Ac-LDL, Biomedical Technologies Inc.) (Voyta et al., 1984; Li et al., 1991).

EC are grown on glass coverslips to near confluency. Since LDL receptors on EC are sensitive to trypsinization, the assay should be done at least 48 h after plating, to allow for recovery. Tissue culture medium is replaced with fresh complete medium containing 10 μg/ml DiI-Ac-LDL, and cells incubated at 37°C for 4 h. (Since LDL can form aggregates, which interfere with its use, they should be removed by spinning in a microfuge for 2 min.) The labeling medium is removed, and cells washed several times with probe-free medium. Cells can be fixed with 3% formaldehyde in PBS for 20 min at room temperature (avoid fixation with methanol or acetone, since DiI is soluble in organic solvents). After rinsing in PBS and distilled water, coverslips can be mounted on glycerol/PBS (9:1) and sealed with cover glass cement (Pfaltz and Bauer, Waterbury, CT; do not use nail polish) or in Gelvatol if a better preservation of fluorescence is desired, and examined under a fluorescence microscope equipped with standard rhodamine excitation/emission filter combinations.

For cell sorting, cells are labeled as before, trypsinized to produce a single cell suspension, and trypsin neutralized in serum-containing medium. Cells and collection tubes are kept on ice prior to and during all manipulations. Cell cytofluorographic analysis is performed using a 514-nm argon laser for excitation, and detecting fluorescence emission above 550 nm. Positive and negative controls (such as bovine aortic EC and skin fibroblasts) should be used to set sample gates.

Angiotensin-converting enzyme activity. The assay measures the conversion of tritiated tripeptide [^{3}H]benzoyl-phenyl-analyl-analyl-proline to [^{3}H]benzoyl-phenyl-analyne and analyl-proline, and is easily performed by using the angiotensin-converting enzyme kit (Ventex Laboratories, Proland, ME). The buffered substrate is added to the endothelial cell monolayer, and incubated at 37°C for different time intervals up to 30 min, terminating the incubation by adding 0.1 N HCl. [^{3}H]benzoyl-phenyl-analyne is separated from the unreacted substrate by extraction with Ventex scintillation cocktail 2, by counting an aliquot of the organic phase in a scintillation counter. Enzyme activity is reported as the percentage of substrate utilized (Chung et al., 1986); 1 unit of angiotensin-converting enzyme activity is defined as the quantity required to hydrolize substrate at an initial rate of 1% per minute at 37°C.

2.3.3.4. Phenotypic modulation of endothelial cells in vitro
Expression of adhesion molecules, growth factors and lytic factors. Adhesive properties of EC are not constant over time, and the normally nonadhesive endothelium becomes capable of binding neutrophils, lymphocytes and monocytes following cytokine-mediated activation after injury or inflammation, thus allowing for migration of these inflammatory cells into sites of pathological processes (Vachula and van Epps, 1992; Imhof and Dunon, 1995).

Similarly, neoplastic cells may take advantage of the host inflammatory response, and thus enhance their metastatic spreading by using cytokine-induced adhesion molecules on endothelium (Bereta et al., 1991). In fact, it has been shown that IL-1α or TNFα treatment of EC in vitro induces the expression of adhesion molecules such as ELAM-1 (E-selectin), INCAM-110, vitronectin and fibronectin receptors, which in turn favor adhesion of carcinoma cells to the activated endothelium (Rice and Bevilacqua, 1989; Lauri et al., 1991; Lafrenie et al., 1992b, 1994). In vivo, it is known that sites of inflammation are more susceptible to cancer cell metastasis, and indeed pretreatment of mice with IL-1 (α or β) enhances the organ lodgement of injected tumor cells (presumably because of increased adhesion to microvascular EC), with the final effect of increasing the metastatic load of target organs (Bani et al., 1991; Bertomeu et al., 1993). Most recently, it has been reported that tumor cell dependent endothelium activation requires the presence of platelets, which release mediators eliciting adhesion molecules expression by EC (Hakomori, 1994).

Beside adhesion, cytokine treatment may also influence paracrine growth interactions between EC and tumor cells, and IL-1β treatment of hepatic sinusoidal cells up-regulates their expression of mannose receptors, and enhances release of paracrine growth factors active on B16 melanoma cells. Furthermore, treatment of mice with an IL-1 antagonist before intrasplenic injection of melanoma cells drastically reduces the amount of liver metastasis, suggesting a significant role for IL-1 in implantation and growth of melanoma cells in the liver (Vidal-Vanaclocha et al., 1994). It is also interesting that some tumor cells may produce their own cytokines, such as TNF (Spriggs et al., 1988) or induce its release from macrophages (Hasday et al., 1990), possibly promoting in that way their adhesion to endothelium.

Taken all together, these data strongly suggest that endothelial activation increases metastasis by increasing cancer cell adhesion, and therefore the effect of cytokine-mediated activation should also be considered when evaluating the results of the adhesion assays in vitro.

However, effects of cytokine activation of EC are not only limited to the induction of adhesion molecules and paracrine growth factors (all favorable events for metastases development). They also appear to include a negative control on metastatic tumor cells, which derives from the fact that many biological properties of EC resemble those of macrophages (Pober et al., 1983; Wagner et al., 1983; Miossec et al., 1986; Beilke, 1989; Sironi et al., 1989). In fact, at least in vitro, tumor cells can be lysed by lung microvascular EC that have been previously activated by a combination of TNFα and interferon γ (Li et al., 1991). This observation might contribute to explain the inability of certain tumor cells to metastatize in the lung, despite their ability to efficiently lodge at this site. For instance, M5076 cells can grow in the lung after intrabronchial but not intravenous injection (Li et al., 1990), although in this latter case they are known to efficiently arrest in the lung (Hart et al., 1981).

Organ-specifically modulated endothelial cells. Endothelial cells, similarly to most of the cells that form a living organism, are in close contact with extracellular matrix (ECM) molecules, which are known to play important roles in regulating their phenotypic expression, as clearly shown in an in vitro system by Madri and Williams (1983). In fact, it is now accepted that adhesive interactions do more than just promote attachment and influence cell shape, since adhesion receptors have been shown to function as transmembrane signalling molecules triggering a series of responses impinging on the regulation of both cell growth and differentiation (Rosales et al., 1995). Along this line, Pauli and Lee (1988) have shown that components of the ECM can modulate EC derived from large vessels (such as aorta or umbilical vein) to assume phenotypic traits of the specific microvascular endothelium of the organ from which the organ-specific biomatrix was prepared. Using monolayers of thus modulated bovine aortic EC (BAEC) in a classical adhesion assay, they show that tumor cells which metastasize to a given organ have a significantly higher binding affinity for BAEC grown on ECMs of the preferred, metastasized organ, than they have for BAEC grown on ECMs of other organs not targeted by these tumor cells. BAEC modulation occurs relatively rapidly (24 h); however, preferential attachment of organ-specific tumor cells is significantly increased if BAEC are grown continuously for 10 generations (3 passages) in dishes coated with organ-specific ECM. Specific cell surface determinants can be extracted from modulated BAEC by 1M urea in PBS, and conserve the ability to mediate organ-specific attachment of tumor cells. Indeed, a lung-specific EC adhesion molecule (Lu-ECAM-1) has been isolated and characterized from the surface of lung-modulated BAEC (Zhu et al., 1991), and appears to be an endothelial chloride channel (Elble et al., 1997).

Preparation of organ-specific biomatrix for EC modulation. (Wicha et al., 1982; Pauli and Lee, 1988; Hutchinson et al., 1989; Reid, 1989). Normal organs from freshly slaughtered 18-month-old bovines are transported on ice to the laboratory and minced into small chunks of tissue, taking care in excluding large blood vessels, fat and other undesired contaminant tissues (of course, organs from other animals, including rats and mice, can be used). Organ pieces are homogenized with a Polytron homogenizer at 4°C in 7–10 volumes of homogenization buffer (3.4 M NaCl, 0.05 M Tris/HCl, pH 7.4, 1 μg/ml leupeptin and 10 μg/ml soybean trypsin inhibitor). The insoluble material is pelleted at 10,000 g for 10 min, collected by filtration on a 160 μm pore size nylon filter, and washed extensively with water containing protease inhibitors as above. The material is further washed with 1 M NaCl and protease inhibitors (2000 g for 30 min) until the supernatant is no longer opaque and contains no detectable proteins. Delipidation is then achieved by layering N-butanol/isopropyl ether 2 : 3 (vol/vol) over an equal volume of water containing the matrix, and shaking for 1 h at room temperature. The matrix material is washed overnight with 1 M NaCl/antiproteases at 4°C, and then treated with DNase at 20 μg/ml and RNase at 100 μg/ml in serum-free medium at 37°C for 1 h with stirring. If desired, the matrix can be stored at this point at −20°C in serum-free medium with 10% (vol/vol) glycerol. When required, the matrix is thawed and repeatedly washed with water until no traces of medium are visible in the supernatant. After the final wash, the biomatrix is pelleted at 10,000 g for 30 min, and pulverized into a fine powder using a Freezer Mill (Spex Industries). Powdered biomatrix is sterilized by γ-irradiation, and can be stored frozen until used.

Molecules responsible for phenotypic modulation of large vessel EC can be extracted from the biomatrix with 10 volumes of 0.5 N acetic acid solution, overnight. at 4°C. After centrifugation in the cold at 15,000 rpm for 30 min, supernatant is collected, and the pellet extracted once more as before. Supernatants are pooled, neutralized with 5 N NaOH, and proteins determined by a colorimetric assay. Neutralized extracts are lyophilized and stored frozen.

Coating of 96 multiwell plates is done with 20 μg of proteins in 0.2 ml of PBS/well overnight at 4°C. After accurate washing of the wells with serum free medium, 5×10^4 EC in 0.2 ml of DMEM containing 10% heat inactivated FCS were plated per each well, and incubated at 37°C until very dense (usually 3–4 days). Monolayers are then rinsed with DMEM supplemented with 0.4% BSA (BSA, Sigma fraction V, treated with periodic acid to remove glycoprotein impurities) and used for the adhesion assay as described below.

2.3.3.5. Attachment assay
Quantitation of tumor cell adhesion on EC monolayers is usually done by adding labeled tumor cells for a short period of time (from 0.5 to 4 h), since this type of interaction during the process of tumor metastasis in vivo is supposed to occur soon after the release of malignant cells into the circulation. We will describe below different methods for labeling and detection of tumor cells to be used in the adhesion assay.

Radioactive labeling. Tumor cells can be metabolically labeled with radioactive agents by incubating with ^{3}H-thymidine (1–5 μCi/ml) in complete culture medium and labeling for 18–24 h. At the end of the assay, adherent tumor cells can be lysed for 30 min at room temperature with 1 M NH$_4$OH (Alby and Auerbach, 1984; Auerbach et al., 1987), or with 1% SDS for 4 h at room temperature (Pauli and Lee, 1988). Lysates are diluted with scintillation cocktails (e.g. Beckman Ready Safe), and counts determined in a beta scintillation counter. If the adhesion assay has been conducted on coverslips, these can be directly placed in 10 ml of Beckman Ready Protein, and radioactivity measured with the beta scintillation counter (Dumont et al., 1992).

^{125}I-deoxyuridine: 0.5–1 μCi/ml in complete culture medium for 24–48 h (Martin-Padura et al., 1991; Lafrenie et al., 1992a, b; Bertomeu et al., 1993; Lafrenie et al., 1994). Na-^{51}Cr: 100 μCi/ml, 30 min at 37°C in serum containing medium (Antonia et al., 1989; Bereta et al., 1991; Haq et al., 1992). Gamma-emitters do not need a scintillation cocktail, and so, if the adhesion assay has been carried out on coverslips, these can be directly placed in scintillation vials and read with a gamma counter. If, however, solubilization is required, it can be achieved by incubation with NaOH (0.5–1.0 N). After transfer of the solubilized material in scintillation vials, radioactivity associated with adherent tumor cells is quantitated with a gamma counter.

Fluorescent labeling. Tumor cells to be labeled with fluorescent dyes are first put in suspension by EDTA treatment (cell concentration ranging from 2 to 5 × 10^6/ml), and then incubated with the chosen dye for 15–60 min.

Hoechst 33342 supravital dye (Sigma B2261 or B4282) is used at 25 mM final concentration, for 30–60 min at 37°C. Adherent tumor cells (with fluorescent nuclei) can then be quantitated with a Fluoroskan II microtiter plate recorder (Flow) at 410 nm (Klepfish et al., 1993).

BCECF-acetoxymethyl ester (Molecular Probes, Eugene, OR) is used at 40 μg/ml final concentration in culture medium, for 15 min at 37°C. After washing, tumor cells are resuspended in HEPES-buffered serum free medium without phenol red (assay medium) at the desired concentration. Confluent monolayers of EC are also rinsed with the assay medium, and a basal autofluorescence (background) is determined with a CytoFluor-2350 system (Millipore) at 485/22 nm excitation with a 530/25-nm emission filter. After addition of labeled tumor cells in assay medium, a second determination with the CytoFluor system is done to estimate the total fluorescence. The actual number of adherent tumor cells is finally calculated after background subtraction as the ratio between the remaining fluorescence and the total fluorescence per each well of sextuplicate wells (Vidal-Vanaclocha et al., 1994).

6-CFDA (Sigma) is used at 40 μg/ml final concentration in serum free medium, 0.1% BSA, pH 6.0, for 30 min at 37°C. At the end of the assay, adherent labeled tumor cells are placed in 0.2% SDS for 30 min at 37°C to release the fluorescent marker. Three volumes of calcium-magnesium-free-PBS are added, and the fluorescence of the cell lysate is measured with a Perkin-Elmer LS-5 luminescence spectrophotometer (excitation maximum 485 nm, emission maximum 538 nm) (Price et al., 1995).

2.3.4. Adhesion to EC monolayers

EC monolayers obtained after growth on either collagen-coated glass coverslips, or tissue culture multiwell plates as previously described (Sections 2.3.3.1 and 2.3.3.2), are rinsed several times in serum free medium, and then incubated 1 h at 37°C with assay medium (usually serum free culture medium supplemented with 0.5–1% BSA), before they are used as substrates for the adhesion assay. Tumor cells are detached from culture dishes either by EDTA or by trypsin treatment. In the latter case, trypsinized cells should be allowed to recover from the enzyme treatment in their growth medium for 30 min at 37°C. After rinsing in serum-free medium, labeled tumor cells are resuspended at the desired concentration in the assay medium and plated onto the EC monolayer at a cell concentration ranging between 10^4 and 2×10^5/cm^2. Incubation is carried out at room temperature or at 37°C for 15–30 min, up to 4 h, either stationary or on a rotating platform at 100 rpm (Auerbach et al., 1987). After the incubation, EC monolayers are rinsed several times with the assay medium in order to remove nonadherent tumor cells. The extent of shear forces exerted by the stream of rinsing medium will of course influence the amount of cells that will be removed, and should be maintained within reasonable physiological limits. We normally let the medium flow through a 10 ml pipette with an electronic pipet aid set at middle power. Finally, adherent labeled tumor cells can be quantitated as described above, depending on the type of label that has been used.

2.3.5. Adhesion to extracellular matrix components

Mammalian organisms are composed by a series of tissue compartments separated from one another by two types of extracellular matrix (ECM): BMs and interstitial stroma (Hay, 1982; Yamada et al., 1985). ECMs consist of three general classes of macromolecules, including collagens, proteoglycans and noncollagenous glycoproteins (such as fibronectin, laminin, entactin and tenascin among others), which are expressed in a tissue-specific fashion. The extracellular matrix (ECM) is a critical component of tissue identity and maintenance of a normal differentiated phenotype (Bissell et al., 1999), and also plays a fundamental role in modulating adhesion of tumor cells and their further response to the microenvironment, and hence their metastatic potential at different organ sites.

Even though sometimes malignant cells arrested in the microcirculation do not migrate further into the organ parenchyma, and grow locally in an expansive fashion until they rupture the vessel wall (Kawaguchi et al., 1983; Crissman et al., 1985; Lapis et al., 1988), in most cases the contact between tumor cells and the endothelium results in EC retraction, with exposure of the underlying BM, followed by invasion of tumor cells in the tissue. Electron microscopy observation on the formation of pulmonary metastasis has shown indeed that tumor cells are often adherent to regions of exposed basal lamina (Chew et al., 1976). The exposed subendothelial matrix is usually a better adhesive substrate for tumor cells than the endothelial cell surface (Vlodavsky et al., 1983; Nicolson et al., 1984; Bastida et al., 1989). The presence of specific adhesion receptors on the membrane of metastatic cells, and the peculiar composition of the extracellular matrix at a given site, will then influence lodging and growth of malignant cells at specific sites (Belloni and Tressler, 1989/90; Rusciano and Burger, 1992). For instance, in the rat 13762 NF mammary carcinoma system the metastatic potential of lung-colonizing cell clones was rather correlated to differential adhesion to lung subendothelial matrix than to microvascular endothelial cells (Lichtner et al., 1989). Similarly, liver-specific colonization of F9 teratocarcinoma cells correlated with their better adhesiveness to fibronectin and liver-derived ECM, than to laminin and lung-derived ECM (Rusciano et al., 1991), in good agreement with the observation that malignant cell that adhere better to laminin than to fibronectin, such as certain mammary carcinoma cells, tend to metastatize predominantly to lung, while those adhering better to fibronectin than to laminin, such as many lymphoma cells, tend to metastatize to liver (Nicolson, 1988b). In a successive study, pre-treatment of melanoma cells with a synthetic peptide containing the laminin-binding domain of the 67-kD laminin receptor inhibited on the one hand tumor cell adhesion to endothelial cells (Castronovo et al., 1991), while on the other hand it increased cancer cell adhesion to laminin and subendothelial matrix in vitro, resulting in vivo in a significant increase in the number of experimental lung metastases (Taraboletti et al., 1993).

Beside adhesion per se, tumor cell interactions with ECM components influence motility, invasiveness and many other important aspects of the metastatic tumor cell phenotype. Tumor cell motility and adhesion are, in fact, closely linked, thus implicating a role for tumor cell adhesion in invasion. Indeed, tumor cell invasion has been proposed to represent a three step process, involving tumor cell adhesion, release of ECM degrading enzymes and tumor cell motility (Liotta et al., 1986a). In order to move through the ECM tumor cells must make firm contacts with matrix molecules, be able to break these adhesive contacts as they move on, and respond to chemotactic molecules which direct their movement. Interactions with the ECM may serve all these scopes, since it is now widely accepted that the ECM does not represent only a passive substrate for cells, but it has also a signaling function either by itself, or through soluble factors retained by its elements. The ECM is in fact a rich deposit of cytokines, growth factors, motility factors, enzymes and enzyme inhibitors (Vlodavsky et al., 1991). Stimulation of the integrin vitronectin receptor ($\alpha_v\beta_3$) on melanoma cells promoted cell invasion by enhancing gelatinase A production (Seftor et al., 1992), and, similarly, laminin A chain peptide fragments induced gelatinase A production, local invasiveness and augmented metastasis formation (Fridman et al., 1991, 1992; Sweeney et al., 1991). Integrin-dependent signaling in normal and tumor cells has been widely explored in recent years, and results show that integrins can hardly be regarded as simple 'sticky' molecules, while they are true receptors capable of generating intracellular signals contributing to control of gene expression, anchorage dependence, cell cycle transit and control of tumor growth (Juliano, 1994; Keely et al., 1998). Moreover, ECM macromolecules may also work as motility attractants, and have been shown, for instance, to stimulate both chemotactic and haptotactic motility in A2058 melanoma cells (Aznavoorian et al., 1990). In fact, both haptotactic and chemotactic responses may be relevant to the metastatic phenotype. Haptotactic migration over insoluble matrix components may occur predominantly during the initial stages of metastatic invasion, while at later stages partially degraded matrix proteins, derived from proteolytic processing of the matrix, could be the major determinant of directed motility. Finally, it has to be considered that some ECM components may actually impede cell adhesion, and thus might influence directional tumor cell motility by promoting the localized detachment of the trailing edge of migrating cells. ECM associated chondroitin sulfate proteoglycans such as decorin (Brennan et al., 1983; Hocking et al., 1998), or the glycoprotein tenascin (Chiquet-Ehrismann et al., 1988; Shrestha et al., 1996), have been suggested to modulate tumor cell adhesion and motility in this way.

Tumor cell adhesion to ECM components can be measured in vitro to either isolated molecules coated on the dish, or to complex ECMs produced by cells in culture or extracted from organs. We will describe first methods to prepare the different ECM coatings on the dish, and next the methodology for the adhesion assay.

2.3.5.1. ECM coating of tissue culture dishes
Nowadays almost all the components that constitute the different types of ECMs are available as purified molecules, which can be used either as single components, or in association with other components, to coat tissue culture dishes and then measure the ability of tumor cells to attach and spread on them. However, a test tube mixture of different components for the coating of a surface will not reconstitute a matrix network, which requires a delicate balance of molecules, assembled in the presence of cells expressing the right receptors (McDonald, 1988). Laminin, collagen type IV, entactin/nidogen and heparan sulfate proteoglycan are the main components of BMs. Collagens type V and VII may not be unique components of the BM, and may extend into the underlying stroma, with anchoring functions. Below the BM, the interstitial stroma contains its own set of collagens (mainly types I and III), and a different set of proteoglycans, glycoproteins and elastins. Fibronectin is present throughout the stroma and is found in some BMs. Both fibronectin and vitronectin are plasma proteins, and their presence in the ECM can result from diffusion from plasma or specific synthesis by epithelial or stromal cells (Liotta et al., 1986a; Yurchenko and Schittny, 1990; Preissner, 1991).

Coating with purified molecules. Once the required adhesion molecule(s) has been put in solution as recommended by the manufacturer (usually a stock solution of 1 mg/ml is prepared), a final dilution is made in sterile PBS so to evenly distribute between 1 and 10 μg/cm^2 per dish. After 1–4 h of incubation at 20 to 37°C, or overnight at 4°C, the dish is rinsed several times with sterile PBS, and then the unoccupied sites on the dish can be saturated with a sterile 1% BSA solution (1 h at 37°C). Collagens are soluble in acidic solutions (0.1/0.5 M acetic acid), and coating is done by letting the solution dry on the dish by overnight incubation under sterile conditions (Murray et al., 1980). If sterilization of the dish is necessary, this can be obtained either during or after the coating by exposure of the dish to u.v. light under a laminar flow hood for few hours (Rusciano et al., 1992). Figure 2.2 shows an adhesion assay on both adhesive (fibronectin) and nonadhesive (tenascin) substrate.

Subendothelial ECM. Cultured EC synthesize an ECM resembling the natural BM of the vessels (Gospodarowicz et al., 1979), that can be used as a substrate to test the adhesiveness of tumor cells to a complex matrix.

Subendothelial ECM is prepared from EC which had been confluent for 5–7 days. There are basically two ways of removing EC from their substrate: either intact, or lysed in a hypotonic solution with a detergent. Both treatments leave the underlying matrix firmly attached to the dish.

Removal of intact EC can be accomplished (after careful rinsing with PBS) by incubation in 2% EGTA in HBSS for 1 h at 37°C, with occasional shaking (Bastida et al., 1989), or by exposure to 2 M urea in medium containing 0.5% FCS, 10–20 min at room temperature (Rogelj et al., 1989; Gospodarowicz et al., 1983). After cell retraction, their removal is accomplished by gentle addition of more DMEM/2 M urea/0.5% FCS. Cells floating in the medium are aspirated. At no time cells are pipetted up and down. This gentle washing process is repeated four times. Once the matrix is free of cells, plates are rinsed with PBS at least three times.

Lytic removal of EC is usually performed in two steps. After rinsing with PBS, cells are first incubated in a hypotonic solution (5 mM Tris-phosphate, pH 7.4, to which 0.05 mM CaCl$_2$ and 0.05% BSA can be added) 15 min at 37°C, and then with detergent (either

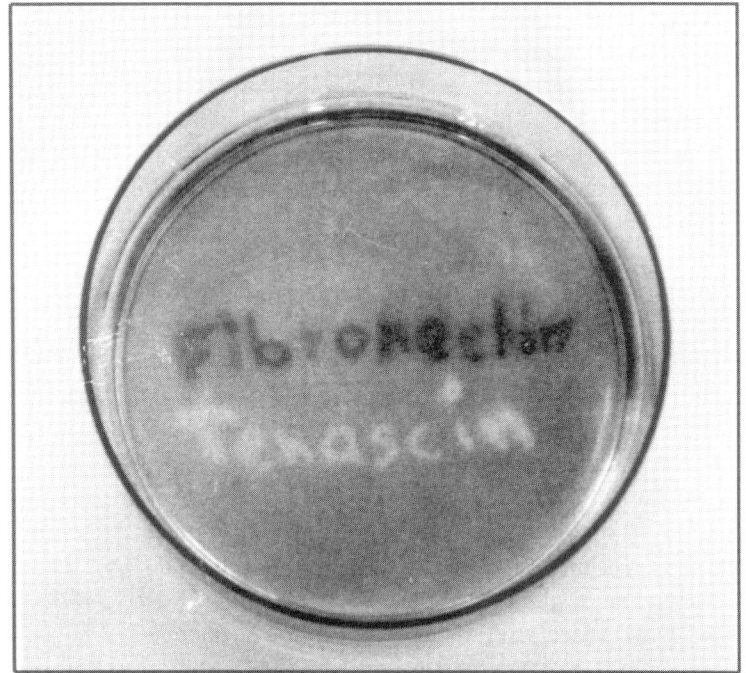

Fig. 2.2. Positive and negative adhesion of fibroblasts on fibronectin and tenascin coating of a tissue culture dish, after saturation with BSA. After 1 h of plating in serum-free medium, cells appear loosely attached on BSA, well spread on fibronectin and not attached on tenascin. Crystal violet staining. (Photograph courtesy of Ruth Chiquet, Friedrich Miescher Institute.)

0.5% Triton X100, 20 mM NH_4OH in PBS for 3 min at 22°C, or 0.2% NP40 in 5 mM Tris-phosphate pH 7.4 for 1 min at 37°C, followed by 2–3 min 0.025 N NH_4OH in PBS to remove remaining nuclei and cytoskeletons) (Vlodavski et al., 1982; Almasio et al., 1984; Bashkin et al., 1989; Lichtner et al., 1989; Rogelj et al., 1989; Taraboletti et al., 1993).

In any case, after EC removal, dishes or coverslips retaining the subendothelial matrix have to be thoroughly rinsed with PBS, and saturated with BSA (1% in PBS, 1 h at 37°C), before they can be used in the adhesion assay.

Finally, it has to be remarked that the composition of the ECM produced by EC can be modulated by culture conditions, and for instance, in the presence of 0.1 mM dexamethasone human umbilical vein EC produce a higher amount of matrix, enriched in fibronectin (Almasio et al., 1984).

Tissue extracts enriched in ECM. Organ-specific ECM can be prepared from organ homogenates as described in Section 2.3.3.4: *Preparation of organ-specific biomatrix for EC modulation* (Wicha et al., 1982; Pauli and Lee, 1988; Hutchinson et al., 1989; Reid, 1989). The powdered biomatrix thus obtained is resuspended in distilled water at 10 mg/ml, and 1 ml aliquots of this suspension are placed in each well of a 24-well tissue culture plate. Coating is achieved by letting the matrix suspension dry under a hood with u.v. light, so to sterilize the matrix at the same time. Here the key to making an evenly coated plate is obtaining a very fine powder by using the Freezer-Mill, and then a very homogeneous suspension. Tissue culture plates with adherent ECM can be stored at 4°C until used. This organ-specific biomatrix can be used to evaluate differential attachment of tumor cells (Hutchinson et al., 1989; Rusciano et al., 1991, 1992), but it has also been shown to influence growth and differentiation (Wicha et al., 1982). Interestingly, it has been reported that for human hepatoma and mammary carcinoma cell lines, tissue specificity in clonal growth on biomatrices correlated with their organ site specificity for metastasis in vivo in immunosuppressed, athymic nude mice, thus suggesting that the ability of tumor cells to colonize specific tissues is, at least partly, influenced by their low density survival and growth on a specific ECM (Doerr et al., 1989).

Another possibility of using a complex ECM for adhesion experiments is given by the use of a reconstituted BM that goes under the name of Matrigel^{TM}. Reconstitution is made starting from components extracted from the BM of the EHS tumor grown subcutaneously in mice. Careful mixing of these soluble components under specified conditions, gives rise to a gel-like structure whose ultrastructure appears as interconnected thin sheets resembling the *lamina densa* zone of BM (Kleinman et al., 1986). Matrigel™ or similar material is commercially available from Collaborative Research/Beckton Dickinson (which provides also a growth factor-depleted Matrigel™), and from Sigma (as ECM-gel). A characteristic of Matrigel™ is that it is liquid at 4°C, while it gels at 37°C. Therefore, Matrigel™ has to be kept on ice, and also further dilutions have to be made in cold buffer to prevent its polymerization. Coating is achieved by layering the required amount of Matrigel™ on a cold dish surface, in order to allow even spreading before polymerization, and then moving the dishes at 37°C for the time necessary to Matrigel™ to fully polymerize and form a continuous structure (4 h to overnight). After rinsing with PBS or serum-free medium, the surface is ready for the attachment assay.

2.3.5.2. Adhesion assay to ECM-coated dishes
Tumor cells, either radioactively labeled, or unlabeled, are detached from the culture dish by EDTA or trypsin treatment (if the adhesion assay is carried out under serum free conditions, trypsin has to be blocked by soybean trypsin inhibitor, and the cells thoroughly rinsed in the assay medium), and resuspended at the desired dilution (1 to 5×10^3 cells/cm^2 give good results) in assay medium (usually serum-free medium which may contain 0.1% BSA), and plated onto coated dishes. Incubation can be either stationary or on a rotating plate if shear forces need to be included, for times ranging from 30 min up to 10 or 20 h, since it has to be considered that the interaction between metastatic cells and the organ ECM is an event more protracted in time than the adhesion to EC. Incubation temperature is usually a physiological 37°C, but incubation at 4°C may give indications about pure mechanical interactions, not involving energy-dependent reactions.

At the end of each time point (a good kinetics should include at least four points), dishes are rinsed three times with PBS, which is gently delivered on the surface through a pipette (take care in using always the same method and force in the rinsing procedure, because application of different shear forces among experiments may alter their reproducibility) with the flow directed against the wall of the dish.

Next, detection and quantitation of tumor cells adherent to specific substrates can be done by either radioactive or nonradioactive methods. For radioactive methods, labeling and detection has been already described in Section 2.3.3.5. Alternatively, tumor cells can also be labeled with fluorescent dyes, and quantitated accordingly (see same section). We will describe below only nonradioactive methods.

Direct staining of tumor cells. If the substrate on which tumor cells have been plated does not interfere too much with the staining procedure, giving an acceptable background (such as in the case when purified ECM molecules are used), then adherent cells can be stained with either crystal violet, or methylene blue.

For crystal violet staining (Gillies et al., 1986, Rusciano et al., 1991, 1992, 1993), adherent cells are first fixed 15 min with 1% glutaraldehyde in PBS, after which time they can be kept hydrated with PBS alone at 4°C until used for staining. Samples are stained for 30 min at room temperature in 0.1% crystal violet in deionized water. After this time, dishes are submerged in a 1-l beaker of deionized water, and destained for 15 min with a continuous, slow stream of deionized water introduced at the bottom of the beaker. Plates are allowed to air dry, and then the crystal violet absorbed onto the cells is solubilized with 0.2% Triton X100 for several hours (usually o.n.) at 37°C. To prevent evaporation, plates are sealed with parafilm. The colored triton solution is finally read with a spectrophotometer at 590 nm. The absorbance is stable at room temperature for weeks, and the resolution of the technique is below 500 cells. Its entire range of accuracy (between 500 and 10^5 cells in 24-well plates) can be achieved by solubilizing the dye in various volumes (0.5–2.0 ml) of the triton solution.

For methylene blue staining (Goldman and Bar-Shavit, 1979), adherent cells are fixed 20 min with 0.5% glutaraldehyde in PBS, and stained for 60 min at room temperature with 1% methylene blue solution in 0.1 M borate buffer, pH 8.5. After thorough rinsing in deionized water, extraction of the color from stained cells is done with 0.1 N HCl for 60 min at 37°C in parafilm-sealed plates, and the o.d./well read at 590 nm. The intensity of the color is proportional to the number of cells.

Enzymatic detection of tumor cells. Coating of dishes with thick layers of organ-derived ECM, or with Matrigel^{TM} gives an elevated background staining with the previous method, which is then not advised in these cases. Possibilities of revealing adherent tumor cells plated on thick layers of ECM include either radioactive labeling (but still, the presence of Matrigel^{TM} may interfere by unspecific retention of radioactivity), or an enzymatic assay based on the ability of active mitochondria in living cells to metabolize and convert a tetrazolium salt derivative into a blue formazan product that can be estimated at wavelengths variable between 490 and 600 nm, depending on which tetrazolium salt derivative has been employed in the assay. For instance, MTT (Mosmann, 1983; Tada et al., 1986; Denizot and Lang, 1986; Gerlier and Thomasset, 1986; Hansen et al., 1989; Vistica et al., 1991) gives an insoluble blue formazan compound, which has to be solubilized, and is better read at 570 nm, whereas XTT (Scudiero et al., 1988; Weislow et al., 1989; Roehm et al., 1991) or MTS (Cory et al., 1991) give a soluble blue formazan product that is directly estimated at 490 nm. This assay is commercialized as a kit by both Promega and Boehringer-Mannheim, and has been used, among other applications, also to estimate adhesion efficiency (Prieto et al., 1993; Klemke et al., 1994). However, since a thick coating of the well interferes with optical reading in a Titertek, the colored medium has to be transferred to another multiwell, or to cuvettes, for o.d. determination.

2.3.6. Adhesion in a dynamic flow system

Shear forces are relevant throughout the adhesion process, since tumor cells are carried in the blood stream, and they must make contact with EC and the subendothelial basement membrane while transported in the flow, similarly again to what has been observed and described for leukocytes (Cohnheim, 1889; Lawrence and Springer, 1991). Therefore, a more accurate evaluation of specific adhesion to EC and ECM should also take into account the effects intrinsic to a dynamic system such as flowing of a liquid on a surface. Several studies have been carried out to elucidate the various steps, and the implicated molecules, of leukocyte adhesion to EC in a dynamic flow system (reviewed in Hammer and Apte, 1992; Toezeren and Ley, 1992; Ley and Tedder, 1995). The model emerging from those studies implicates selectins at first stages of interactions, as the molecules responsible for slowing down leukocytes carried in the blood stream, bringing them to a slow rolling on the endothelial surface. P- and L-selectins, respectively expressed on endothelium or leukocytes, can mediate the initial capture of leukocytes flowing in the blood, and in synergy with E-selectin present on EC they also mediate the subsequent rolling phase. Leukocyte integrins and their ligands are required for firm adhesion, but may also mediate rolling in cooperation with E-selectin, which, in turn, might also be required for firm adhesion (Ley and Tedder, 1995). Both selectins and integrins have been implicated in the adhesive interactions of tumor cells with EC under physiological flow conditions (Giavazzi et al., 1993; Toezeren et al., 1995), in a model that by analogy to what happens for leukocytes, has been named 'docking and locking' (Honn and Tang, 1992; Honn et al., 1994). Tumor cells entering the microvasculature may initially attach only loosely to EC, and this docking step is thought to be mediated by relatively weak and transient adhesion mechanisms involving carbohydrate-carbohydrate (such as Le^x-Le^x and glycosphingolipid-glycosphingolipid), carbohydrate-selectin (such as s-Le^x-E-selectin), and protein-protein (such as

PECAM-1-PECAM-1) recognition (Honn and Tang, 1992). This initial cell attachment event will lead to the activation of tumor and/or EC, finally resulting in increased surface expression and/or functional maturation of other adhesion molecules, typically integrins, and finally to the second phase (locking) of tighter tumor cell adhesion (Tang and Honn, 1994/95).

The methodologies used to measure rolling and adhesion in a dynamic system appear to be quite lengthy and troublesome to set up, and will not be described in this book. We will refer, instead, the interested reader to papers where relevant methods have been described.

Intravital microscopy has been, and still is, the main technique to observe the rolling behavior (Atherton and Born, 1973; House and Lipowski, 1987; Mayrovitz, 1992; Mayadas et al., 1993). Simulation in vitro of a dynamic interaction with an EC monolayer has been obtained with a perfusion chamber (Sakariassen et al., 1983, Bastida et al., 1989), and observation of the rolling phenomena became possible with parallel plate-flow chambers (Lawrence et al., 1987, 1990). The use of these chambers was initially limited to monitoring of neutrophils rolling on cultured EC monolayers, but more recently they have been modified to show that neutrophils can also roll on surfaces coated with purified E- and P-selectins (Lawrence and Springer, 1991, 1993). Moreover, studies by Kojima and colleagues (1992) suggested that under flow conditions glycosphingolipid homologous interactions between tumor and endothelial cells may predominate over lectin- or integrin-based mechanisms. Most recently, a capillary tube method has been described, which allows to measure the rolling behavior in fused silica capillary tubes coated with selectins (Schmuke and Welply, 1995). Finally, it might be worth mentioning that Cellco Inc., (Germantown, MD 20874, USA) has developed several Artificial Capillary Cell Culture Modules (CELLMAX) specifically designed for the culture and study of EC under flow, and which could also be exploited to study adhesion under dynamic flow conditions.

2.3.7. Adhesion strength

Tumor cells that have the ability to attach to some defined substrates with the same apparent efficiency, as measured by the adhesion assays described above, may nonetheless adhere to the different substrates not with the same strength. Therefore, even though each substrate appears to be able to support cell adhesion, the tenaciousness characterizing each interaction may be different, and it may have an influence on the metastatic ability of tumor cells (Leung-Tack et al., 1988).

Several types of assays have been described, which aim at measuring the adhesion strength between test cells and a substrate (which can be either a cell monolayer, or a matrix coating). We will describe one of these assays in more detail, and mention the others for reference.

Minimal shear force. The method has been developed to prevent removal of weakly adherent cells during washing, detaching non-adherent cells by gravity and gentle shear force, rather than by aspiration and washing (St. John et al., 1994). The adhesion assay is carried out in a 96-well microtiter plate as usual, taking care to fill the wells up to the top (300 μl) in order to avoid bubbles during the following step. After the desired incubation time(s), the microtiter plate is gently immersed face up, but oblique, into a large plastic container filled with 2.5 l of 0.85% NaCl solution. The plate is gently inverted while submerged in the fluid, to avoid trapping air bubbles in the bottom of the wells. The plate is then raised and allowed to float on the surface facing downward. The saline solution is stirred with a magnetic bar at 300 rpm for 7 min at room temperature. The plate is then resubmerged, turned upward and shifted to a similar container filled with 100% methanol by using the same precautions as before. The plate is kept submerged and rotated several times in the methanol to allow the saline to mix with methanol, without exposing the cells to air. After 5 min the plate is removed right side up with the wells filled with methanol, and kept 60 min more at room temperature. Methanol can be finally shaken from the wells, and adherent cells quantitated by staining or label counting if radioactively or fluorescent labeled cells have been used (Section 2.3.3.5).

This basic assay is susceptible to be modified in order to measure the ability of adherent cells to withstand increasing shear forces, which can be obtained by increasing the rate of mixing. Using this approach, the strength of cell adhesion to different substrates, or the strength of adhesion of different cell types to the same substrate, can be compared.

Centrifugation. The method is obviously based on the application of a centrifugal force to adherent cells, in order to find out which force is required to dislodge most of the cells from the tested substrates (McClay et al., 1981; Hertl et al., 1984; Lotz et al., 1989). The main advantage of this method is that it allows a precise quantitation of the force per cell required for dislodgement, which can be calculated according to the equations:

$$F_D = (\rho_{cell} - \rho_{medium}) \times V_{cell} \times RCF$$

in which F_D is the dislodgement force per cell, ρ_{cell} is specific cell density (1.07 g/cm^3), ρ_{medium} is the specific density of the medium (1.00 g/cm^3), V_{cell} is cell volume (to be calculated from the average cell diameter, which can in turn be estimated either at the microscope, or with a Coulter Counter), and RCF is expressed in gravities (g). The resulting force is expressed in dyne/cell, and is usually in the range 10^{-5}, 10^{-6} (McClay et al., 1981).

Capillary glass tubes. With this method, radioactively labeled tumor cells are allowed to adhere to the inner surface of glass capillary tubes either uncoated, or coated with different adhesion molecules. After incubation, adherent cells are subjected to variable shear forces by application of variable flows through an automatic pipette connected to the tube. By varying the temperature and the viscosity of the washing medium, shear forces in the range of 16–130 N/m^2 can be obtained (Mege et al., 1986; Leung-Tack et al., 1988).

CHAPTER 3

Motility, deformability and metastasis

Up until the 1970s, cancer invasion and metastasis was mostly seen as a mechanical process, resulting from passive growth pressure coupled with low tumor cell cohesion (Eaves, 1973). It was not until recently that it was recognized that growth pressure alone is insufficient for invasion (Meyvish et al., 1983; Thorgeirsson et al., 1984), and cannot account for the difference in invasive behavior between many rapidly proliferating benign and malignant neoplasms. Cancer invasion is in fact an active process, which requires tumor cell locomotion. Escape from the primary site and entrance into the circulatory system, and next extravasation at a secondary site and local invasion and growth, are all steps of the metastatic cascade involving the ability of metastatic cells to attach and walk on 'biological' surfaces, and withstand a certain degree of deformation.

3.1. Motility and metastasis

Several lines of evidence indicate that active motility of tumor cells is required in the metastatic process. The metastatic potential of Dunning R-3327 rat prostatic adenocarcinoma sublines best correlates with their ability to move on a substrate, as assessed either by a visual grading system of time-lapse videomicroscopy, or by a computerized mathematical system using spatial temporal Fourier analysis (Mohler, 1993). Similarly, the high metastatic capacity of a v-fos-transfected rat fibrosarcoma cell line correlated with an increase in invasiveness associated with enhanced cell motility (Taniguchi et al., 1989). Finally, intraperitoneal injection in Sprague-Dawley rats of a C5-derived chemotactic factor locally enhanced the metastatic burden after intravenous injection of Walker carcinosarcoma cells (Lam et al., 1981).

The relevance of motility in metastatic behavior is not only limited to animal models, and expression levels of a Motility Related Protein (MRP-1/CD9) in human cancer cells appear to be related to the prognostic outcome of breast cancer patients (Miyake et al., 1996), as well as of colon cancer (Mori et al., 1998), or oesophageal squamous cell carcinoma (Uchida et al., 1999) patients. Cell motility might also be a possible therapeutic target. Experiments conducted on murine melanoma cell lines (where metastatic ability may depend on actin isoforms influencing cell motility (Sadano et al., 1994)), have shown that treatment with low levels of adriamycin, not inhibiting tumor cell growth, but blocking their migratory response to haptotactic agents, was also able to prevent their invading through a reconstituted basement membrane (BM) in vitro (Repesh et al., 1993).

Directed cell migration is a highly coordinated multifactorial process, requiring at least three different types of mechanisms to operate simultaneously for cell locomotion to occur: (a) intracellular force-generation mechanisms involving the cytoskeleton; (b) polarity-determining mechanisms involving the insertion of new membrane mass into the leading edge; and (c) adhesion mechanisms involving interactions of the cell surface with the substrate. However, a detailed analysis of the physiology of cell movement is beyond the scopes of this chapter, and the interested reader is referred to some excellent reviews published on this theme (Singer and Kupfer, 1986; Condeelis, 1993; Stossel, 1993; Parent and Devreotes, 1999). Cells initiate movement in response to surface stimulation, usually in the form of soluble or substrate-bound extracellular molecules that bind to specific transmembrane receptors, thus triggering a cascade of signalling events that finally activate the complex machinery of movement (Stossel, 1989; Brundage et al., 1991; Hahn et al., 1992; Ridley and Hall, 1992; Ridley et al., 1992). There are basically three different types of motility response for a cell to a specific stimulation: (a) chemokinesis indicates induction of random cell locomotion; (b) chemotaxis; and (c) haptotaxis represent the response to signals orienting cell movement along a defined gradient path. A chemotactic response is elicited by soluble ligands, whereas haptotaxis describes the directed migration of cells along a gradient of anchored molecules. Most of the extracellular matrix (ECM) molecules may exhibit both haptotactic and chemotactic activities towards migrating cells. In fact, in order to move through the very tight matrix meshwork, migrating cells have to be able to carve a corridor through it by cleaving the components of such a net, and in so doing they will release chemoactive fragments that will further influence their motility behavior (Nabeshima et al., 1986). Fibronectin, laminin, collagens and elastin promote the haptotactic migration of several tumor cell types (Zetter and Brightman, 1990), while thrombospondin, which is released by activated platelets, stimulates both their chemotactic and haptotactic movements (Taraboletti et al., 1987).

Beside ECM components, a group of secreted cytokines specifically inducing cell motility has been identified, which can be either autocrine or paracrine in action. Migration stimulating factor (MSF) is produced by fetal and tumor-derived fibroblasts, and stimulated the rapid migration of producer cells into collagen matrices (Schor et al., 1988, 1993). Autocrine motility factor (AMF) is secreted by mouse and human melanoma cells, and stimulates both random and directed cell motility by interacting with its cognate gp78 cell surface receptor (Silletti et al., 1991; Watanabe et al., 1991a). Transformed cells both secrete and respond to AMF, which, on the other hand, is capable of stimulating motility of untransformed cells as well (Liotta et al., 1986b). AMF could thus play a paracrine role for normal cells, whereas tumor cells could stimulate their own motility in an autocrine fashion. In different tumor model systems, such as mouse melanoma (Watanabe et al., 1991b) and fibrosarcoma (Watanabe et al., 1993), or human prostatic carcinoma (Silletti et al., 1995), the autocrine interaction between AMF and its receptor has been shown to influence the metastatic ability of malignant cells. AMF might belong to a family of autocrine motility cytokines acting through G-protein-coupled receptors, since another similar factor, named autotaxin, has been identified in the conditioned medium of human melanoma cells (Stracke et al., 1992). Scatter factor (SF) is the prototype of paracrine motility factors, being secreted by mesenchymal cells and active on epithelial cells (Stoker et al., 1987; Gherardi et al., 1989), on which it can induce both random and directed cell motility (Stoker, 1989). Scatter factor has been found to be identical to hepatocyte growth factor (Weidner et al., 1991; Furlong et al., 1991: hence the new name HGF/SF), which was initially discovered as a growth factor involved in liver regeneration (Nakamura et al., 1984). Beside its effects on growth and motility, HGF/SF has also been shown to be a broad morphogenic (Brinkmann et al., 1995; Zarnegar and Michalopoulos, 1995), and a potent angiogenic factor (Bussolino et al., 1992). The pleiotropic response induced by HGF/SF is fully mediated by a single transmembrane receptor, the product of the c-met protooncogene (Bottaro et al., 1991; Weidner et al., 1993a). The inappropriate activation of c-met signal transduction pathway has been linked to malignant progression in several different tumors both in vitro and in vivo (Prat et al., 1995) and to their metastatic ability (Rong et al., 1994; Rusciano et al., 1995, 1998; Lin et al., 1998). Most recently, other molecules have been identified, which possess scattering activity. A scatter factor-like factor (SFL) has been found to be produced by bladder carcinoma cells (Bellusci et al., 1994), and monocyte-conditioned medium also contains a novel scattering activity (Jiang et al., 1993). All these factors may represent a family of cytokines whose regulated expression induces cell motility in physiological situations such as wound healing, organ regeneration and embryogenesis, and whose constitutive autocrine expression may induce or enhance the metastatic ability of tumor cells.

Beside HGF/SF, several other growth factors (such as PDGF, TGFα and β, EGF, FGF, IGF, ECGF, G-CSF and GM-CSF, IL-6, TNFα and IFNγ, and bombesin) have been found to promote cell motility (Stoker and Gherardi, 1991) and some also possess the ability to induce colony scattering (Warn, 1994). Carcinoma cells can be scattered by acidic FGF (Valles et al., 1990; Jouanneau, 1991) or EGF (Shibamoto, 1994), and c-src activation seems to be required for the scattering response (Rodier, et al., 1995).

Active cell movement requires a traction force exerted by moving cells, which thus must be adherent to a relatively immobile substrate, either ECM or neighboring cells. Adhesive interactions have to be of the right strength, because if too weak they may not support efficiently the traction, while if too strong they may block it. Therefore, expression levels of specific cell adhesion molecules will influence the motile and invasive ability of migrating cells. Overexpression, for instance, of the integrin receptors $\alpha 5\beta 1$ in CHO cells (Giancotti and Ruoslahti, 1990), or $\alpha 4\beta 1$ in mouse melanoma cells (Qian et al., 1994) inhibits respectively motility or invasion and suppresses either the tumorigenic or the metastatic phenotype of recipient cells. ECM components may dramatically alter the motile behavior of cells, as shown by a rat prostatic carcinoma subline which, although highly motile on plastic, failed to move efficiently on elastin, and also to metastasize in the lung with the same efficiency of other highly motile sublines able to migrate both on plastic and on elastin (Mohler et al., 1991). Finally, a direct link has been shown between AMF-induced cell motility and integrin expression. Stimulation of gp78 (AMF-receptor) in murine melanoma cells, beside inducing cell motility, enhanced their adhesion and spreading on fibronectin, and invasiveness through Matrigel^{TM}, by an increased translocation of $\alpha 4\beta 1$ and $\alpha IIb\beta 3$ from the cytoplasm to the cell surface (Timar et al., 1996).

3.1.1. Assessment of motile behavior of tumor cells

Most of the time cell motility is intended as the translation of a cell from one place to another, while the term also includes all the motile manifestations of a cell, such as ruffling (rapid rhythmic movements of short segments of cell membrane), undulation (slow rhythmic movements of long segments of cell contour), pseudopodal extension (nonrhythmic extension and retraction of narrow based segments of cell contour over comparatively longer distances) and translation, which can be vectorial (displacement of cell center in a straight line), or irregular (when displacement deviates from a straight line) (Mohler et al., 1987a).

There are basically two different ways of estimating cell motility in vitro: direct observation of moving cells through a video-microscope (time-lapse videomicroscopy), and quantitation of the amount of cells that moved in response to a specific stimulus after a certain time of induction (Boyden chamber, artificial wound and colony scattering, chemotaxis under agarose, phagokinetic tracks). While the latter methods can only estimate cell translation, the former one allows the evaluation of all the above parameters describing cell motility. However, since setting up of videomicroscopy methods and the interpolation of data thus obtained would require a book on its own to be adequately described (for instance: Shinya Inoue and Kenneth R. Spring, 1997 *Video Microscopy*, 2nd edn., Plenum Press, New York), we will only mention some fundamental aspects of videomicroscopy, while describe in more details the indirect methods for the estimation of cell locomotion.

3.1.1.1. Time-lapse videomicroscopy
Time-lapse cinematography, originally pioneered by Abercrombie (1953, 1957), who first suggested that locomotory behavior of malignant cells could differ from that of normal cells, has evolved and has been replaced by the more sophisticated time-lapse videomicroscopy. This technique allowed a more precise description of cell motility, and suggested a correlation between cancer cell motility

and metastatic potential (Fulton, 1984). Soon after, a visual grading technique for the assessment of cell motility by time-lapse videomicroscopy was developed, and applied to the characterization of clonal sublines of the Dunning R3327 rat prostatic cancer of different metastatic potential (Mohler et al., 1987a, b; Doyle et al., 1992). These studies established that membrane ruffling, pseudopodal extension and cellular translation correlated best with metastatic potential, and their grading could be combined to give a motility index, which neatly discriminated high metastatic cells (index grade >6) from low metastatic cells (index grade <4). However, this visual motility grading system could merely establish whether, or not, a cell has moved, without giving any quantitative details of this movement. This has been obviated by the combined use of time-lapse videomicroscopy, image analysis techniques, and a new spatial-temporal two-dimensional Fourier analysis of cell motility (Partin et al., 1989), confirming again the correlation between the motility index and the metastatic behavior of rat prostatic cancer cell lines (Partin et al., 1992).

The basic requirements for time-lapse videomicroscopy are: an inverted microscope, connected to a high resolution video camera and a time-lapse video recorder. Cells are plated at low density in filming flasks (Falcon) one day before observation to allow complete spreading in the desired medium. Images, magnified 400 times, are recorded every 15 sec for at least 2 h. The film is then played back at 24 frames per second to allow visualization of motile parameters. Visual grading is performed by a double blind test, with the aid of independent observer with no knowledge of the identity of the specimens. Membrane ruffling, pseudopodal extension and cell translocation are graded from 0 (no motility) to 5 (excessive motility) by each observer. The grades are summed to yield final scores of zero to 10 for each motility parameter. Averaged motility grades relative to the three considered parameters can be summed and divided by three to give a motility index for each cell under observation (Mohler et al., 1987b). Evaluation of at least 10 cells is required to characterize within 95% confidence intervals a cell line (Doyle et al., 1992). This grading system, however, completely depends on subjective evaluation of the observers, and does not give any objective quantitative information. Since motility manifestations of individual cells resemble a wave-form motion that is possible to analyze by mathematical techniques, and the complex shape of a cell can be decomposed and mathematically represented by a Fourier transform, Partin et al. (1989) developed a mathematical, computerized image analysis technique utilizing a spatial temporal Fourier analysis that is capable of analyzing time-lapse data, and providing quantitative accurate information on the various types of cell motility. When comparing visual grading with the Fourier analysis, correlation coefficients ranged from 0.71 to 0.77 for the considered motility parameters, indicating that the technique can be applied to the evaluation of motility. By increasing the resolution of video microscopy to better approximate that of the human eye, so to render more precise the mathematical interpolation of the data, this method should be able to give quantitative results in a very reliable way.

Alternatively, another method to quantitate cell motility independently from human judgement has been developed by Tatsuka et al. (1989). With this system, quantitative estimation of cell motility, after video recording, is done by counting the total intensity of the cumulative trace image in a window on a TV monitor. A trace image is obtained by subtracting a digital image for cells in any one video frame from a digital image for the same cells in the subsequent frame, and trace images for these particular cells are accumulated at intervals corresponding to the video frames (cumulative trace image). However, to use this method to compare different cell lines, which may appear with different intensities in a video frame if phase-contrast microscopy is used, the Allen video-enhanced contrast-differential interference contrast (AVEC-DIC) microscopy must be used. The trace image thus generated corresponds to cell motility, including shape changes such as ruffling and other motions which accompany cell locomotion (vectorial translation) and/or occur at a fixed position without vectorial translation. With this method the intensity of the trace image is higher in transformed cells than in nontransformed cells (Tatsuka et al., 1989).

Drawbacks of the method: Specialized equipment is required to guarantee constant culture conditions during long term microscopic observation of cells, and for digitalization of recorded images for further mathematical interpolation of data. Moreover, this technology only allows simultaneous analysis of a small number of cells, and might thus be selective, so that validation of data by other biological and/or biochemical methods is necessary. Finally, quantitative estimation of chemotaxis is hardly allowed by time-lapse videomicroscopy, and other methods, easier and cheaper to set up in the lab, are available to this purpose.

3.1.1.2. Filter membrane assay
The extensively used filter micropore assay, originally developed by Boyden (1962) to study leukocyte chemotaxis, is based on a chamber separated in two compartments by a filter with pores that allow active migration, but not passive diffusion of cells (usually 8 μm pores are used for adherent mammalian cells). Cellulose membranes (150 μm thick) have now been replaced by polycarbonate membranes (10 μm thick), that allow detection of cells that have crossed the filter, and not just migrated into it, thus improving the reliability of the quantitative analysis (Horwitz and Garrett, 1971). Modified Boyden chambers are now commercially available (Costar Transwell, Nunc and Falcon Inserts) with a variety of membranes of different pore size. Cells are plated in the upper compartment, and the chemotactic agent is added to the lower compartment, from which it diffuses through the filter to form a gradient that induces cell migration. After an appropriate incubation time, filters are fixed and stained, and the amount of cells migrated to the lower side of the membrane is determined. Chemotaxis can be distinguished from chemokinesis by exposing the cells to the attractant without a concentration gradient, and to a positive or negative gradient of the chemotactic agent (checkerboard analysis).

Materials
Two compartments Boyden chambers (commercially available as transwell filter plates). Fixative and cell staining solutions as described below;
microscope for visual and semiquantitative analysis;
titertek multiwell reader, or spectrophotometer for quantitative analysis.

Procedure
Cells are detached by EDTA or trypsin treatment (in the latter case a time of recovery in serum-containing medium might be needed), and resuspended at the desired concentration in the plating medium. If scope of the experiment is the identification of chemotactic agents, the use of serum-free medium containing BSA 0.5–1.0% will facilitate the further isolation of the attractant. Serum-containing medium, on the other hand, will improve attachment and spreading on the membrane, and might act synergistically with exogenous chemoattractants to enhance migratory effects. Serum concentration, of course, must be the same in the two chambers, to avoid chemotactic effects due to the serum itself. Cell density on the membrane has to be high enough to allow a detectable number of cells to migrate towards the lower face of the membrane, however not too high because, mostly in the case of cells of epithelial origin, intercellular interactions might prevent migration through the filter. In this case, a bell-shaped curve is expected when evaluating the amount of migrated cells vs. cell density.

Different dilutions of the chemoattractant in plating medium are placed into the lower compartment, and the chamber is incubated at $37°C$ for a period of time that usually ranges between 4 and 24 h. The assay is also suitable for the evaluation of haptotactic stimulation. In this case the attractant (usually extracellular membrane molecules) is insolubilized on the bottom surface of the filter by letting the filter float on the surface of a solution (either PBS or carbonate buffer, 50–100 mM, pH 9.6) containing the desired molecule at concentrations ranging between 10 and 120 μg/ml for 2–18 h at

37°C. Filters can be dried and stored until used, or directly rinsed with PBS and used in the assay. Vitronectin and collagen (Klemke et al., 1994), laminin (McCarthy et al., 1983) and thrombospondin (Taraboletti et al., 1987) have been used successfully in the assay.

After incubation with the cells, filters are gently rinsed with PBS, fixed and stained. Diff-Quik (available from American Scientific Products, McGraw Park, IL, USA, or from VWR Scientific Products, or from Dade Behring) contains both a dedicated fixative and two different dyes for staining and counterstaining. Otherwise methanol fixation (100% methanol, 5′ at r.t.) and hematoxylin/eosin staining (Terranova et al., 1989), or glutaraldehyde fixation (1% glutaraldehyde in PBS, 15′ at r.t.) and crystal violet staining (0.1% c.v. in water, 30 min at r.t.) (Rusciano et al., 1995) can be used. Other dyes can also be employed, as long as they stain cells and not the membrane.

Quantitation of migrated cells can be done microscopically or colorimetrically. If a microscope is going to be used, the filter is removed from its support, placed on a microscope slide bottom side down, and cells which have not migrated to the lower surface of the filter are removed by gently rubbing with a cotton swab. Cells which have migrated to the lower surface are visualized on a microscope at 400× magnification, and either several random fields are counted and averaged, or quantitation can be done with the aid of an image analysis system (such as Optomax Image System IV, Optomax Inc., Hollis, NH, USA) (Lester et al., 1991; Evans et al., 1991; Repesh et al., 1993). Alternatively, indirect quantitation can be obtained by solubilization of the dye (crystal violet staining is advisable) incorporated by cells migrated to the lower surface. In this case, after removal of cells from the top of the filter with the cotton swab, the filter is placed in a small volume of 0.2% Triton X100 in water overnight, and the blue color is then read at 590 nm (Rusciano et al., 1995).

Checkerboard analysis. The filter assay does not allow one to directly observe migration of cells, and track them to establish whether their locomotion is random or directed. Therefore, to distinguish whether a factor is chemotactic, chemokinetic or a combination of the two, the checkerboard assay was introduced (Zigmond and Hirsch, 1972). The method consists in exposing the cells to a range of attractant concentrations (I) in the absence of any gradient (i.e. the same concentration is in the two compartments of the chamber), and to a (II) positive (lower compartment > upper compartment) or (III) negative (lower compartment < upper compartment) gradient of the chemotactic molecule (Fig. 3.1).

Increasing values along the diagonal suggest positive chemokinesis; elevated values in II and no variation from control in III

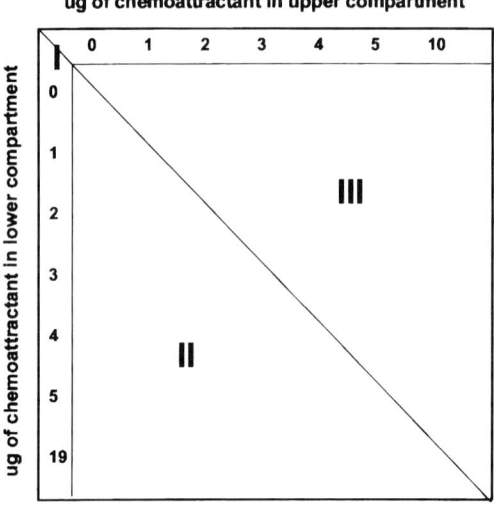

Checkerboard analysis of directed motility

Fig. 3.1. A scheme illustrating checkerboard analysis of directed cell motility. Three fields are identified in the scheme: (1) wells in field I contain an equal amount of chemoattractant in the upper and lower compartments; (2) wells in field II contain a higher concentration of chemoattractant in the lower than in the upper compartments; and (3) the reverse happens for wells in field III.

indicate positive chemotaxis (Terranova et al., 1989). Results, in fact, may show a higher complexity, since some agents can be both chemotactic and chemokinetic, and because some chemotactic agents at high concentrations can be inhibitory of locomotion.

Haptotaxis is properly defined as the directed migration of cells on an adhesive substratum gradient. Coating of one surface of a polycarbonate filter (which is 10 μm thick) with a solution of an adhesive molecule or peptide should theoretically generate a gradient through the filter, although this has not been directly measured. Moreover, molecules adsorbed to the lower surface of the filter might desorb during the course of the experiment (Grinnel, 1986), generating thus a chemotactic, beside the haptotactic gradient, further complicating the interpretation of the results.

An alternative method to evaluate haptotaxis has been developed (Brandley and Schnaar, 1989; Brandley et al., 1990). The method is based on the covalent immobilization of adhesive peptides on an inert, hydrated polyacrilamide gel surface. Adhesive peptides are covalently immobilized via stable, uncharged amide bonds, thus effectively eliminating desorption. Slope of the gradient (dc/dx), concentration of the immobilized molecule, and shape of the gradient (linear or exponential) can be defined through the use of commercially available gradient makers. A variety of ligands containing a primary amine (including peptides, proteins, glycoproteins, glycosides and polysaccharides) can be stably immobilized, and gradients can be quantitated either chemically or radiochemically. However, preparation of the immobilized ligand requires some chemical expertise, and might result tricky for a lab with no adequate background, so as microscopical or biochemical quantitation of the cells along the gradient, and the interpretation of the relative data. In fact, this method has not been widely used in the study of the haptotactic response, and laboratories still prefer using the microfilter assay.

3.1.1.3. 'Agarose droplet' and 'under agarose' assay

Although the presence of agarose in the test accommodates the two methods, which derive from a test originally described by Carpenter (1963), they differ in that the 'agarose droplet' measures outwards random migration of cells embedded into an agarose drop, whereas the 'under agarose' assay is based on the migration of cells under an agarose gel, and allows the distinction between random and directed migration stimulated by the presence of a chemoattractant.

Agarose droplet. A 2% stock solution of low melting point agarose (e.g. Bethesda Research Laboratories) is dissolved and sterilized by autoclaving. At the moment it is required for the experiment, the agarose solution is liquefied by boiling, and then cooled to 38°C in a water bath to keep it liquid. Cells are detached from tissue culture plates as usual, and resuspended at high density (3.33×10^7/ml) in their growth medium (with FCS, if required), however buffered with 25 mM Hepes pH 7.4 instead of sodium bicarbonate. Cell suspension and agarose stock solution are mixed together at a ratio 9:1 (e.g. 900 μl cells + 100 μl agarose) so to attain a final concentration of agarose of 0.2%, and immediately a 1 μl drop is placed on the bottom of a 96 multiwell plate for tissue culture. The test is performed at least in quadruplicate. Agarose drops are fully gelled by incubation of the plate in a refrigerator for 5–10 min, after which time 100–200 ml of cooled, regular growth medium, which may contain dilutions of factors to be tested for their chemokinetic effects, are added with care to avoid detaching the agarose droplet. Microtiter plates are then placed in a CO_2 incubator at 37°C, and daily inspected by phase contrast microscopy using an inverted tissue culture microscope. Time zero, no migration control plates must be considered after a few hours from the plating. When the extent of cell migration from the agarose droplet in the test plates is considered to be enough for the experimental purposes (usually between 2 and 5 days), a better visualization of migration can be obtained by fixing and staining the cells. In this case plates are treated with 4% formaldehyde in PBS, 30–60 min at r.t., and stained

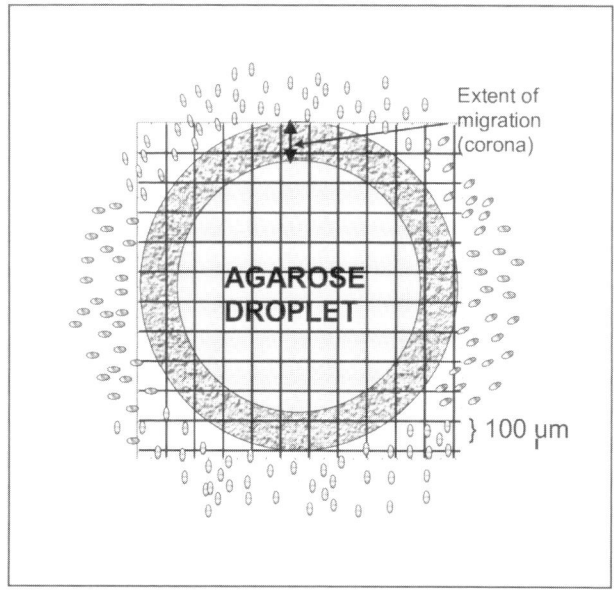

Fig. 3.2. The agarose droplet assay. The figure illustrates a hypothetical result of the assay, showing an almost homogeneous migration of cells outside the original droplet, that can be quantitated by measuring the amplitude of the corona. Some cells have migrated even further, but they are not taken into account for quantitation.

with 1% crystal violet in 95% ethanol for 30 min at r.t., destained under tap water, and air dried. Quantitation of the migration extent can be done visually at the microscope (Varani et al., 1978), using a calibrated grid in the eyepiece (Fig. 3.2). Area of migration within the corona can be estimated by aligning the grid in the eyepiece with the agarose droplet, and counting the number of grid spaces occupied by the corona. In addition, the number of isolated cells which have migrated beyond the edge of the corona can be counted as well.

Alternatively, wells can be photographed by transmitted light under a dissecting microscope. Total areas of cell outgrowth of both time zero control and test wells are digitized using a Sigma-Scan measurement system (Jandel Scientific, Corte Madera, CA, USA). After subtraction of time zero control areas from total outgrowth of test wells, net areas of outgrowth can be calculated (Akiyama et al., 1989).

Under agarose. This assay was originally developed to study leukocyte migration (Nelson et al., 1975); however, it has also been applied to the study of chemotactic and chemokinetic effects of FGF on endothelial cells (Stokes et al., 1990). Cells are allowed to migrate under an agarose gel in which a chemoattractant (or a control solution) forms a diffusion gradient. The differential migration of cells toward the chemoattractant is taken as a measure of its chemotactic activity.

A 1% stock solution of agarose (e.g. Difco) is dissolved and sterilized by autoclaving. After cooling to 45°C in a water bath, it is diluted 1 : 1 with prewarmed, twice concentrated complete growth medium, and 5 ml of this agarose medium are plated in a 60×15 mm tissue culture dish. After the agarose is solidified, a better hardening is obtained by keeping the dish in the refrigerator for 30–60 min, so that cutting of the wells is facilitated. Six series of three wells, each 2.4 mm in diameter, and spaced 2.4 mm apart can be cut in each plate using a sterile plexiglass template and a sterile stainless steel punch (Fig. 3.3A). Agarose plugs are removed with a needle, since use of vacuum may result in distortion of the wells. The use of a template is also recommended (Nelson et al., 1975) in that it allows reproducible alignment of the wells, and prevents alteration of the agarose-plastic interface, which can affect cell migration. Punching of the wells has to be done taking care not to scratch the plastic, since this could also interfere with cell migration. Alternatively, a different template can be made, which includes 18 punches arranged so to obtain the desired series of holes in the agarose dish.

A cell suspension is prepared in growth medium at a density of 1 to 3×10^7/ml, and 10 μl (1 to 3×10^5 cells) are plated in each center well of a triad. Outer wells receive 10 μl of serial dilutions of the chemoattractant, while the inner wells receive 10 μl of con-

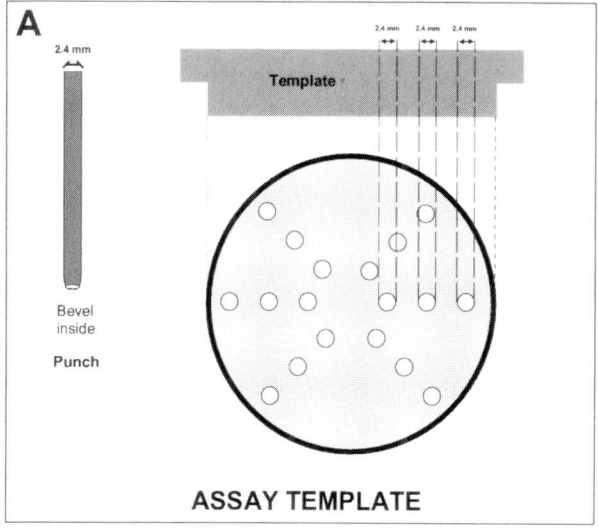
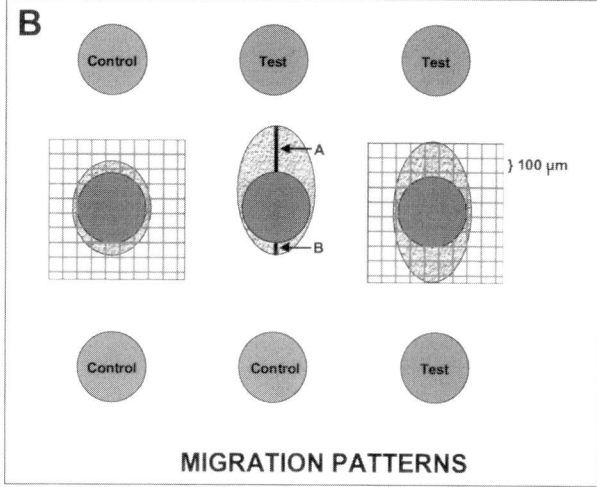

Fig. 3.3. The under agarose assay. (**A**) The illustration shows how to make a template in an agarose plate, that allows testing of six different combinations of chemoattractants. (**B**) The expected results from the under agarose assay: the extension of migration of the cells from the central well can be quantitated as for the droplet assay by using a calibrated grid under the microscope.

trol medium. The dish is then incubated at 37°C in a humidified incubator with 5–10% CO_2 for an appropriate length of time (from 18 to 72 h), after which cells are fixed with the agarose in place by adding 3 ml of methanol first, and 3 ml of 47% formaldehyde in PBS next, each for 30 min at r.t. If required by the timing of the experiment, cells can be left in methanol o.n. in the refrigerator. After fixation, the agarose gel is removed in one piece from the dish, and cells can be stained with hematoxylin/eosin, or crystal violet (see above), and air dried. Quantitation of migration can be done either microscopically, using a reticulum in the eyepiece, or on a photograph, by calculating both the ratio and the difference between the linear distance the cells have covered from the margin of the well toward the chemotactic factor (A) and the linear distance cells have covered toward the control well (B). A/B (chemotactic index) and A–B (chemotactic differential) are calculated per each replicate, and mean values and standard deviations are determine per each value. Quantitation of cell migration might also be expressed as the effective migration area, also estimated with a reticulum in the eyepiece, subtracting the area of migration toward the control well from the area of migration toward the chemoattractant. A better calculation can be obtained by digital image analysis of photographs, as illustrated before (agarose droplet assay). In this case the area of migration of cells between two control wells is subtracted from the area of migration of cells between one control well and one well containing the factor, or cells between two factor-containing wells (Fig. 3.3B).

The assay in this configuration, however, does not allow a clear distinction between chemotaxis and chemokinesis. The evaluation of a chemokinetic effect can be done by including the attractant in the agarose itself, so to see its effects on locomotion in the absence of a gradient (the same concentration of chemoattractant has to be present, of course, also in the cell plating medium). Moreover, a mathematical model to interpret the linear migration patterns of leukocytes in the under agarose assay has been developed by Lauffenburger et al. (1983), and adapted to endothelial cell migration by Stokes et al. (1990). The method has been designed to provide a rigorous quantitative assessment of the chemokinetic and chemotactic responses, where migration is evaluated in terms of the random motility and chemotaxis coefficients, μ and χ, which are defined in a mathematical model. The interested reader is referred to this paper and papers thereafter, for an exhaustive explanation of the mathematics underlying the model.

Finally, since under physiological conditions cells migrate not in agarose, but within an extracellular matrix meshwork, an experimental assay has been designed, similar to the agarose assays, but utilizing a collagen matrix instead of agarose (Yamada et al., 1990).

3.1.1.4. Migration track assays
There are two possible ways of detecting migration tracks produced by a moving cell: the phagokinetic track assay (Albrecht-Buehler, 1977a, b) and the extracellular matrix (ECM)-track assay (Bade and Nitzgen, 1985). Phagokinetic tracks are negative tracks, generated by moving cells after phagocytotic clearing of a surface homogeneously coated with protein-colloidal gold particles. In contrast, the positive ECM-track assay is based on the immunohistochemical detection of matrix proteins deposited by migrating cells onto the culture substrate.

Phagokinetic track assay. The procedure developed by Albrecht-Buehler (1977a, b) is described (a more detailed procedure appears in D. L. Spector, R. D. Goldman and L. A. Leinwand (eds.) 1998, *Cells: A Laboratory Manual*, CSHL Press, pp. 77.1–77.10).
1. Coverslips preparation. Cleaned glass coverslips are coated with a protein solution, necessary for the retention of gold particles. A solution of BSA (10 g/l) in double distilled water is prepared, filtered through 0.2 μm filters, and stored in the refrigerator. Coverslips are coated by a short immersion (few seconds) in the BSA solution, after which the excess BSA is drained by touching the edge on a paper towel. The coverslip is then immersed in high grade absolute ethanol, dried under the hot stream of an air dryer, and

placed in a suitable dish (depending on the size of the coverslip, 48 or 24 multiwell plates can be appropriate), where it can be stored until the colloidal gold solution is ready. The protein coating can also be done with other molecules, such as ECM proteins, if their effects on cell migration has to be considered (Volk et al., 1984). In this case the ethanol fixation step is omitted.

2. Gold particle preparation. To 11 ml of double distilled water stirred in a beaker on a heated magnetic stirrer, 1.8 ml of 14.5 mM $AuCl_4H$ and 6 ml of 36.5 mM Na_2CO_3 (both in double distilled water) are sequentially added, and the solution is immediately heated to the boiling point. As soon as boiling, 1.8 ml of a 0.1% formaldehyde solution in water are quickly added, and the heating is turned off. Gold particles form immediately, producing a brownish solution, which appears clear blue in transmitted light.

3. Coating of coverslips. This is done by layering the still hot (80–90°C) gold particle solution on top of the coverslip (2 ml for a 22 mm^2 coverslip), followed by 45 min incubation. After rinsing the coverslip 4–5 times with serum-containing growth medium, they are placed at the bottom of tissue culture dishes, sealing them with one drop of medium (if ECM coating of the coverslip has been used, serum should be omitted in the growth medium). No further sterilization of the coverslip appears to be required for up to 2 weeks of cell growth on particle-coated coverslips, most likely because the hot particle suspension sterilizes the test coverslips sufficiently (Albrecht-Buehler, 1977a).

4. Cell plating. Cells are plated at low density (100 cells/cm^2) in the desired culture medium, and let to grow in the incubator for one or more days, as required. Ingested particles do not seem to interfere with cell division, and they seem to be evenly split between daughter cells (Albrecht-Buehler, 1977a). At the end of each time point, the medium is removed from the well, and replaced with a 3.7% solution of formaldehyde in PBS, for 30 min at r.t., to fix cells. Coverslips are finally mounted on glass slides using an appropriate embedding medium (e.g. Elvanol, Du-Pont, Wilmington, Delaware).

5. *Microscopical examination.* Coverslips can be observed under dark field illumination. Tracks are best visible at low magnification (50–100X), with the gold particle coating appearing as a finely granulated light-grey background, whereas particle-free tracks appear black. Cells appear as bright white clusters, in consequence of particle clusters accumulated inside, or on the dorsal surface of cells. The size of the clusters, however, does not coincide with the cell size, which, under phase-contrast microscopy, appears larger than the cluster. The nucleus is always free of ingested particles, and thus may appear as a dark circular spot within the bright area. Further analysis of the tracks can be done on photographs, or directly at the microscope, with a connected image analyzer (e.g. Artek model 982, Farmingdale, NY) which allows to calculate geometrical parameters of the tracks (Mignatti et al., 1991).

ECM migration tracks. The assay is based on the immunohistochemical detection of ECM components deposited on the substratum by the cells while migrating (Bade and Nitzgen, 1985; Bade and Feindler, 1988). Laminin (Bade and Nitzgen, 1985), fibronectin (Bade and Feindler, 1988; Seebacher et al., 1988), plasminogen activator inhibitor-1 (PAI-1: Seebacher et al., 1992) have been used as ECM track markers.

Cells are plated either on coverslips or directly onto tissue culture dishes (e.g. Nunc SlideFlasks, in which the bottom of the flask is a removable microscope slide with a surface growth area of 9 cm^2) at low density (100–1000 cells/cm^2 depending on cell type and size), preferentially in serum-free medium, so to avoid the background given by ECM protein present in serum, which stick to the plating surface. After a suitable time of incubation at 37°C in a humidified CO_2 incubator (usually 2 days), cells are rinsed with PBS, and fixed (3.5% formaldehyde in PBS, 15 min at r.t.) on their support. After further rinsing with PBS, a permeabilization step (0.05% Triton X100, 5 min at r.t.) can be performed, if also intracellular antigens have to be evaluated. After rinsing again in PBS, antibodies against the protein(s) of choice to be detected in the migration tracks are added according to the manufacturer suggestion, incubating usually 30–60 min at r.t. The primary antibody is washed out by thorough rinsing in PBS, and a labeled secondary antibody (rhodamine and/or FITC can be used for double staining: in this case, of course, primary antibodies against the chosen antigens have to be raised in different animals) is added for microscopical detection. Stained preparations are sealed in a mounting medium (e.g. Gelvatol minimizes fluorescence decay), and stored at 4°C or −20°C (depending on how long they have to wait) until they are examined at the microscope. Quantitative analysis of migration is done by scoring at least 1000 cells per preparation. This can be done directly at the microscope, either visually, or with the aid of an image analyzer, or it can be done by using photographs of serial fields. Migration can be expressed as the percentage of cells unambiguously connected with a well defined migration track, and the average length of migration tracks under different conditions can be estimated as well with the aid of a reticulum in the eyepiece, by image analysis or on photographs.

3.1.1.5. In vitro wound assay
This assay is based on the ability of some cell types, under certain circumstances, to migrate into the empty space produced in a cell monolayer by an appropriate cell scraper (Lipton et al., 1971; Bürk, 1973). It allows evaluation of chemokinetics, or at most haptotactic effects if the denuded surface retains ECM molecules.

Cells to be tested are grown in a tissue culture dish (not a flask, which would be difficult to scrape) until confluent. Most of the growth medium is removed, and an artificial linear wound is made in the monolayer, by using, e.g. a sterile pipet tip, or a 'rubber policeman'. Care has to be taken to clean all the wounded area from cells, because in case an uneven scraper is used, which leaves behind some attached cells, their presence at the end of the experiment may complicate the interpretation of the results. The plate is then rinsed several times with PBS to remove all the debris, and the test medium (containing dilutions of the supposed migration-stimulating agent) is finally added to the wounded culture, which is then placed back in the incubator. The refilling of the wound can be checked at the microscope at different times, and usually the experiment is terminated at 24 h. The dish is rinsed with PBS, and cells are fixed (e.g. 1% glutaraldehyde in PBS for 15 min), and stained (e.g. 0.1% c.v. in water, 30 min at r.t.). If a thin wound was produced (e.g. by scraping with a 100 μl plastic tip), it may be expected to see it either unchanged or completely filled by cells respectively kept under nonmigrating, or migrating conditions. If a thicker wound was produced with a rubber policeman, the number of cells migrating into the empty space may be counted in a series of randomly selected areas, so to obtain quantitative results. Alternatively, photographs of the wounded area can be taken at different times during the 24 h incubation period, and the distance between the two boundaries of cells measured on pictures of identical magnification. Cell migration is then expressed as the distance cells have moved into the acellular area over time (Chen et al., 1994), as schematically illustrated in Fig. 3.4.

Cell growth during the time of the assay might influence the results, in the sense that the filling of the artificial wound could be contributed from both cell replication and movement. In order to avoid this problem, the assay should be conducted in serum-free or at least low-serum conditions, to limit cell growth. Alternatively, inhibitors of DNA synthesis (mitomycin-C or aphidicolin at 0.5 μg/ml), can be included in the culture medium, since it has been shown that active migration may depend on protein, but not DNA synthesis (Geimer and Bade, 1991; Chen et al., 1994).

3.1.1.6. Scatter assay
Colony scattering is probably the easiest assay for detecting factors able to induce cell movement. It was originally described by Stoker and Perryman (1985), to detect a scattering activity for MDCK cells present in fibroblast conditioned medium.

Cells suitable to be used in the scatter assay are those that have a tendency to grow in packed colonies, such as epithelial cells (e.g.

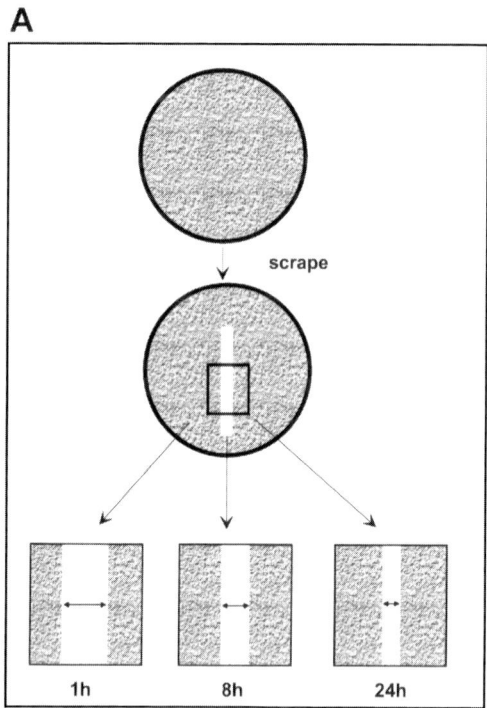

Fig. 3.4. The scrape motility assay. (**A**) An artificial wound is made in a confluent petri dish, that cleans a defined area from cells. The kinetics of recolonization of the area is recorded for the next 24 h, and gives an estimate of motility for the cell population under consideration. (**B**) A circular wound has been produced in a sheet of epithelial cells, in triplicate wells, and repopulation of the wounded area has been recorded for the next 48 h with a microscope equipped with a Hamamatsu CCD camera with a Sigma 28–70 mm lens, and images digitalized on a connected microcomputer. (**C**) Images were processed and analyzed using Image-1 (Universal Imaging Corporation), and calibration was done by using the circular groove cut into the well during wounding, which has a known diameter, as an internal standard for each image. (Photograph courtesy of Dr Michael Berman, CibaVision, Basel.)

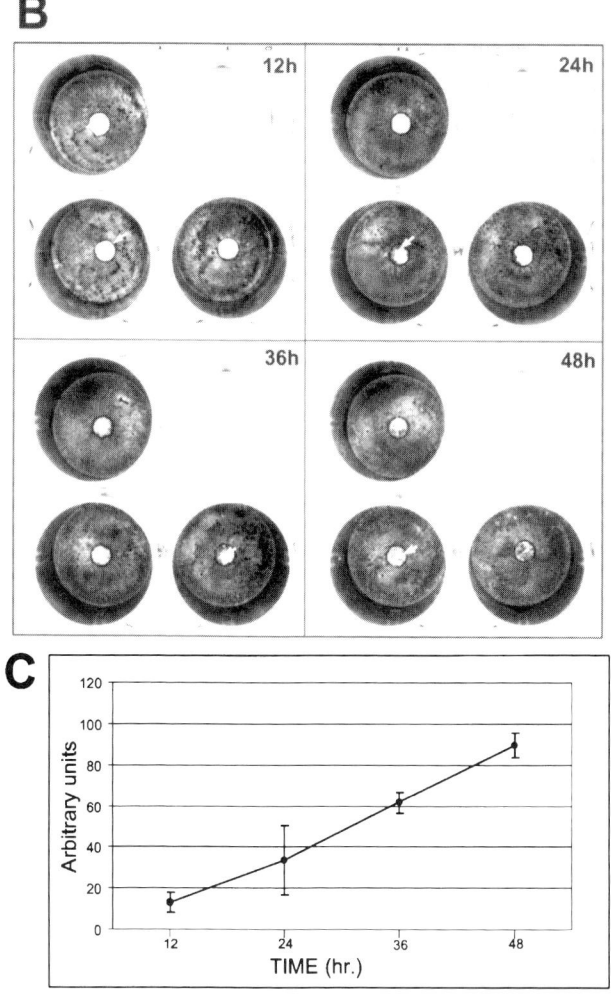

Fig. 3.4. (**B**) and (**C**).

MDCK cells are typically used as test cells in this assay), although also other cell types, e.g. melanoma cells (Rusciano et al., 1995) can be used, provided they are responsive to the tested factor.

In a typical assay, cells are plated at low density (100 or 200/cm^2) in 96 or 48 multiwell plates, in order to allow them to grow as isolated colonies. At this point, molecule(s) to be tested for their scattering activity are added to the wells at different dilutions, and the plates returned to the incubator. The scattering effect is usually visible about 6 h later, and is manifested as a dispersion of the colony, with the cells at the periphery of the colony migrating out and taking a fibroblastic phenotype. Quantitative analysis of scattering can be done by comparing the size of some preselected colonies at different time points (e.g. 0, 6, 12 and 24 h). Areas occupied by the same colony during these times are measured on photographs. Colony scatter is then defined as the increase in colony area over time (Chen et al., 1994). Alternatively, solely for qualitative results, cells plated at low density, can be treated soon after attachment to the dish with the scatter factor, so that colonies developing in the absence or in the presence of the factor can be compared for their morphology after a few days of growth (Fig. 3.5) (Rusciano et al., 1995).

3.2. The role of active and passive deformability in invasion and resistance to shear stress forces in the blood stream

Both active and passive cell deformations are required for successful migration of metastatic cells from the primary tumor into target organ(s). In fact, tumor cells in the vasculature must be resistant to passive deformation derived by frictional forces arising at contact points with vessel walls (Weiss and Dimitrov, 1984), and by shear stress caused on the one hand by turbulent flow (Brooks, 1984), and on the other hand by their transit through capillaries which generally are rigid and smaller in diameter than tumor cells (Zeidman, 1961; Weiss, 1976). Besides, active deformability is required when

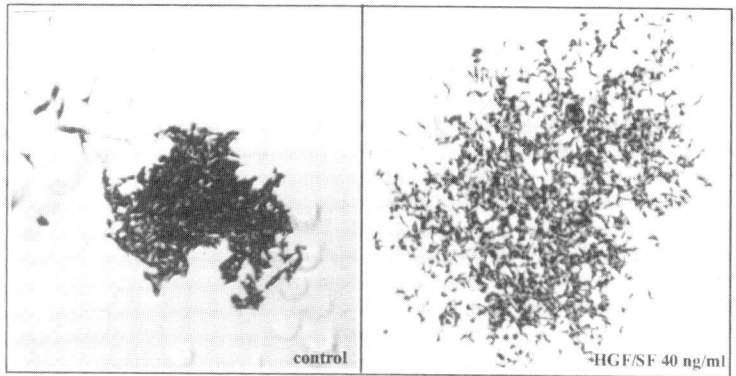

Scatter Assay

Fig. 3.5. Scatter assay. In the illustrated example, B16 melanoma cell colonies have been allowed to grow from few cells either in the absence (control) or in the presence (40 ng/ml) of HGF/SF. After fixation and staining with crystal violet, representative colonies have been photographed.

metastatic cells have to migrate outside the capillaries and within the organ tissue.

Furthermore, tumor cells (at least those derived from solid tumors) are not designed to withstand the circulatory stress, and cancer cells delivered to various target organs via the blood stream may be lethally damaged during their passage through the microvasculature (Weiss, 1990). It might thus be expected that highly deformable tumor cells could be more metastatic than their less deformable counterpart, and that treatments influencing cell deformability would also affect their metastatic potential. Moreover, since the initial arrest of circulating cancer cells in the target organ occurs at least in part by mechanical trapping in the microcirculation, different levels of deformability might also influence the recirculation of such tumor cells, and determine thus a different pattern of organ metastasis.

Furthering such considerations, Tullberg and Burger (1985) selected B16 melanoma tumor cell variants able to penetrate filter pores 10 times smaller than their own diameter. These size-selected tumor cells showed cytoskeletal alterations and an increased metastatic capacity after subcutaneous implantation into mice. When the deformability characteristics of these cell variants were estimated (Ochalek et al., 1988), it was found that they were related to the metastatic potential, i.e. the more metastatic cells were also the more deformable. Similar results had already been reported by Sato et al. (1977) with rat ascite hepatoma cell lines, and most recently Mittelman et al. (1994) made similar findings by using murine lymphoma cell lines.

The involvement of the cytoskeleton in cell deformability and metastatic potential is suggested by other studies in which either the cytoskeleton was disrupted by drug treatment (Hagmar and Ryd, 1977; Hart et al., 1980), or cells were selected on the basis of their degree of actin organization (Raz and Ben-Ze'ev, 1987). In either case, a correlation was found between metastatic ability and cytoskeletal organization, indicating that a loose cytoskeleton could favor crossing of the first capillary bed and seeding of viable cells to downstream organs (Hagmar and Ryd, 1977).

3.2.1. Evaluation of cell deformability

Passive deformability. Two methods were devised to measure cell properties that are related to their deformability. The micropipette aspiration method (Sato and Suzuki, 1976) consists in measuring the negative pressure required for a cell or a portion thereof to enter a pipet the opening of which is smaller than the diameter of the cell. This method is laborious and requires a skilled operator. Cells have to be measured one at a time, and since cell and capillary diameters must have a constant ratio, the sampling of a population is time consuming and requires a set of different pipettes. As a consequence, this method is not widely used, and will not be further described.

A more common alternative are filtration methods generally relying on the time it takes for a given volume of cell suspension to pass through micropore membranes at constant pressure (Sato et al., 1977). Although this method only gives average information on the cell population, and individual properties of the cell, such as fluid mechanical parameters cannot be deduced, it is simple, economical and well suited to address the question of deformability in relation to metastatic ability (Ochalek et al., 1988; Mittelman et al., 1994).

Materials

A 13-mm holder for Nucleopore filters equipped with a dual inlet valve (Nucleopore Co., Cambridge, MA);
nucleopore filters, 13 mm, 10 μm pore size;
fraction collector;
syringes (1 and 50 ml);
Coulter counter;
hemocytometer.

Procedure

Cell filterability is influenced by a variety of biological and technological factors (Nordt, 1983). Thus, in order to be reliable and reproducible, Nucleopore filtration techniques must fulfil certain criteria, and namely: (1) the most part of the input cells must be recovered in the filtrate; (2) cellular aggregation should be minimized by choosing conditions which permit relatively short filtration times; (3) cell viability should be high and not lost on filtration; (4) cell size distribution should not be influenced by filtration; and (5) differences in cell to filter and cell to cell adhesion of different cell lines should not be responsible for differences in filterability. In order to fulfil these criteria, experimental parameters such as cell to pore ratio, filtration pressure and cell culture conditions have to be standardized. The following optimal conditions have been established for filtration of B16 melanoma cells (mean cell diameter 17.4 $\pm$ 0.21 μm, mean diameter of cell nuclei 9.8 $\pm$ 0.27 μm): 20 cm H_2O driving pressure, cell-to-pore ratio 1 : 1, temperature 22°C (Ochalek et al., 1988). Care has to be taken to derive tumor cells from similar culture conditions, since cell density has been found to influence filterability.

Nucleopore filters from the same lot should be used throughout the experiments. While pore diameter is uniform, there might be considerable variability in pore density. Therefore, filters should be checked before use by passing a certain amount of medium (20–50 ml), and measuring the time it takes to pass through: this is proportional to the total pore area. Only filters giving similar flow values should be used in the experiments.

Cells are harvested from the dish by EDTA treatment, resuspended and rinsed in PBS-CMF, and finally resuspended at the desired concentration in serum free medium buffered at pH 7.2 with Hepes and filtered through 0.22 μm Millipore membranes. Cell viability is checked by trypan blue dye exclusion test, and aggregates, if present, can be removed by filtration through cell strainers (Costar).

A 20-ml suspension with an appropriate cell concentration is prepared. Cells are counted and sized at the Coulter counter. Nucleopore filters are placed in the filter holder (the active filtration area of a 13 mm filter is 19.6 mm^2, corresponding to roughly 1.96×10^4 pores), and the chamber of the holder is filled via the lateral inlet with 1 ml of the cell suspension. Soon after, the valve is connected to the reservoir containing 20–50 ml of filtered medium, and pressure is kept constant during filtration by dropwise addition of fresh medium to the reservoir (Fig. 3.6A). Fractions of filtrate are collected in siliconized tubes at timed intervals (around 2 sec) using a fraction collector, and cell numbers immediately determined. A cumulated percentage curve as a function of time for each cell line to be analyzed is constructed from the fraction counts. Viability of the cells in filtrates is checked by combining fractions, centrifuging and resuspending the pellet in 0.3 ml of 0.2% trypan blue in PBS. Control cells, not subjected to filtration, are treated similarly, and the results compared.

The above conditions allow 90% recovery of the input cells, and differences in filterability of cells among different lines appear both in filtration time and filtration kinetics. All the filtration curves exhibit steep slopes in the beginning and then decrease, so that 50%

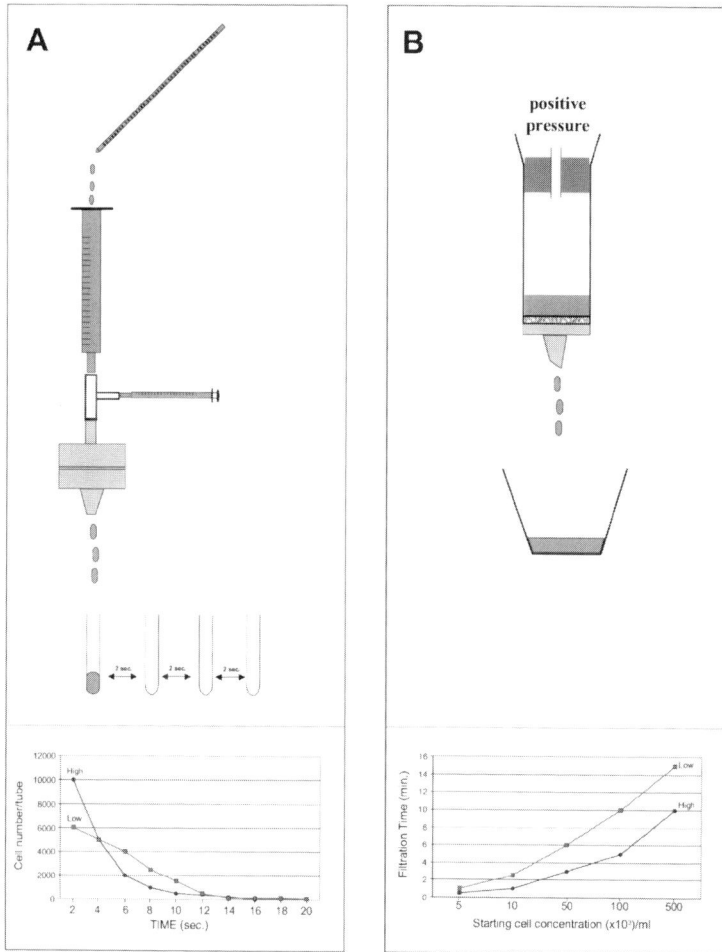

Fig. 3.6. Deformability measured as filterability. (**A**) Cells passing through the pores of the filter are collected in siliconized tubes at 2 sec intervals. Cells in each tube are counted, and a curve generated of cell N vs. time. The steeper the curve, the more deformable the cells. (**B**) The time for a defined cell concentration to pass through the filter at constant pressure is recorded, and a curve is generated as a function of variable cell concentrations. HIGH and LOW in the graph legend refer to more and less deformable cells.

of the cell input passes through the filter after 10–50 sec, i.e. more than 10 times shorter than the total filtration time (Ochalek et al., 1988).

Alternatively, measurement of cell filterability can be estimated by recording the passage time of cell suspensions at different concentrations, generating curves of filterability vs. cell concentration (Sato et al., 1977; Mittelman et al., 1994). The set up in this case consists of a Nucleopore filter holder which is loaded with the cell suspension, and to which a constant positive pressure (between 5 and 24 cm H_2O) is applied (Fig. 3.6B). Filterability is then defined as the ratio between the filtration rate of the cell suspension (V_c) and the filtration rate of the same volume of medium alone (V_m). Since the input volume for different cell suspensions is the same, then filterability is proportional to the filtration time of the cell suspension (t_c) vs. medium alone (t_m):

$$\text{Filterability} = V_c/V_m = t_c/t_m$$

A rigidity index (RI) can also be estimated, which is identical to the filterability index of red blood cells (Hanss, 1983):

$$RI = (t_c - t_m)/t_m \times (1/C)$$

where C is the concentration of cells in suspension.

Since the filtration rate through a Nucleopore filter is sensitive to the ratio cell diameter/pore diameter (Kikuchi, 1991), such ratio should be maintained constant when comparing different cell lines which differ in size.

Active deformability. If the capability to squeeze through tissues is an important step for most metastatic cells, then a selection based on the ability of cells to cross filter holes about ten times smaller than their cellular diameter should theoretically yield cells with a higher metastatic capability (Tullberg and Burger, 1985). The selection procedure is described in Chapter 6.

CHAPTER 4

Extracellular matrix degradation and invasion

In this chapter, an account is given of well established and new methods that allow insight into the capacity of cells (tumor, stromal or immune) to degrade components of the extracellular matrix (ECM) and then overcome the anatomical barriers they face during invasion.

Key factors in the complex process of invasion and metastasis are proteolysis of the local connective tissue stroma and locomotion into the region of the matrix modified by proteolysis.

Hydrolysis of ECM molecules has been identified in invasive tumors, and a variety of enzymatic activities have been studied that may be involved in the degradation of ECM: a few examples are gelatinases, urokinases, glycosidases, and cathepsins. (A number of enzymes active on ECM are illustrated in Table 4.2.) While all may play a role in matrix lysis, for some a correlation has been demonstrated between their level of expression and the invasive and metastatic aggressiveness of tumor cells. Measuring markers of potential invasion should then offer crucial insight in pharmacological and clinical screening and monitoring.

However, only further analysis can give a measure of the actual invasion, since a series of additional steps must be completed by the cells in order for them to eventually arrest and grow in an appropriate site, often different and distant from the primary tumor (metastasis). Examples are: cycles of attachment to matrix molecules or fragments—traction on them, release of contacts, intravasation into the blood stream, evasion of immune attacks, arrest onto the endothelium, extravasation, proliferation, etc.

A number of methods are available today for investigating the various aspects of this process of matrix degradation and tissue invasion. Study of degradation aims to identify and quantitate at mRNA or protein level a series of markers involved in the disruption of ECM molecules. These markers—hydrolases, their activators, inhibitors, receptors, degradation products, etc.—may be found to be involved or even crucial to invasive and metastatic aggressiveness.

Techniques for studying invasion—measurement of cell invasion in vitro and/or in vivo, and metastatic potential in syngenic or immunodeficient animals—are designed to verify the steps critical to the invasive phenotype, and to quantitate the eventual invasive and metastatic aggressiveness.

Obviously, care must be taken when choosing a method and interpreting the results: for example, a high level of specific enzyme can imply an increased invasion in vivo, but the enzyme might be counterbalanced by overexpression of a specific inhibitor, lack of activation, or increased disruption. For this reason, a complete study of potential tumor aggressiveness should incorporate examination of a very wide range of related molecules (receptors, ligands, activators, proteases, inhibitors, etc.), in parallel with in vitro migration and in vivo invasion assays. Tailoring these latter assays allows study of specific steps involved in invasion, such as those listed above.

In the following sections, we will describe in detail some of the methods to study both invasion-related properties of cells, and their actual invasive behavior.

4.1. Degradation

A number of methods are available today to investigate the various aspects of this process of matrix degradation and tissue invasion. A first line of analysis addresses questions such as:

- which hydrolases are involved in the invasive process?
- do these hydrolases correlate with tumor aggressiveness?
- what cells are producing these hydrolases?
- are the hydrolases active, are they blocked by inhibitors?
- may other hydrolases act on the same substrate?

Answers to these questions are obtained through measurement of expression of specific molecules at mRNA or protein level.

4.1.1. Substrate-capture ELISA

This is a low complexity assay, which allows one to detect and quantitate diluted antigens (proteases) capable of binding to specific substrates. However, it cannot give information about whether, or not, the antigen is intact and functional.

To quantify an antigen related to invasion (hydrolase, inhibitor, activator, etc.), commonly used procedures are Western blotting and ELISA. The only requirement is the availability of a specific antibody. However, a positive reaction does not necessarily indicate that the antigenic molecule is intact, able to bind to its substrate and be active on it. In fact, while both procedures may also reveal impaired antigens, the second does not allow distinction between complete and fragmented antigens.

ELISA is also of limited application when the antigen is extremely diluted, such that other molecules interfere with the performance of the technique. One approach in this case is the so-called 'substrate-capture ELISA'. Here, a substrate for a protease, for example, is bound to the plastic plate in order to absorb and concentrate the specific enzyme for detection by immune reaction. The first modification of a standard capture assay technique in which a metalloprotease substrate is used to capture the enzyme of interest was described by Wacher et al. in 1990, and is summarized in generalized form below. The substrate-capture ELISA 'greatly simplifies the mixture in which the enzyme is detected and removes potentially interfering substances, thus avoiding some of the difficulties inherent in the usual assay protocol requiring coating of antigen or antibody directly onto the solid phase'. This assay has been used to detect type IV collagenase in lung cancer patients (Garbisa et al., 1992), and can be used (with appropriate choice of substrate) both in basic research and for screening and clinical follow-up for other antigens related to the invasive phenotype. This method can detect as little as 50 ng of type IV collagenase corresponding to 0.03 ng/μl (given a well loading of 150 μl).

Materials
Gelatin;
TSE buffer: 0.05 M Tris-HCl, 0.2 M NaCl, pH 7.6, 10 μM EDTA;
antibody buffer: 3% BSA in PBS, 0.1% Tween-20, pH 7.3;
control samples: hrMMP-2 or hrMMP-9.

Procedure
Gelatin is dissolved to 1% in the TSE buffer by warming to 55°C in a water bath, and allowed to cool to room temperature. 300 μl of gelatin solution are dispensed into microtiter plate wells, and allowed to incubate overnight at r.t. The next day wells are emptied by inverting the plate, which is then chilled to 4°C. Wells are rinsed twice with TSE buffer, and then 150 μl of antigen (enzyme) solution are added to each well (serial dilutions in the TSE buffer). Binding of the enzyme to its substrate is allowed to occur for 1 h at 4°C. The wells are emptied by inversion and washed twice with TSE buffer. A dilution (according to the titre) of the antibody specific for the protease is added in 150 μl (if using monoclonal antibodies, these do not have to be directed against epitopes involved in binding to the substrate), and incubation carried out for 3 h in the cold, after which time the plate is rinsed twice with TSE buffer. Next, 150 μl of 0.5 μg/ml solution of the second antibody-peroxidase conjugate are added, and incubation continued for 3 h at 4°C. The plate is then washed twice with TSE buffer and soaked for 5 min in the same buffer. After emptying the plate, and removing all the liquid

by gentle tapping on a paper towel, color development reagents are added, and developed until a reasonable amount of color is detectable (usually 10 min should suffice). The plate is then scanned at a microplate reader at the required wavelength.

4.1.2. Hydrolase assays

These are a series of low complexity assays that are used for: (a) detection and quantitation of enzymatic activities (extracted from tissues or cell layers, or released in culture medium); (b) quantitation of level of enzyme activation (in some cases); (c) assessment of enzyme inhibition by new compounds; and (d) measurement of tissue inhibitors in a given sample. They do not allow one to see whether, or not, the enzymatic activities are blocked by inhibitors.

4.1.2.1. Zymography

Zymography easily overcomes the limitations of the ELISA technique concerning the functionality of the detected molecules. When enzymes (without prior reduction and boiling) are electrophoresed in gels where their substrate is co-polymerized with polyacrylamide, they leave traces of their activity under suitable conditions. The removal of SDS used for electrophoresis by nonionic detergent and the use of appropriate gel-incubation buffer allows the enzymatic molecules to recover their function and degrade the substrate; after staining, the gel shows clear digestion bands against the background.

Heussen and Dowdle in 1980 first described this method for detecting plasminogen activator (PA) through the use of polyacrylamide gels containing co-polymerized plasminogen and gelatin: PA releases plasmin which degrades gelatin. This technique can be used to detect as little as 1 mU of uPA, while omission of plasminogen from the gel allows detection of plasminogen-independent gelatinolytic activities. A variety of suitable substrates (e.g. for elastase) may be co-polymerized into the gel.

In some cases, such as that of gelatinases, the method not only reveals the activated form of the enzyme but also the zymogen, giving rise to a closely spaced doublet of digestion. This counterintuitive outcome results from the electrophoresis conditions, which may expose binding and active sites of the enzyme without loss of the propeptide. Densitometric analysis of each band of the doublet may then be used to measure the activation level of a protease.

If after electrophoresis the gel is incubated in a suitable buffer containing a known or unknown inhibitor, zymographic analysis may also prove useful for assessing the inhibitor's activity, and for determining its IC50. In this case, allow full penetration of the inhibitor into the gel at 4°C before bringing the temperature to 37°C.

However, since the SDS-PAGE mostly dissociates enzyme-inhibitor complexes, the results of zymography fail to give information on the level of the true enzymatic activity of the sample. An example of application of this technique is given below for detection of gelatinolytic activities (Fig. 4.1).

Materials
Enzyme substrate: 1% gelatin (wt/vol) in distilled water;
Enzyme buffer: 50 mM Tris-HCl, pH 7.4, 0.02 M NaCl, 5 mM $CaCl_2$, 0.02% (wt/vol) Brij-35;
Control sample: 5–20 μl serum-free medium conditioned by HT1080 human fibrosarcoma, SK-N-BE human neuroblastoma, or other suitable cell lines.

Procedure
Prepare a 9% polyacrylamide separating gel containing 100 μl/ml gelatin with 3% (wt/vol) stacking gel. Apply samples in sample buffer, without prior heating or reduction. Remove SDS from the gel by incubation in 2.5% (vol/vol) Triton X100 for 30 min. Incubate the gel at 37°C overnight in enzyme buffer. Stain the gel for 30 min in staining solution and de-stain in the same solution without dye.

Enzymatic (gelatinolytic) activities will appear as clear bands against the blue background of stained substrate (gelatin). For sam-

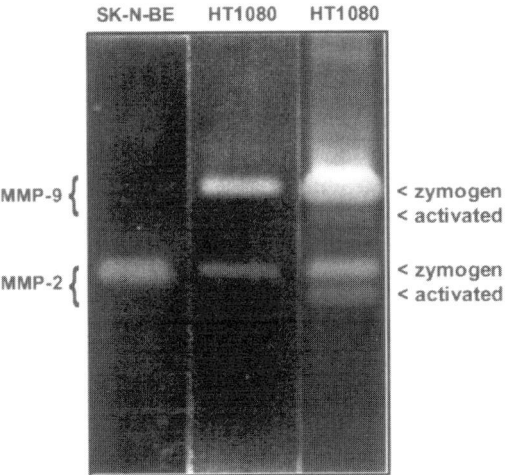

Fig. 4.1. Gelatin zymography of 30 μl culture medium conditioned by human HT1080 fibrosarcoma and SK-N-BE neuroblastoma cells. Top bands correspond to MMP-9 gelatinase, lower bands to MMP-2 gelatinase. HT108 conditioned medium was collected after 6 h (central lane), or 24 h (right lane) of incubation: a longer incubation time results in activation of the pro-enzyme (unpublished, Garbisa et al.).

ples within the linear response range, the bands can be measured and compared using an image analyser with densitometric software.

4.1.2.2. Inverse zymography

Zymography may also be used to study presence and amount of enzyme inhibitors in a sample. In this case, after electrophoresis, the gel is incubated at 37°C in a buffer containing the proteolytic enzyme for which a blocking inhibitor is looked for. The enzyme will digest the co-polymerized substrate, apart from where the band containing the inhibitor has migrated. However, failure of this method may result for several reasons, but mainly because of the masking effects of other proteins migrating with the inhibitor.

4.1.2.3. Solid-phase substrate assay for glycosidases

Invasive cells have to confront several extracellular matrix components including proteoglycans, such as heparan sulphate proteoglycan (HSPG). Degradation of HSPG is involved in tumor cell invasion: heparanase can specifically cleave the HS side chains of HSPGs. Several studies have shown that highly metastatic cells produce high amounts of heparanase compared to the nonmetastatic counterparts, and that invasion can be prevented by specific inhibitors of heparanase.

Various biochemical methods are available for measuring heparanase and other glycosidase activities. The former can be assayed by a method that makes use of a solid-phase substrate for heparanase, such as chemically-modified heparan sulphate coupled covalently at its reducing terminal saccharide to agarose gel beads. The reader is referred to Nakagima et al. (1986).

4.1.3. Northern blotting and RT-PCR

Northern blotting permits detection and quantitation of mRNA for invasion-related molecules. It does not give information, though, about the efficiency of mRNA transcription into protein, and the activation level of enzymatic proteins.

Northern blotting can be applied to reveal mRNA levels in treated vs. untreated cell cultures, as well as in tumor specimens vs. normal homologous counterparts. Overexpression of proteases and their inhibitors has been demonstrated using Northern in transformed tissues (Caenazzo et al., 1998) and treated tumor cell cultures (Negro et al., 1997), respectively.

Given the incomplete information obtainable with this technique, the analysis should involve an extensive number of invasion-related antigens (proteases, receptors, activators, inhibitors).

Useful probes for some invasion-related-molecules mRNA can be obtained by PCR procedure from specific cDNAs (a list of primers with the expected size of the amplified region is given in Table 4.1).

4.1.4. Degradation of immobilized matrix component by living cells in vitro

The assays described so far only allows one to tell whether a certain enzymatic activity is present or not in a given cell type, but do not reveal anything about its real involvement in the matrix degradation process. Therefore, they need to be complemented by another set of low complexity assays, designed to investigate the actual substrate degradation ability of living cells in culture. As a drawback, this type of assay does not yield the exact identity of the proteolytic enzymes involved, unless specific inhibitors are used and shown to prevent the digestion of substrate.

A first assessment of a cell's capacity to degrade a specific ECM molecule can be made by measuring the amount of degradation products released from a radiolabeled substrate attached to plastic culture wells. Living tumor cells are then inoculated into the coated wells and incubated at 37°C. Fragments may be excised from attached labeled molecules under the action of cell proteolytic activities: the level of degradation of the substrate is given by measuring the soluble radioactive fragments in the media.

While this approach can give immediate indication of whether, or not, a specific ECM component may be degraded by living cells, the use of noncytotoxic inhibitors during incubation is required in order to be able to assign the activity to a specific class (metallo-, serin-, cystin-, aspartic and glyco-) (Table 4.2).

The procedure to measure type IV collagen degradation is detailed below (Garbisa et al., 1980a). The technique can be easily adapted to other matrix components, single or in combination.

TABLE 4.1

α1(IV)COLL (Soininen et al., 1989)	
sense	5′-TTTGCATCACGAAATGACTAC-3′
antisense	5′-AAGGTGGACGGCGTAGGCTTC-3′
dimension/annealing/amplif.	413 bp/56°C for 60 s/25 cycles
MMP-1 (Smith et al., 1990)	
sense	5′-ACCCCAAGGACATCTACAGC-3′
antisense	5′-CACCTTCTTTGGACTCACACC-3′
dimension/annealing/amplif.	381 bp/56°C for 60 s/32 cycles
MMP-2 (Collier et al., 1988)	
sense	5′-ACCTGGATGCCGTCGTGGAC-3′
antisense	5′-TGTGGCAGCACCAGGGCAGC-3′
dimension/annealing/amplif.	447 bp/62°C for 30 s/28 cycles
MMP-9 (Wilhelm et al., 1989)	
sense	5′-GGTCCCCCCACTGCTGGCCCTTCTACGGCC-3′
antisense	5′-CCTTTCCCTCCTCACCTCCAC-3′
dimension/annealing/amplif.	762 bp/56°C for 60 s/32 cycles
TIMP-1 (Carmichael et al., 1986)	
sense	5′-TGCACCTGTGTCCCACCCCACCCACAGACG-3′
antisense	5′-GGCTATCTGGGACCGCAGGGACTGCCAGGT-3′
dimension/annealing/amplif.	552 bp/56°C for 60 s/32 cycles
TIMP-2 (Stetler et al., 1990)	
sense	5′-TGCAGCTGCTCCCCGGTGCAC-3′
antisense	5′-TTATGGGTCCTCGATGTCGAG-3′
dimension/annealing/amplif.	590 bp/62°C for 30 s/27 cycles
MT1-MMP (Sato et al., 1994)	
sense	5′-CAGCAACTTTATGGGGGTGAGTC-3′
antisense	5′-GGTTCTACCTTCAGGTTCTGG-3′
dimension/annealing/amplif.	647 bp/62°C for 60 s/27 cycles
uPA (Duffy et al., 1996)	
sense	5′-GTCGTGGACTACATCGTCTACCTG-3′
antisense	5′-CCATTCTCTTCCTTGGTGTGACTG-3′
dimension/annealing/amplif.	564 bp/62°C for 30 s/30 cycles
TGF-β1 (Derynck et al., 1985)	
sense	5′-GCCCTGGACACCAACTATTGCT-3′
antisense	5′-AGGCTCCAAATGTAGGGGCAGG-3′
dimension/annealing/amplif.	161 bp/60°C for 45 s/35 cycles
GAPDH (Ercolani et al., 1988)	
sense	5′-ACCACAGTCCATGCCATCAC-3′
antisense	5′-TCCACCACCCTGTTGCTGTA-3′
dimension/annealing/amplif.	450 bp/60°C for 30 s/25 cycles

TABLE 4.2

Enzymatic activities on extracellular matrix molecules

Class	Zymogen kDa		Current name (alph. order)	Abbrev.	Collagens	Substrates Others	Inhibitors
Met	57	105	bacterial collagenase	BC	all native		
		55	interst. collagenase	MMP-1	I II III VII VIII X OF		
		72	gelatinase A	MMP-2	gel I II III IV V VII X XI	Ag CS El En Fn Ga La PG Vn	
		57	stromelysin-1, transin-1	MMP-3	nh II III IV IX	Ag Fn La PG	
		28	pump-1/matrilysin	MMP-7	nh IV X	Ag El Fn La PG	α2-M.Globulin
		75	PMN collagenase	MMP-8	I II III V VII VIII X		cysteine
		92	gelatmase B	MMP-9	gel IV V VII X XIV OF	Ag El En Fn Ga La PG	doxorubicin
		57	stromelysin-2, transin-2	MMP-10	nh III IV V	Ag El Fn La PG	DTT
		51	stromelysin-3	MMP-11		Ag Fn La α1-antitrypsin	EDTA
		54	MΦ-met.-elastase (MME)	MMP-12	gel IV	El Fi > Fn PG	EGTA
		60	collagenase-3	MMP-13	gel I II III IV	Ag	mercaptoethanol
		66	memb-type 1 MMP(-14)	MT1-MMP	gel I II III	Ag El Fn La MMP-2 MMP-13	o-penanthroline
		72	memb-type 2 MMP(-15)	MT2-MMP	gel	Fn La MMP-2	phosphoramidon
		64	memb-type 3 MMP(-16)	MT3-MMP		MMP-2	SC44463
		57	memb-type 4 MMP(-17)	MT4-MMP			TIMP-1, -2, -3, -4
		70	Xenopus collagenas-4	MMP-18			
		54	RASI-1	MMP-19			
		54	enamelysin	MMP-20		Amelogenin	
	97	170	gelatinase C (seprase)		gel		
	85	340	meprin		gel IV	Fn La	
Ser	30	36	cathepsin G	cathG	I	El Fn La PG	α1-prot. inhibitor
			chym.-like neu.protease		II V	El	α2-M.globulin
		110	dipeptidyl-peptidase IV	DPPIV	X-Pro-		α2-plasmin inhib.

TABLE 4.2
(continued)

Class	Zymogen kDa	Current name (alph. order)	Abbrev.	Collagens	Substrates Others	Inhibitors
Ser	33	elastase	El.ase	gel IV	El MMP-2 MMP-3, uPA	anti-TH III
	36	fibroblast activ.prot.α	FAPα			aprotinin
	95	guanidino-benzoatase	GBase	gel		benzamidin
	55	plasmin	PL	V	Fi Fn La PG	hirudin, leupeptin
	85	seprase		gel		PAI-1, -2, -3
	97	serin-like Fn.ase				PMP-C, PMSF
	120	thrombin	TH		Fn	
	37	tissue-type pl.^{gen} activ.	tPA	gel	Fn La	protease nexin I
	28	try.like neu.protease		I	Fi Pg	SBTI, serpins
	70	urok.-type pl.^{gen} activ.	uPA			squash inhib.
	47				Fi Pg	TFPI-2, TPCK
Cys	25+	cathepsin B	cath B	I II IV V IX XI	Fn La PG	α2-M,globulin
	5	cathepsin B-like	cath B$_{like}$			antipain
		dipeptidyl peptidase I	cath C		serin pro- enzymes	cystatins A, B* (I)
		cathepsin F	cath F			cystatin C (II)
	28	cathepsin H	cath H	I II IV V IX XI	Fn La PG	E64D
		cathepsin K	cath K		El	kininogens (III)
	21+5	cathepsin L	cath L	I II IV V IX XI	El Fn La PG	LHVS
		cathepsin N	cath N	N-ter I II III		NEM
		cathepsin O	cath O			serpin Crm A
		cathepsin S	cath S		El	serpin SSCA
	170	cystin-like gelatinase		gel		
Asp	34	cathepsin D	cath D		Fn HG La PG	neutral pH
	34	pepsin				pepstatin

TABLE 4.2
(continued)

Class	Zymogen kDa	Current name (alph. order)	Abbrev.	Substrates Collagens	Others	Inhibitors
Gly	250	acetyl-glucosaminidase			HA HE HS	α2-M. globulin
	100	heparinase			He HS	Ca^{++}, Fe^{++}, Fe^{+++} Zn^{++}
	70	galactosidase			KS	carrageenans
	270	glucuronidase			CS DS HA HE HS	dextran sulfate
	60	hyaluronidase			CS HA	fucoidan
	82	iduronidase			DS HE HS	heparin derivatives
	120	chondroitinase ABC			CS DS	serum
	40	chondroitinase AC			HA CS DS	sulfated chitin
		chondroitinase B			DS	suramin
	43	heparinase			HE	
	55	hyaluronidase			HA AMPS	

Enzy. abbrev.: cheym. = chymotrypsin; memb. = membrane; MΦ-met. = macrophage-metallo; MMP = matrix metallo-proteinase; pl.^{gen} = plasminogen; PMN = polymorpho-nucleates; try. = trypsin.
Substr. abbrev.: Ag = aggrecan, AMPS = acid mucopoly-saccharides, CS = chondroitin sulfate (s.), DS = dermatan (s.), El = elastin, En = entactin, Fi = fibrin, Fi> = produces fibrin, Fn = fibronectin, Ga = galectin-3, gel = gelatin, HA = hyaluroic acid, HE = heparin, HG = hemoglobin, HS = heparan (s.), KS = keratan (s.), La = laminin, nh = nonhelican region of …, N-ter = amino-terminal of …, OF = oncofoetal/laminin-binding (Pucci-Minafra), Pg = plasminogen, PG = proteoglycans, Vn = vitronectin.
Inhib. abbrev.: DTT = dithioghreitol; E64D = epoxysuccinyl-L-leucyl-amido-3-methyl-butane ethyl ester; EDTA = ethylene-diamine-tetraacetic acid; EGTA = ethylen-glycol-tetra-acetic acid; LHVS = morpholinurea-leucine-homophenyl-alanine-vinylsulfone-phenyl; NEM = N-ethyl-maleimide; PAI = plasminogen activator inhib.; PMSF = phenil-methyl-sulfonil fluoride; SBTI = soybean trypsin inhib.; TIMP = tissue inhib. of metalloproteinases; TPCK = tosyl-L-phenyl-alanyl-chloro-methane; * Also termed *stephins*.
[By Garbisa according to "*Enzyme & Protein* 49, June '96", "*Ann. Rev. Physiol.* 59, 63, '97", "*Extracellular Matrix Proteases* Calbiochem-Oncogene 1, April '99".]

Procedure

Well coating. Prepare the required number of wells at least 1 day prior to the experiment, by coating clusters of 24 tissue culture wells (16 mm diameter) as follows: first, with unlabeled type IV collagen (10 μg in 200 μl of 0.1 N acetic acid) to ensure a linear rate of degradation over a short period of time. Allow to evaporate overnight in a laminar flow hood. Then, with [^{14}C]proline-labeled type IV collagen (in 200 μl of 0.1 N acetic acid): 2000 cpm in the case of 12–24 h incubation with high number of cells; 10,000 cpm in the case of 2–4 h incubation, or a few cells. Allow to evaporate overnight in a laminar flow hood. Sterilize the collagen-coated wells by exposure to ultraviolet light for 20 min, in laminar flow hood; avoid longer exposure.

Cell seeding. Harvest cells in log phase of growth by incubating with 0.1% EDTA for 3–5 min. Wash the cells once with 1 ml of complete medium (with serum and antibiotics) to remove EDTA, and once with 1 ml of serum free medium. Determine the cell number and viability. Prior to the actual plating of the cells, wash the tissue culture clusters with 1 ml of serum-free medium for 10 min (serum contains gelatinases, though these are blocked by inhibitors), to remove loosely adherent collagen and fragments. Measure the radioactivity of randomly chosen washes, to assess the final availability of labeled collagen (usually >75%). Inoculate the cells into each well, in concentration ranging from 1×10^3 to 1×10^5 cells/ml, and add 1 ml of serum-free medium. Serum-supplemented medium can be used to prevent cell starvation and block of proliferation, but note that serum inhibitors might restrain collagen degradation. Set control wells containing: 1 ml of serum-free medium without cells ($\sim$10% counts released, as background: B); 1 ml of bacterial collagenase (10 μg/ml) dissolved in serum-free medium ($\sim$90% counts released, as maximum digestion level: D_{max}). Incubate the clusters at 37°C for the desired amount of time (usually overnight).

Measurement of degradation. At the end of the incubation period, harvest 0.8 ml of medium from each well and mix this with 10 ml of appropriate solution for scintillation counting aqueous samples. For cells exhibiting degradation activity, the counts released into the medium are proportional to cell number. Subtract B, and express degradation as percentage of D_{max}. Compare the relative activity of cell lines using the linear portion of the dilution curve of activity vs. viable cell number.

The above method allows degrading activity for a specific substrate by living cells to be measured. If, instead, measurement is required on a cell extract or a conditioned medium, approaches such as zymography (see Section 4.2 above) and in-solution assay are available. The latter can make use of radiolabeled substrate, and after incubation measures the fragments produced and separated from undigested molecules by TCA (trichloroacetic acid) precipitation. For details, refer to an example method described for type IV collagen substrate by Garbisa et al. (1980b).

4.1.5. Co-polymerized matrix component degradation by living cells in vitro

Another approach for studying a cell's capacity to degrade a specific substrate lies midway between zymography and the immobilized component method described above. Here, the cells are seeded on the bottom of a dish, which is then overlaid with soft agar containing a substrate for proteolytic activity. After incubation under appropriate conditions, lithic plaques appear within the agar, indicative of enzymatic activity. The assay is described below to evidence urokinase-plasminogen activator (uPA) activity.

Materials

Purified agar: 2.5% in distilled water, dissolved and sterilized by autoclaving;
Plasminogen: 1 mg/ml in PBS (filter sterilized);

Milk (carnation nonfat dry milk): 8% (wt/vol) in distilled water. Heat in boiling water for 30 min. Centrifuge and take supernatant. Aliquot and store frozen at $-20°C$;
2× concentrated tissue culture medium.

Procedure

Cells to be analyzed have to be plated at clonal density, so that they grow in isolated colonies. Once colonies are present in the dish, growth medium is removed, and the plate rinsed very well with PBS to eliminate any trace of serum. In the meantime, the agar is melted by heating to boiling, and cooled down at 45°C in a water bath. Also tissue culture medium and milk are brought to the same temperature. A mix is then prepared, by adding in the following order: 0.2 ml 2.5% agar; 0.125 ml 8% milk; 0.25 ml 2× medium; 0.025 ml plasminogen (final concentration 50 μg/ml). Then, 0.35 ml of this mixture are evenly layered onto a 60-mm dish containing the cell colonies, and allowed to cool at r.t. until semisolid (10–30 min). The dish is then carefully transferred to the incubator at 37°C. After 18–24 h lithic plaques can be seen on a black box transilluminator (the milk solution is whitish, and lithic plaques appear as clear spots on a opaque background—see Fig. 4.2).

4.2. Invasion

In the group of techniques designed to study invasion, focusing on cell invasion in vitro and/or in vivo, and metastatic potential in syngenic or immunodeficient animals, gives direct answers to questions such as:

- are these cells aggressive?
- to what extent are they invasive and/or metastatic?
- which cell-cell and cell-matrix interactions take place during invasion?
- how important is immune surveillance?

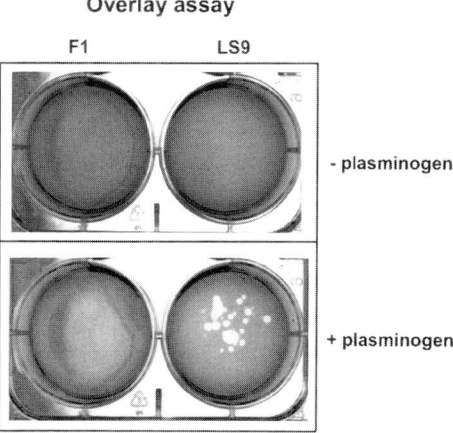

uPA activity in B16 melanoma cells

Fig. 4.2. Agar overlay assay for plasminogen activator. B16-F1 and -LS9 cells grown as isolated colonies have been overlaid with agar/milk (see text) in the presence (lower wells) or absence (upper wells) of plasminogen, to detect cell surface associated uPA activity. B16-LS9 cells clearly show a higher activity as compared to B16-F1, in which the activity is undetectable under this assay conditions (Rusciano et al., 1998a).

- which body district or organ will the tumor cells preferentially target?
- what role do growth factors and hormones play in spreading and targeting of tumor cells?

4.2.1. Amnion invasion assay

The assay allows quantitation of invasion through a basement membrane (BM). As for the degradation assays, it is not immediate the identification of the proteolytic enzymes involved.

The finding that metastatic tumor cells can secrete enzymes that degrade BM collagen has provided indirect evidence that these cells have the capability to damage the structural component of the BM

(Garbisa et al., 1980a). Tumor cells have been shown to cross the chorioallantoic membrane, mouse urinary bladder and canine blood vessels in vitro (Poste et al., 1980; Vlaeminck et al., 1972). Verification of active penetration of BM requires a direct interaction of cells with isolated intact BM.

Large intact surfaces of whole human amnion BM for in vitro studies can be easily prepared from fresh human placentas, and their integrity may be verified by permeability and morphology studies (Liotta et al., 1980b). Microscopic examination of invasive cells cultured on amnion surfaces may demonstrate production of local discontinuities.

Procedure

Obtain fresh human normal-term placentas within 1 h of delivery, and inspect them in a biohazard hood. If necessary, match placenta and cell species. Aseptically, peel the amnion membrane away from the chorion by hand and blunt dissection, and rinse it with PBS supplemented with antibiotics (100 U/ml penicillin, 100 μg/ml streptomycin, 0.5 μg/ml Fungizone). Use only the outer portions of the amnion (>600 cm^2, <0.5 mm thick) not directly adjacent to the umbilical cord and decidua; transfer them into DMEM with antibiotics.

For cold amnion. remove the epithelial layer by submerging the membranes in 1 M urea for 30 min at 25°C and then by gentle wiping with sterile gauze; rinse several times with PBS with antibiotics; use immediately or store in DMEM with PBS at 37°C.

For labeled amnion. incubate in proline/glutamine-free DMEM containing 100 μg/ml ascorbic acid, 100 μg/ml BAPN, 20% dialyzed foetal calf serum and 2 μCi/ml L-[^{14}C]proline; incubate at 37°C on a gyratory shaker for 36 h, wash the membrane exhaustively with PBS (to remove nonincorporated radioactivity) and denude the amnion as above.

Amnion mounting. Use a 24-well two-compartment chemotaxis chamber apparatus, with 1.6 cm well diameter. Stretch and clamp the denuded BM, either labeled or nonlabeled, within the holders, thus dividing the upper and lower compartments (Fig. 4.3). For invasion studies, sandwich a Millipore filter (8 μm pore size) in direct contact with the amnion stromal surface.

4.2.1.1. Analysis of invasion
Cold. Detach cells from plastic flasks with 5 min treatment with trypsin-EDTA solution, centrifuge 5 min at 500 g, and resuspend in DMEM with 0.15% fetuin and antibiotics. Apply cell suspension to the amnion BM at a concentration of $3 \times 10^5/1.5$ ml in each individual amnion holder. Replace the medium in the upper compartment every 24 h and incubate the holder at 37°C. At one day intervals, submerge the holders in 10% formalin and leave overnight to fix the membranes. Discard the amnion membrane, stain the filters using the standard haematoxylin-eosin procedure and mount on glass slides. Cells that migrated through the whole thickness of the membrane remain trapped within the Millipore filter: count with a light microscope at 400×.

Labeled. Seed 3×10^5 cells/well/3 ml of serum-free DMEM and incubate at 37°C for 70 h. Cells usually attach within 2 h. Take aliquots of medium from the upper compartments at 2, 22, 46 and 70 h of incubation; mix with 5–10 ml of suitable solution and measure in the scintillation counter. After 70 h take a sample also from the lower compartment and count. Incubate control wells for 12 h at 37°C with 2 ml of 0.1% of bacterial collagenase (BC) dissolved in 0.5 M Tris-HCl, 50 mM $CaCl_2$, 0.2 M NaCl, pH 7.4. Express the results as percentage of maximum release (BC) of collagen-incorporated radioactivity (Russo et al., 1986).

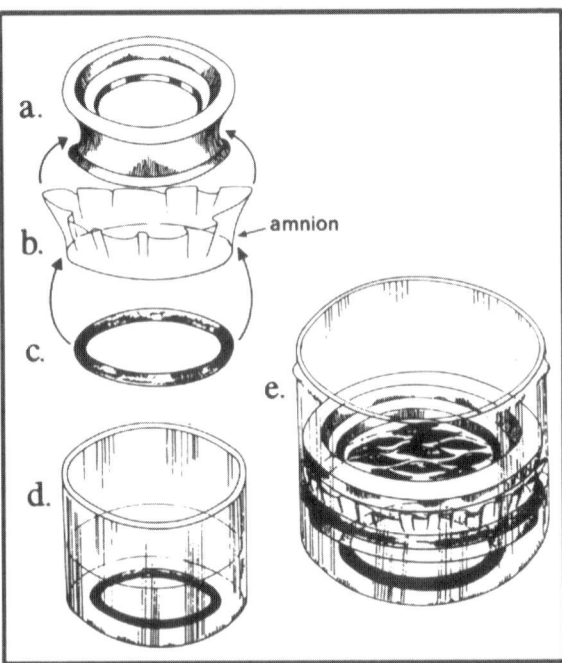

Fig. 4.3. (**A**) Diagram of the amnion invasion assay. The invasion chamber represents a cylindrical well produced by a Teflon ring (a) to which epithelium-free amnion (b) is fastened with the aid of a viton ring (c), to face the BM side up and stromal side down. A smaller lower chamber is created by a silicone rubber ring support attached to the bottom of a 35-mm tissue culture well (d) with silicone grease, and filled with medium. The (upper) invasion chamber is placed on this support, and medium with or without additives (to be tested for invasion-blocking or stimulating ability) is added to this chamber 1 h prior to the addition of labeled cells to be tested for invasive ability. Medium is then added to the tissue culture well (d) outside these chambers to bring the fluids inside and outside the Teflon ring to the same level: (e) represents a well that includes the complete invasion chamber seeded with cells on the BM. (Reproduced from Yagel et al., 1989.) (**B**) (a) Human amnion. Epithelium (EP), basement membrane (BM), connective tissue stroma (ST). Haematoxylin-eosin PAS stain. (b) Denuded human amnion membrane. Basement membrane (BM), connective tissue stroma (ST), Millipore filter (F). Haematoxylin-eosin, PAS stain. (Reproduced from Russo, 1986.)

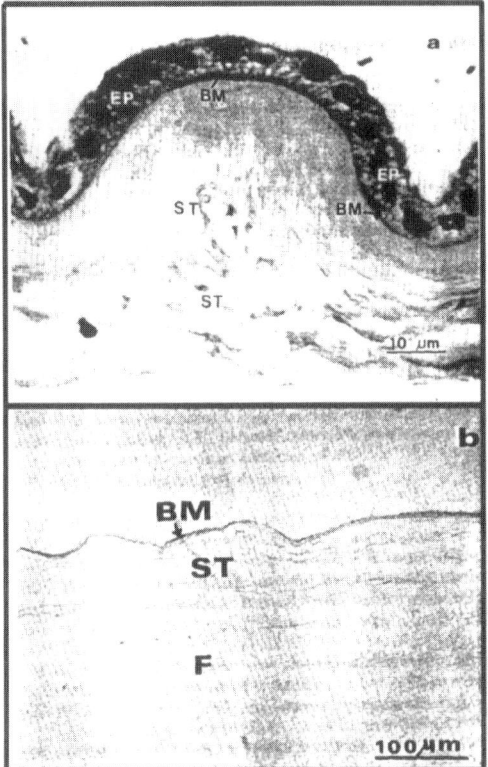

Fig. 4.3. (continued).

4.2.2. Chemoinvasion (Matrigel^{TM})

The Boyden-chamber chemoinvasion assay, developed by Albini et al. (1987) using Matrigel^{TM}, is an alternative simplified model system for measuring cell capacity to attach to extracellular matrix, degrade it, and migrate towards a chemoattractant. Matrigel^{TM} is a commercially available mixture of BM components extracted from the EHS murine sarcoma; it is used as a coating on the top of a porous filter, where it represents a matrix barrier to traversal by invasive cells. The optimal amount of Matrigel^{TM} and the use of chemoattractants allow the assay to be completed within 5 h; if the lower compartment is filled with medium without chemoattractants, the assay may require as long as 2–5 days (Hendrix et al. 1987).

Following cell seeding and incubation, nonmigrated cells on the upper surface of the filter are removed, and the cells that actively migrated to the under surface of the filter are counted. This assay—as well the amnion assay—may also be used to investigate the ability of hydrolase inhibitors or growth factors to restrain or augment the invasive activity respectively, as well as measure the capacity of a chemoattractant to accelerate the movement of invasive cells.

Materials

Matrigel^{TM} (Collaborative Res., Beckton Dickinson);
blind well chemotaxis chambers with 13-mm diameter filters;
polyvinylpyrrolidone-free polycarbonate filters, 8 or 12 μm pore size (depending on the cell size).

Procedure

Dilute Matrigel^{TM} in cold distilled water to 50 μg in 50–100 μl. Apply 25–50 μg Matrigel^{TM} to the filters, dry under hood, and reconstitute with serum-free medium. Place a solution of chemoattractant in the lower compartment of the Boyden chamber (in the absence of chemoattractant, very little cell migration occurs over a 6–h period). Place the coated filters in Boyden chambers, and close the chamber. Add to the upper chamber 2 to 3×10^5 cells in appropriate medium containing 0.1% BSA. Incubate at 37°C in 5% CO_2 for 4–6 h. Remove the cells from the upper surface of the filter by wiping with a cotton swab or by passing them on tissue paper. Fix the filter (methanol or ethanol) and stain (haematoxylin-eosin or toluidine blue). Count the cells from various areas of the lower filter surface. Alternatively, migrated cells can be solubilized in order to release the dye, which can then be quantitated at the spectrophotometer, as described in migration assays (see Chapter 3).

4.2.3. Invasion through cell monolayers

This type of assay is even closer to the real situation in vivo, since it allows to address interactions between the tumor cells and a continuous layer of cells, and thus takes into account in the invasion assay also parameters such as adhesion, retraction and penetration. Of course, being an in vitro assay, there is no guarantee that it is predictive of in vivo invasive behavior.

During intravasation and extravasation, tumor cells must traverse not only the BM barrier but also the continuous layer of endothelial cells which lie on the BM. Adhesion to, separation of and intrusion into endothelial (or epithelial) cells are as crucial to tumor spread as degradation of the extracellular matrix. Crissman et al. (1985) estimate that the endothelial cell layer may delay tumor cells from contacting the BM in vivo for up to 2 days.

These invasion steps are most easily studied on amnion complete with epithelium, or in bioengineered cell monolayers. These can be easily produced by seeding vascular endothelial cells (EC) onto reconstituted extracellular matrix (Matrigel^{TM}), or corneal EC onto bovine lens capsule mounted in a combi-ring dish.

Following incubation (w/wo chemoattractant), the interaction of cells with the monolayer is examined by scanning and transmission electron microscopy. Using this approach, Boxberger et al. (1989) documented that invasive cells force retraction of endothelial cells, and subsequently undermine the monolayer. Instead, noninvasive cells remained mostly in a typical roundish morphology, and only very few attached to and invaded into the endothelial monolayer. It should, however, be noted that the authors also report a number of counter-intuitive results (invasive behavior by nonmetastasing cells, lack of adhesion of highly metastatic cells): these should be borne in mind when drawing conclusions.

4.2.4. Organ culture assays

Where tumor cell interactions are to be studied with host tissues, in vitro invasion systems may be used with cultured tissue fragments. In some cases, normal tissue architecture has been preserved. In this case, even in absence of host immunologic or nonimmunologic cellular reactions and hormonal or neural influences, the patterns of invasion by malignant cells or tumors are similar to those found in vivo. The abilities of tumor cells to invade at the cut edges of normal organ tissues (liver, lung, skin, heart and kidney) have been examined. Although normal tissues undergo degeneration and cellular rearrangement with time, such techniques may prove useful in short-term experiments on the tissue specificity of binding and invasion of tumor cells with different organ colonization patterns. The reader is referred to Nicolson et al. (1985) for details.

4.2.5. Chick chorioallantoic membrane

This is an invasion assay of greater significance for the real invasive ability of malignant tumor cells, since it measures their intravasation ability (i.e. the ability to penetrate blood vessels).

Ex vivo models, such as mouse bladder (Poste et al., 1980) or human amnion (Russo et al., 1986) either denuded of epithelium and/or reseeded with endothelial cells (Foltz et al., 1982) may be considered too simplified systems. According to Kim et al. (1998), these 'recapitulate poorly the structure of the blood vessels and, in particular, small vessels ... where most of the cancer cell invasion is believed to take place'. These authors describe an interesting alternative where cells are inoculated on the chick chorioallantoic membrane (CAM) of an artificially created air sac in chick embryo. To detect and quantitate tumor cells actively penetrated in the ventral 'lower CAM', genomic DNA is extracted and used as a template for human Alu sequence identification by PCR. These sequences, unique to human and higher primate DNA, are repetitive sequences of about 300 bp. Present in high abundance (ca 5%) in the DNA of the explanted human tumor cells, they are absent from the host chicken cells. PCR amplicons produce a radiolabeled band with intensity proportional to the amount of human DNA present, from as few as 50 cells. This method allows insight into the kinetics of intravasation and cellular properties crucial for completion of this process.

An analogous procedure based on injection of GFP-transfected cells (see below) and monitoring by image analysis may give information on the pathways and growth of invasive cells.

4.2.6. In vivo monitoring of invasion pathways and metastases of GFP-tagged tumor cells

The invasion pathways, arrest and clearance of tumor cells in various tissues can be studied by injecting tumor cells into experimental animal models. In this case, a way of tracking single tumor cells within the circulatory system and the organ microenvironment is required.

Recent advances now allow cells to be tagged by transfection with fluorescent markers derived from green fluorescent protein (GFP). This protein is expressed by the jellyfish *Aequorea Victoriae*, and is now being increasingly used in visual marking. The use of permanently GFP-transfected cells offers a nonradioactive alternative tool for studying tumor invasion and metastatic growth. GFP-fusion proteins may also be used in vitro to study molecular events instrumental to invasion. For example, GFP-tagged receptor for a complement component allows direct recording of receptor-distribution at the leading edge of living neutrophils crawling towards a chemoattractant (Servant et al., 1999).

GFP tagging is described in more details in Chapter 7.

CHAPTER 5

The role of growth interactions in cancer metastasis

Sustained, uncontrolled cell proliferation is a necessary requirement for a tumor (either primary, or at a secondary metastatic site) to arise, in the first place, and then to become life-threatening. Although progression towards malignancy requires the acquisition of other properties, such as motility and invasion, if growth remains limited and in check, it is most likely that no overt disease will ever become detectable. Similarly, during metastatic spread, the mere presence of viable tumor cells in an organ is a necessary but not a sufficient condition for metastasis development. The last decisive step will be the proliferation of the cells that lodged and survived in the organ parenchyma into a clinically detectable colony. Metastatic cells that migrate from an organ to another have to deal with a new environment, which may not be ideal to support their survival and growth. Therefore, the ability to respond to autocrine or paracrine growth factors will be a distinctive advantage for those cells endowed with such properties. Indeed, the increased responsiveness of some cancer cells to paracrine growth factors expressed differentially at particular metastatic sites could explain why certain cancers show a preference for metastatic growth at certain sites, as demonstrated by the seminal work carried out in the lab of Max Burger (Burger and Madnick, 1983; Sargent et al., 1988; Rusciano et al., 1993). Accordingly, a variety of oncogenes, activated by different mechanisms, frequently have been shown to encode growth factors, receptor tyrosine kinases, or other enzymes that participate in mitogenic signaling (Bishop, 1991; Aaronson, 1991). Beside its relevance for tumor growth itself, an enhanced proliferation rate also serves the scope of generating intratumoral heterogeneity. It is in fact only through multiple divisions that mutations can arise and become fixed in the cell progeny, thus contributing to both tumor progression and to its chances of surviving the adverse conditions it will find during the phase of distant spreading (metastasis). Tumor cells within an established metastasis may even lose—due to genetic instability—those properties enabling them to be invasive and metastatic, but they cannot lose their net proliferation ability for clinically relevant metastases to appear.

Many different factors are known to control cell differentiation and proliferation. Depending on their biochemical properties and mode of action, growth factors are divided into lipophilic and hydrophilic. Lipophilic hormones—typically steroids—are passively transported inside the cell, and interact with cytosolic or nuclear receptors. They induce slow, long-lasting responses (hours or days) in their target cells. Water soluble signaling molecules include derivatives of amino acids (epinephrine, norepinephrine and histamine) and arachidonic acid (prostaglandins), peptide hormones (such as insulin, glucagon, ACTH, FSH . . .) and growth/differentiation factors (such as NGF, EGF, FGF . . .). Hydrophilic hormones interact with cell surface receptors, triggering immediate response (milliseconds to seconds), with effects that, however, may extend over days. Depending on the distance over which the signal must act, growth factors can be classified as endocrine (when target cells are distant from the source of ligand), paracrine (target cells are close to signaling cells), autocrine (signaling and target cells are one and the same: i.e. cells respond to substances that they themselves produce) and juxtacrine (signaling molecules are bound to the surface of signaling cells).

Autocrine growth stimulation. Progressive independence from exogenous growth factor stimulation is a common hallmark of most malignant tumors. Loss of hormone responsiveness correlates with increased malignancy and decreased survival rates in both breast (Sheikh et al., 1994–1995) and prostate (Dorkin and Neal, 1997) cancer. In general, as tumors progress to more malignant or more metastatic phenotypes, they become less dependent on serum-derived growth factors for their growth in vitro (Kath et al., 1990; Herlyn et al., 1990; Wright and Huang, 1996). Autocrine growth stimulation can be achieved in different ways: a cell may express a growth factor receptor which is constitutively active also in the absence of the cognate ligand (either because of activating mutations in the receptor gene, or because of overexpression of the gene product), or the cell can produce at the same time both the receptor and its ligand, thus triggering continuous activation of the receptor. These autocrine loop mechanisms can be produced either by extracellular release of the growth factor and its binding to an appropriate extracellular receptor on the same cell, or by ligand-receptor interaction inside the cell (intracrine stimulation) (Lang and Burgess, 1990; Browder et al., 1989; Stachowiak et al., 1997). Autocrine production of multiple cytokines has been shown during melanoma progression (Mattei et al., 1994), while overexpression of erbB-2 and elevation of MAP kinase activity have been associated with progression to estrogen independence in breast cancer (Tang et al., 1996; Coutts and Murphy, 1998). Similarly, both autocrine and paracrine pathways are up-regulated in androgen-independent prostatic tumors, and may thus replace androgens as primary growth stimulatory factors in cancer progression (Russel et al., 1998). Multiple factors, beside activation of oncogenes and inactivation of tumor suppressor genes, seem to be involved in the acquisition of growth independence by colon carcinoma cells (Markowitz et al., 1994). Finally, activation of autocrine loops by transfection of a secreted form of a growth factor (Rogelj et al., 1988; Egan et al., 1990), or its receptor (Cortner et al., 1995), may induce a fully malignant phenotype in NIH 3T3 cells.

The autocrine production of growth factors may give metastatic tumor cells an intrinsic advantage to establish a growth at most of the organs in which they can efficiently lodge. No correlations have been reported so far between organ-specific colonization and autocrine stimulation, but the possibility cannot be ruled out that specific interactions between tumor cells and the local microenvironment may modulate the expression of autocrine growth factors, or their receptors, thus leading to organ specific effects on the metastatic ability of tumor cells.

Paracrine growth stimulation. Primary tumors are often dependent, at least at their initial stages, on endocrine, and/or paracrine growth factors released from surrounding host cells. Typically, tumors of the female reproductive tract (breast, ovary and uterus) are estrogen dependent (Murphy, 1998), while tumors of the prostate in males are androgen dependent (Wilding, 1995). The same type of tumors can also be dependent on paracrine factors, such as IGF II or LIF in the case of breast cancer (Yee et al., 1988; Dhingra et al., 1998), or prostate-specific factors in the case of prostate cancer (Gleave et al., 1991). After progression to the metastatic phenotype, however, their growth factor responses often change, and while dependence on steroid hormones is usually lost, the requirements of at least a fraction of metastatic cells become more compatible with the cytokines present at the secondary sites (Nicolson, 1993; Radinsky and Fidler, 1992; Rusciano and Burger, 1992). Experimental metastatic models have amply demonstrated that the organ preference of metastasis is correlated with the enhanced mitogenic response of the metastatic cells to cytokines released from the target organ (Burger and Madnick, 1983; Horak et al., 1986; Nicolson and Dulski, 1986). Organ-specific growth factors can be soluble, as shown in several model systems, such as the B16 melanoma (Nicolson and Dulski, 1986), RAW117 large cell lymphoma (Nicolson, 1988a), 13762NF mammary adenocarcinoma (Nicolson, 1988a), Walker 256 mammary carcinosarcoma (Manishen et al., 1986), PC3 prostatic adenocarcinoma (Chackal Roy et al., 1989). Some growth factors may also exist in membrane-bound form (Massagué and Pandiella, 1993; Bosenberg and Massagué, 1993), and they may function both as intercellular adhesion molecules and as true growth factors (Anklesaria et al., 1990; Steele, 1989). Direct cell contact between host hepatocytes and tumor cells has been shown to be required for paracrine (juxtacrine) growth stimulation of liver-specific B16 melanoma cells (Sargent et al., 1988; Rusciano et al., 1993). Intriguing enough was the finding that when identification of the paracrine factor(s) responsible for lung, liver or bone colonization in the different systems was attempted, in each case the same factor was found: transferrin (Cavanaugh and Nicolson, 1991; Rossi and Zetter, 1992; Rusciano et al., 1994). Moreover, transferrin was found to give a selective advantage also to the growth of brain metastases of B16 melanoma (Nicolson et al., 1990) and mammary carcinoma (Inoue et al., 1993). Since transferrin is not an organ-specific growth factor, rather an ubiquitously present nutritional factor, the most likely explanation is that it is a necessary, yet not sufficient factor for metastasis development. More recently, overexpression of the EGF receptor has been reported to be involved in liver colonization by human colon carcinoma cells (Radinsky et al., 1995; Singh et al., 1997; Parker et al., 1998).

Beside soluble growth factors released by organ and tissue cells, an additional source of organ-derived growth promoting molecules is the extracellular matrix (ECM) or basement membrane (BM), where growth factors can be bound to ECM proteoglycans or glycoproteins (Taipale and Keski-Oja, 1997). ECM-bound growth factors are usually inactive, but they may be released in their active form after cleavage of the proteoglycan acting as a linker (Ruoslahti and Yamaguchi, 1991). Interestingly, heparanase activity has been shown to correlate with the metastatic potential of various tumor cells, while heparanase inhibitors reduced the incidence of lung metastasis in experimental animals (Vlodavsky et al., 1990). Most intriguingly, Doerr and collaborators (1989) reported that hepatoma and mammary carcinoma cell lines had a selective growth advantage, and formed more colonies when plated at clonal density on ECM derived from their preferred target organs in vivo. Finally, besides acting as a storage compartment for growth factors, the ECM directly modulates differentiation (Aharoni et al., 1997), survival (Meredith et al., 1993; Frisch and Ruoslahti, 1997; Ilic et al., 1998), growth and growth responses (Levesque et al., 1991; Nathan and Sporn, 1991; Thiery and Boyer, 1992; Howe et al., 1998) of normal and tumor cells. Hence, specific interactions between tumor cells and the ECM can influence their tumorigenic and metastatic behavior (Imhof et al., 1996; Sy et al., 1996; Holzmann et al., 1998; Zvibel et al., 1998; Boudreu and Bissell, 1998). In fact, co-injection of tumor cells with a natural ECM derivative (Matrigel^{TM}) can enhance, or even make possible, the establishment of primary tumors in vivo from otherwise scarcely tumorigenic cell lines (Fridman et al., 1991; Topley et al., 1993).

Inhibitory growth interactions. Autocrine and paracrine growth interactions can also be inhibitory, as illustrated by the negative effects on growth exerted by kidney- and liver-conditioned media, which mainly inhibited the growth of those malignant cells unable of kidney or liver metastatic colonization (Nicolson and Dulski, 1986; Nicolson, 1988a, b). The responsible factors appeared to belong to the TGFβ family (Tucker et al., 1984). TGFβ is a potent inhibitor of epithelial cells, but it might turn beneficial to tumor cells. Prostatic cancer cells may autocrinally produce TGFβ, which despite its negative effects on tumor cell growth under routinary in vitro culture conditions, can enhance prostate tumor growth and metastasis in vivo (Barrack, 1997). Interestingly, prostate cancer cells may produce other autocrine inhibitory factor, such as spermine (Smith et al., 1995), thus offering an explanation to the slow growth rate of prostatic tumors at their primary site. However, when prostatic cancer cells metastasize to the bone, paracrine growth factors locally present may revert their growth rate, triggering rapid proliferation and development of metastatic colonies. Other factors can also trigger different effects on normal and tumor cells, such as IL-6, which is growth inhibitory for melanocytes and noninvasive melanomas, while becomes growth stimulatory for invasive melanomas (Lu and Kerbel, 1993). Similarly, hepatocyte growth factor/scatter factor (HGF/SF) is a potent mitogen for hepatocytes, keratinocytes or melanocytes, whereas it inhibits growth of their transformed counterparts (Mizuno and Nakamura, 1993). Interestingly, it appears that growth inhibitory effects exerted by HGF/SF on B16 melanoma cells could contribute to the liver-specific colonization ability of B16-LS9, a selected sub-line of B16. B16-LS9 cells overexpress c-met, the cellular receptor for HGF/SF (Rusciano et al., 1995), thus causing its constitutive activation (Rusciano et al., 1996), and making the cells increasingly sensitive to HGF/SF effects, both in terms of enhanced motility and invasion, and growth inhibitory response (Rusciano et al., 1998a). However, while HGF/SF-induced growth inhibition was competed in vitro by liver-derived extracts, it was not by lung-derived extracts. Since both lung and liver contain HGF/SF (the lung even more than the liver: Tashiro et al., 1990; Rusciano, unpublished), growth of metastatic colonies is then allowed in the liver and not in the lung (Rusciano et al., 1998a).

In conclusion, net cellular growth is the final result of contrasting forces, such as growth stimulation and growth inhibition, survival and apoptosis/anoikis. We will consider each of these elements in the dynamics of metastatic growth, providing the relevant assays to establish the nature of the driving forces of growth.

5.1. Methods to evaluate growth interactions in vitro

This section will be divided in two main parts, describing first protocols to evaluate effects on cell growth, i.e. how to estimate cell numbers, or the fraction of growing cells, and next methods to determine the presence of autocrine and paracrine growth effects.

5.1.1. Estimation of cell growth

Cell cultures maintained in vitro contain at least three populations of cells: actively growing, resting, and dying. The balance between these three compartments will determine the net result: active growth, stationary growth, growth decrease. In order to evaluate the effects of factors affecting growth behavior, methods are required that allow to estimate the total amount of cells, and those in active proliferation.

5.1.1.1. Cell count
The most direct way of assessing the amount of cells present in a culture is counting them. This requires, of course, that enough cells are available to obtain statistically significant counts. If cells are growing in suspension, they are immediately available for counting, either under the microscope, with a hemocytometer, or by automatic cell counting, with an electronic cell counter. If cells are growing adherent on a substrate, they have to be first detached, and then resuspended in an appropriate medium. In any case, for reliable counts, the cell suspension should consist as much as possible of single cells. Counting cells with the hemocytometer (e.g. Neubauer chamber), requires at least a concentration of 10^5 cells/ml, which will result in an average of 10 cells per square of the chamber. Also, in order to obtain reliable counts, the cell concentration should not be too high (over 5×10^6/ml: 50 cells per square). Count at least 10 squares (the 4 corner squares plus the central squares in the two fields of the chamber). Counting cells with the hemocytometer allows a direct visualization of the cells, their shape and the presence of aggregates, and if done in the presence of trypan blue (0.2% final concentration) also allows to tell viable (those excluding the dye) from dead (colored in blue) cells. The use of an electronic cell counter (such as the Zeta series of Beckman Coulter) allows to count more diluted cell suspensions, and with higher precision, since it counts the number of cells in larger volumes (0.1–1.0 ml) than the hemocytometer (0.1 μl per square). Since cells are not visible with this instrument, good care has to be taken to use well mixed single cell suspensions, and calibrate the cell size window, so not to count cell fragments or aggregates. The use of an electronic counter will allow not only to count cells in a population, but also to evaluate

their size distribution, which can be important when studying the metastatic properties of the cells (Rusciano et al., 1992). Neither instrument (electronic or manual) permits to tell the proliferating fraction among the total number.

Materials

Counting device (manual or electronic).

Procedure (anchorage dependent assay)

Cells are randomly plated at day 0 in multiple wells or dishes, so to consent at least triplicate determinations for each time point. The initial plating density will determine the successive time points: if cells have been plated at medium to high density (about 10^3–$10^4/cm^2$), cell numbers can be evaluated starting 18–24 h after plating (at which time a cell number close to the actual number of plated cells is expected), and each day until confluence is attained. Population doublings (PD) can then be calculated using the formula (Kath et al., 1990):

$$PD/day = [LN \text{ (fold increase)}/LN2] \times [1/\text{days of incubation}]$$

So, if for instance on the fifth day of culture starting from 10^4 cells, a total cell number of 3.2×10^5 cells is reached, then PD/day = LN 32/LN2 × 1/5 = 1.0

In case of plating at clonal density ($10^2/cm^2$ or less), a lag phase is expected before growth can be assessed, and determinations can usually start 5–7 days after plating. In this case, growth of cells is expected to occur in well defined colonies, so that an alternative way of evaluating growth would be to fix and stain the cells, and count the number of colonies and their average content of cells; if cells within the colonies are not countable, an approximate number can be inferred by establishing a correlation between the o.d. readings of the solubilized dye and cell number (see below for further details about fixation, staining and indirect estimation of cell number). In this case PD and colony-forming efficiency (CFE) can be calculated from the following formulae (Herlyn et al., 1987):

$$PD/day = [LN \text{ average N cells per colony}/LN2] \times [1/\text{days of incubation}]$$

where at least nine randomly selected colonies (>4 cells per colony) per duplicate dishes have to be considered for cell count; and

$$CFE\% = [\text{average N. colonies per dish}/\text{N. of cells initially plated}] \times 100$$

So, if for instance, at the 10th day of growth, starting from 100 cells plated per dish, an average number of 70 colonies per dish is counted, with a mean number of 16 cells per colony, then:

$$PD/day = (LN16/LN2) \times 1/10 = 0.4 \quad \text{and} \quad CFE = 70\%.$$

5.1.1.2. Incorporation of labeled DNA precursors

All the proliferating cells have to make new DNA in order to undergo mitosis and sustain their growth. Therefore, a labeled nucleotide (either radioactive or nonradioactive) added to the culture medium will be incorporated in the DNA of all those cells going through their S-phase during the time of labeling. Hence, a pulse-labeling (0.5–2 h) will label only the actual proliferating fraction of cells, and is thus indicated to estimate mitogenic effects in the short term; a longer labeling time (24–40 h) will be likely to label all the growing cells in the dish (as long as they do not have extremely long doubling times), and thus gives a better estimate of the real amount of proliferating cells.

Radioactive labeling. The incorporation of [^{3}H]thymidine into acid-insoluble material has been widely used to detect growth stimulatory or inhibitory effects on many different cells. However, the incorporation of exogenous thymidine into the DNA also depends on its rate of transport across the plasma membrane, and its phosphorylation by thymidine kinase, which has to occur before the nucleoside is incorporated into DNA. Therefore, beside factors affecting growth, also factors affecting transport or thymidine kinase activity, will influence the amount of thymidine incorporated into DNA. If there is evidence, or a strong suspect, that transport may be responsible for the observed results (usually a decrease in incorporation), then increasing the concentration of thymidine in the experiment (up to 1 or 10 mM) should overcome the transport effects, since at high concentration the nucleoside can passively diffuse inside the cells.

Materials
[methyl-^{3}H]thymidine;
trichloroacetic acid (TCA) 10 and 5%;
0.5 M NaOH;
1.0 M HCl;
scintillation fluid (e.g. Ready Safe, Beckman)

Procedure
Cells are labeled in tissue culture (the use of multiwell plates is convenient for this type of experiments; at least triplicates per each point of the experiment are prepared) by adding the desired amount of radioactivity, for the required length of time: 0.5–1.0 μCi/ml for long-term (24–40 h) labeling, 2.0–5.0 μCi/ml for short (0.5–2 h) pulse labeling. At the end of the labeling time, the radioactive medium is removed, and the cell monolayers rinsed three times with ice-cold PBS. Both the medium and the washes have to be disposed of according to biosafety rules. Cell monolayers are then incubated at 4°C with ice-cold 10% TCA, for 15–30 min, to remove radioactivity present in the cells, not incorporated into DNA. Next, they are rinsed once with ice-cold 5% TCA, and finally they are solubilized with 0.5 M NaOH for several hours. The solubilized material is then transferred into scintillation vials, neutralized by the addition of HCl, diluted with at least 10 volumes of the scintillation cocktail,

and counted in a beta counter (Sargent et al., 1988; Rusciano et al., 1991). 10% SDS can also be used to solubilize cells, but its compatibility with the scintillation cocktail used has to be verified.

^{3}H-thymidine can be substituted by ^{3}H-BrdU, or by ^{125}I-thyimidine. In this latter case, scintillation counting is not required, and the solubilized cells—after TCA extraction—can be directly counted in a gamma counter. Of course, the use of ^{125}I as a label poses higher health hazards if not properly handled.

Bromodeoxyuridine labeling. A valid, nonradioactive, alternative to thymidine labeling, is the use of 5-bromo-2′-deoxy-uridine (BrdU), which is also incorporated into DNA of cells during S-phase. Cells that have incorporated BrdU into DNA are then detected using a specific antibody against BrdU and an enzyme- or fluorochrome-conjugated secondary antibody. Nuclear immunostaining is readily visible, usually with uniform intensity and no background. S-phase positive nuclei show uniform staining of the entire nucleus, or its periphery in the form of a ring or irregular coalescing granulations only. The difference most likely depends on the duration of DNA uptake of BrdU: partial labeling during S-phase will result in an uneven distribution of the label (Risio et al., 1986). The accuracy of the BrdU immunohistochemical method is comparable to ^{3}H-thymidine autoradiography (Lacy et al., 1991).

Cell proliferation by BrdU staining can be detected either at the single cell, or at population level, depending on the label on the secondary antibody used, and the method of detection. If a fluorochrome-conjugated secondary antibody is used (FITC or rhodamine labeled), detection will be done by microscopical observation, and counting of the fraction of positive nuclei among the total (visualized by propidium iodide staining). Alternatively, FACS analysis of the labeled cells might give cumulative data, but requires that the cells are brought in suspension after labeling. Use of an enzyme labeled secondary antibody (peroxidase or alkaline phosphatase) will allow either detection of single labeled cells by microscopic analysis, if an insoluble dye product is used for revelation (e.g. NBT/X-phosphate for alkaline phosphatase: AP), or evaluation of whole populations if a soluble dye is used (e.g. ABTS or TMB for peroxidase: POD).

Materials

BrdU stock solution (1000×): 10 mM in PBS;
anti-BrdU antibody, and secondary labeled antibody;
PBS and PBS/0.1% BSA for antibody dilution and washings;
chromogenic substrates for POD or AP, if enzyme-labeled secondary antibodies are used;
cell proliferation kits based on BrdU incorporation are commercially available (e.g. by Roche Molecular Biochemicals), which allow both ELISA-based determination, and single cell analysis at the microscope;
DAPI (Molecular Probes), 1 mM in ethanol;
a mounting medium for glass coverslips, if single cell detection by microscopy observation is performed (e.g. Vectashield by Vector Laboratories, or Fluoroguard by BioRad Laboratories are commercially available antiquenching mounting media, if a fluorochrome is used); for more general use 1 part of PBS and 9 parts of glycerol can be employed;
spectrophotometer, or multiwell plate reader (e.g. Spectramax Plus, Molecular Dynamics) if an ELISA method has been chosen; Inverted microscope equipped with epifluorescence illumination if single cell reading has been selected.

Procedure

Cells are plated at the desired density, either on sterile glass coverslips for microscopic examination, or directly into multiwell plates (prepare each point at least in triplicate) for population analysis by ELISA. After treatment of the cells with the desired growth agents, BrdU is added, following the same criteria already discussed for thymidine labeling (short pulse, or long term). Cells are then washed free of residual BrdU by rinsing with PBS, and fixed in ethanol. Fixed cells can be stored at 4°C for further analysis. BrdU incorporated into DNA has to be made available to antibody binding, so either denaturation of DNA (by alkaline or acidic treatment) or digestion with DNases (primary anti-BrdU antibodies available with staining kits already contain an optimized mix of nucleases) is required. Primary anti-BrdU and secondary labeled antibodies are used sequentially, according to the manufacturer's instructions. If microscopic examination of individual cells is done, then DAPI (1 : 10,000; 0.1 μM final concentration) can be added with the secondary antibody labeling, in order to make all the nuclei (of both replicating and nonreplicating cells) visible. In case of enzyme labeling of the secondary antibody, a chromogenic substrate is added following the producer's recommendations, to obtain either an insoluble or a soluble coloration. If coverslips have to be analyzed, these are finally mounted on glasses, using an appropriate mounting medium. ELISA plates are quantitated by o.d. reading.

Results after microscopy are expressed as the ratio between total number of positive nuclei (BrdU staining) and total number of nuclei (DAPI staining) counted in 10 or more randomly selected fields, averaged on the number of replicates per each experimental point. This kind of analysis thus allows to express a proliferative index for the cell population. Results after ELISA detection are simply expressed as o.d. values, vs. an untreated control. A calibration curve can be prepared, to correlate the actual o.d. readings with the number of cells per well, so that cell numbers can be extrapolated on the linear portion of the curve.

5.1.1.3. *Enzymatic conversion of tetrazolium dyes*

In general, assays using tetrazolium dyes (such as MTT, XTT and MTS) measure the cellular conversion of the dye into a formazan product by the action of NADH-generating mitochondrial dehydrogenases found in living, metabolically active cells. The conversion of the tetrazolium compound to a formazan product is monitored by a shift in absorbance, and easily measured with an ELISA plate reader (Mosmann, 1983). The formazan product resulting from conversion of MTT is insoluble, and has to be dissolved by detergents or organic solvents before optical reading; it is, however, quite stable, allowing storage of plates for up to 7 days before reading. On the contrary, MTS is converted into a formazan product that is soluble in tissue culture medium, thus allowing immediate reading. MTS formazan products also allow for extended color development after the initial reading, until the reaction is stopped by the addition of SDS for final determination. XTT has a limited solubility, and is not stable in solution, requiring daily preparations of fresh solutions using prewarmed (37°C) or hot (60°C) culture medium to produce a 1 mg/ml solution. The formazan products it yields are, however, soluble in tissue culture medium, similarly to MTS. Both MTS and XTT require an additional electron transport reagent (PMS: phenazine methosulfate) in order to be converted into formazan products.

The use of either MTT or MTS shows a linear relationship between cell number and the amount of formazan product produced. However, the ability to convert MTT or MTS to a formazan product may vary with cell type, and depends on a cell's metabolic capacity. Most eukaryotic cells yield significant level of formazan product from MTT or MTS at low cell numbers (Smail et al., 1992). The method for MTT staining is described below. It is however advisable and more practical to use the ready-to-use cell proliferation kits from Promega, which are based on either MTT or MTS staining.

Materials

Microplate reader (e.g. Spectramax Plus, Molecular Dynamics) with filter at 570 nm;

MTT (Sigma, M2128): stock solution 5 mg/ml in PBS. Aliquot and store frozen in the dark at $-20°C$;

solubilization buffer: SDS 10% in HCl 0.01 M (alternatively, acid isopropanol: 100 μl 0.04N HCl in isopropanol can be used).

Procedure

Cells are plated at the desired density (usually 10^3 to 5×10^3/well) in 96 multiwell plates, in 90 μl of medium. At the chosen time, 10 μl/well from MTT stock solution are added, for 4 h at 37°C. Formazan crystals are then dissolved by incubation with 100 μl/well of solubilization buffer. If an SDS-based one is used, solubilization has to be overnight at room temperature; sealing the plates with parafilm between the lid and the plate itself is advisable to prevent evaporation. If acid isopropanol is used, it will take only few minutes to dissolve the formazan crystals, but in this case o.d. readings have to be made within 1 h, and care has to be taken in preventing differential evaporation from the wells, thus resulting in high variability of readings among the replicates. The solubilized color is measured by its absorbance at a wavelength ranging between 550 and 600 nm.

5.1.1.4. Fluorescence cell viability assay

The assay is based on the membrane-permeant fluorogenic substrate calcein-AM, which is hydrolyzed by intracellular esterases to yield a green fluorescent product (calcein) which is easily determined by fluorometry (ex/em: ~495/~515 nm). Thus, green fluorescence is an indicator of cells that have esterase activity as well as an intact membrane to retain the esterase products. Since different cell lines may have different esterase activity on calcein-AM, a calibration curve should be prepared per each cell line, in which a known number of cells is correlated to the amount of fluorescence read by the instrument.

Materials

Calcein-AM (molecular probes);
microplate reader equipped with a fluorometer (e.g. Cytofluor 2300, Millipore).

Procedure

Cells are plated in 96 multiwell plates at the desired density in 100–200 μl of medium, and treated with the desired growth factors. At the end of each time point, each well (at least triplicates are required) is exposed to 1.17 μM calcein-AM (prepared as a 1000×

stock solution in DMSO, and conserved frozen at $-20°C$ inside a sealed bag containing a desiccant and protected from light) for 45 min at $37°C$. After thorough rinsing of the wells with Ca^{++}/Mg^{++}-free PBS (with no pH indicator: CMF-PBS), each well is left in 100 μl of CMF-PBS, and the fluorescence determined at the fluorometer (Marchetti et al., 1993).

This same procedure can also be used in microscopy or FACS analysis: coupled with a nuclear dye such as ethidium homodimer-1, a red fluorescent nucleic acid stain (ex/em: $\sim$495/$\sim$635 nm) that is only able to pass through the compromised membranes of dead cells, and indicates the proportion of live vs. dead cells. (The viability assay kit including calcein-AM and ethidium homodimer-1 is sold by Molecular Probes, Cat. # L-3224.)

Some tumor cell lines may express high levels of the multidrug resistance proteins, which expel calcein-AM from the cells (Essodaigui et al., 1998), thus lowering the sensitivity of the assay.

5.1.1.5. Crystal violet staining of cell nuclei

All the procedures mentioned so far require cells in an active metabolic state, so that the labels are incorporated and/or converted into detectable products. This implies that not only effects on growth, but also changes in the metabolic state of the cells may influence the determination that is being done. An alternative method is passive staining of the cells, which is only function of the number of cells present in the dish, and solubilization of the dye for optical reading. Crystal violet (c.v.: N-hexamethylpararosaniline) is a basic dye which stains cell nuclei (Dutt, 1980). A method has been developed (Gillies et al., 1986), and further improved (Kueng et al., 1989) to stain cells in tissue culture with c.v., and estimate the number of cells in the dish, based on a calibration curve that should be run per each considered cell line. The method is economic and sensitive enough to allow determinations in multiwell plates from 6 to 96 wells per plate.

Materials
    Glutaraldehyde 11%;
    crystal violet 0.1% in buffered solution;
    10% acetic acid;
    microplate shaker;
    microplate reader (e.g. Spectramax Plus, Molecular Dynamics).

Procedure
Cells are plated as desired, and treated with growth factors. At the end of each time point, cells in replicate wells are fixed by addition of 11% glutaraldehyde to the culture medium (final concentration of glutaraldehyde has to be 1%: so, add 10 to 100 μl of medium in 96 multiwell plates, or 100 μl to 1 ml in 24 multiwell plates, and so on) for 15 min at room temperature, with mild shaking of the plate (about 500 rpm). Plates are then rinsed three times by sequential submersion in deionized water, and air dried. Air drying is important, because c.v. staining is less efficient on wet cells (Kueng et al., 1989). Also, the intensity of staining with c.v. is dependent on the pH of the c.v. solution. Crystal violet dissolved at 0.1% respectively in 0.2 M solutions of phosphoric acid (pH 2.5), formic acid (pH 3.5), water (pH 5.0), Mes (pH 6.0), or boric acid (pH 9.0) had a progressively increasing staining ability, and conserved a linear relationship with the amount of cells in the plate (Kueng et al., 1989). Crystal violet in boric acid, however, has to be prepared fresh each time, because the solution is unstable at that pH. Staining is for 20 min at room temperature with shaking, after which time the plates are rinsed extensively under deionized water, and air dried before dye solubilization in 10% acetic acid. After 15 min with shaking, the solubilized dye is measured by o.d. reading at 590 nm, directly in the plate against a blank of unstained cells, or even a blank with an empty well, since absorbance of an unstained cell monolayer is equivalent to that of an empty well.

    There are some advantages of this method. First, all time points after fixation can be stored dry, and processed all at the same time, thus eliminating some intraexperimental variability. Then, in case of overstaining because of a highly elevated cell number (o.d. readings >2.0), plates can be decolored by acetic acid, and restained at a lower pH, so to decrease dye uptake by the cells.

5.1.1.6. Colony formation ability in soft agar

Anchorage independent growth, such as growth in soft agar, is a hallmark of transformation for those mammalian cells that usually require a substrate to which adhere in order to proliferate. Moreover, growth factors may differentially affect cell growth depending on whether cells are grown as adherent monolayers, or as colonies in a three-dimensional array. For instance, A431 carcinoma cells, which overexpress both the hepatocyte growth factor/scatter factor (HGF/SF) receptor, c-met (Tajima et al., 1992) and the epidermal growth factor (EGF) receptor (Gill and Lazar, 1981), are growth stimulated by HGF/SF, but inhibited by EGF when grown as monolayers. However, when A431 cells were plated in soft agar, or grown as subcutaneous tumors in vivo, EGF treatment promoted their growth under each condition (Lee et al., 1987; Ozawa et al., 1987). On the other hand, liver-specific B16 melanoma cells, also overexpressing c-met (Rusciano et al., 1995) were growth inhibited by HGF/SF both as monolayer cultures and as colonies in soft agar (Rusciano et al., 1998a).

Materials

1 and 0.7% Agarose (DNA grade) in distilled water. Dissolve and sterilize by autoclaving;

$2\times$ tissue culture medium with $2\times$ FCS (e.g. 20%, if 10% is routinary used for culture).

Procedure

The two-layer soft agar culture system, originally introduced by Hamburger and Salmon (1980), consists of a bottom layer of hard agar, and a soft layer of agar in which cells are dispersed.

Bottom layer: Melt the 1% agar solution in boiling water or in a microwave oven, and cool it to 40°C in a water bath set at that temperature. In the same water bath, warm the concentrated ($2\times$) medium. Mix equal volumes of the two solutions to obtain 0.5% agar in $1\times$ culture medium, $1\times$ (e.g. 10%) FCS. Add enough volume to make a thick layer on the bottom of the well (typically, 1.0 ml/well is used for 24 multiwell plates, 1.5 ml/well is used for 6 multiwell plates), and let it set at room temperature until it is hardened. The plates can now be stored at 4°C for up to 1 week.

Top agar: melt 0.7% agar and cool it at 40°C, while warming up to the same temperature the concentrated ($2\times$) medium, as above. A higher temperature could be fatal to cells. In the meantime, plates with the bottom agar have to be brought to room temperature, and cells to be tested can be detached (it is important to have a single cell suspension) and appropriately diluted (usually, 5,000 cells are plated per each well of a 6 multiwell plate; in this case a cell suspension at 2×10^5 cells/ml is required). Then, 0.1 ml of cell suspension (enough for 4 wells) are distributed in 10 ml sterile tubes, resuspended with 3.0 ml of $2\times$ medium at 40°C, and finally diluted (by gentle mixing) with 3.0 ml of 0.7% agar at the same temperature. Then, 1.5 ml/well are distributed on top of the base agar layer and allowed to set, before moving the plates to the incubator for 10–14 days. Some liquid tissue culture medium can be added on top of the top agar, and changed weekly until the end of the experiment. Each experimental point should be made at least in triplicate, and the experiment should contain a positive control of anchorage independent growth (e.g. ras-transformed cells). The growth factors to be tested can be added to any or every of the three layers (bottom, top, liquid phase) (Fiebig et al., 1987; Hermann et al., 1987). To make colonies more visible, cells can be stained either with crystal violet (0.5 ml of 0.005% c.v. for at least 1 h), or by adding MTT (1.0 mg/ml) to the upper liquid phase of tissue culture medium 24 h before the final detection.

5.1.2. Detection of growth factor effects

Paracrine growth interactions between tumor and host cells occurring in vivo can be addressed in vitro using either co-culture techniques, or tissue-specific extracts in the growth assays described above. Autocrine growth stimulation, by definition, would not require cooperating cells; however, it cannot be ruled out that growth of cancer cells in a specific organ microenvironment might influence autocrine expression of certain growth factors, which the cells would not produce under routinary culture conditions. Therefore, a more thorough evaluation of autocrine growth effects would require to test the cells also in presence of tissue specific elements.

5.1.2.1. Autocrine growth effects

Activation of growth stimulatory autocrine loops is a frequent event during cancer progression, and may give an obvious advantage to those cells able to express both a growth factor and its receptor (Kath et al., 1990). As a consequence, tumor cells that activate autocrinally their growth factor receptors, may have decreased requirements for growth in vitro in tissue culture (Quinn et al., 1996). Therefore, measuring growth kinetics at decreasing FCS concentrations (from 10 to 0.1%) might give useful hints about the presence of autocrine growth stimulatory events: a better growth under serum-limiting conditions may suggest activation of some autocrine circuit. Keeping a small amount of FCS (at least 0.1%) in the tissue culture medium is useful because allows cell spreading, and contributes that minimal amount of exocrine (externally added) growth factors which might be necessary for the cells to be able to survive and respond to their own autocrinally produced factors. Of course, activation of an autocrine loop does not necessarily mean de novo secretion of a bioactive ligand, it might also imply new expression of a growth factor receptor in presence of a secreted ligand, as observed, for instance, in sarcomas, where production of c-met, HGF/SF receptor, is switched on (Kuhnen et al., 1998; Ferracini et al., 1995; Cortner et al., 1995). Moreover, a result similar to autocrine stimulation is also given by constitutive activation of a signaling receptor, either by activating mutations, such as in ras (Billadeau et al., 1997) or in tpr-met (Ramesh et al., 1998), or by overexpression, such as in erbB2 (Hynes and Stern, 1994), or c-met (Rusciano et al., 1996). Only identification of the responsible factor(s), and a detailed molecular analysis at both the protein and the DNA level can clarify the mechanisms involved.

However, if secretion of an autocrine ligand is suspected, its presence in the conditioned medium might be revealed by testing the growth effects of the medium on the same cells.

Materials

Set up for tissue culture.

Procedure

Source cells are usually plated at high density, to maximize the yield of secreted factors. Conditioned medium (c.m.) can be produced either at low serum (between 1 and 0.1%) in order to keep a minimal amount of exocrine growth factors, or under serum-free (SF) conditions. In this case, precoating of tissue culture plastic with adhesion factors (e.g. fibronectin, laminin or collagen IV, usually at 1 μg/cm^2) should be done. Alternatively, cells can be plated at low serum, and, once spread out, switched to SF-medium. In this case, before collecting SF-c.m., one or two changes of SF-medium (each of at least 4 h) have to be made, to allow shedding of serum particles from cells and dish, which might influence the result of the experiment. Conditioned medium to be tested can be collected after 24 h of incubation, and a second collection (on addition of fresh medium) can be done after further 24 h. Of course, cells have to be examined at the microscope, and look healthy. If considerable cell lysis occurs under SF conditions, proteins and lytic enzymes released by the dying cells could be toxic to test cells, and mask any effects of secreted growth factors. Conditioned medium must be cleared by centrifugation (3000 rpm, at 4°C) to eliminate cellular debris. Further ultracentrifugation at 100,000 g is necessary to ensure that only soluble factors, and no membrane-bound factors, are present in the c.m. If necessary, c.m. can be filtered through 0.22 μm for sterilization, but care has to be taken to minimize the surface of filtering, which could unspecifically bind proteins present in the medium. Once prepared, c.m. can also be aliquoted and stored frozen for some time (usually a month or two).

Test cells (the same as source cells, if autocrine growth effects have to be shown) are plated at very low density (<100 cells/cm^2), in the same kind of medium used to produce the c.m. Once attached (for 4–8 h after plating), their medium is replaced with c.m. at different dilutions (100, 50, 10, 1 and 0%). Growth has to proceed for at least one week, so one change of medium can be done after 3–4 days. At the end of the experiment, cells can be quantitated by any of the methods described above.

Plating cells at low density has the scope of minimizing the effects of freshly produced autocrine factors, while maximizing the effects of factors contained in c.m. Conditioned medium, by its own nature, is partially exhausted, therefore dilution with fresh medium should restore its ability to support cell metabolism. Best effects can be expected at dilutions between 50 and 10%. The method should be validated in the lab by using a cell line in which autocrine growth stimulation has been demonstrated. However, it should be always considered that the absence of a response is no proof that there are no autocrine growth effects, because the amount of factor produced could be below the threshold of detection, or because its stability in solution is low. A way to obviate to this last problem, could be to plate the experiment in transwells, where the same cells serve again as source of factor (plated at high density in the bottom well), and as tester (plated in the transwell, at low density). To prevent overgrowth of source cells, the transwell should be shifted to freshly plated source cells every 48 h, until the end of the experiment.

5.1.2.2. Paracrine growth effects
Malignant tumor cells lodging at distant organ sites have to cope with a new microenvironment, not necessarily well suited to sup-

port their survival and growth. Therefore, those cells endowed with autocrine growth stimulatory ability, or the possibility to respond to paracrine growth factors, will have a growth advantage which might result critical for the establishment of successful metastases (Rusciano and Burger, 1992). Paracrine growth factors can derive from different sources, such as parenchymal cells (Sargent et al., 1988), fibroblasts (Okumura et al., 1992), endothelial cells (Hamada et al., 1992), mast cells and macrophages, beside noncellular sources such as the interstitial extracellular matrix or the BM (Doerr et al., 1989). If a tissue- or organ-specific effect is suspected, the relative factors could be recovered in organ extracts, organ conditioned medium, or organ tissue culture.

Organ extracts. Whole organs to use as source of factors are easily obtained from laboratory animals (usually, mice or rats). The possible species-specificity of candidate factors has to be taken into account. Animals are anesthetized by pentobarbital i.p. injection, and the blood removed by whole body vascular perfusion (Alterman et al., 1989). A 23 gauge needle is inserted into the left ventricle of the heart, while a venous vent is opened into the right atrium to provide adequate drainage and prevent overfilling of the systemic vascular system. Ten ml of PBS containing 1% heparin, and prewarmed at 37°C are circulated through the organs at a constant perfusion pressure of 25 mmHg. Organs are then removed from the animal, and cleaned as much as possible from surrounding extraneous tissues, such as large blood vessels and adipose tissue. All the subsequent steps are performed at 4°C. After further rinsing in PBS/heparin, organs are cut in small pieces, which are rinsed again in PBS/heparin, and then homogenized (e.g. with a Polytron homogenizer) in PBS (between 2 and 5 volumes) with antiproteases (a cocktail of 2 μg/ml aprotinin, 1 μg/ml leupeptin and 1 mM AEBSF can be used). Nuclei are removed from the homogenate by centrifugation at 2000 g ($\sim$3000 rpm in a table-top centrifuge) for 15 min, and the supernatant further centrifuged at 100,000 g ($\sim$37,000 rpm in a regular ultracentrifuge) for 1 h. The supernatant is saved as the soluble organ fraction (after sterilization by filtration and protein determination, it can be stored frozen at $-70°C$); the pellet is resuspended by homogenization in PBS/antiproteases, and centrifuged again as before. The super is discarded, and the pellet (representing the insoluble fraction, containing membranes, organelles and the organ matrix) can now be extracted. Extraction with high salts (e.g. 2 M NaCl, 5 mM EDTA in 30 mM Tris-HCl, pH 7.3 plus antiproteases as above; overnight in the cold) (Rusciano et al., 1991, 1992) will release surface-adsorbed factors, whereas extraction with a detergent (e.g. 25 mM octylglucoside in PBS, plus antiproteases: 1 h at 4°C) (Rusciano et al., 1993) will remove also transmembrane proteins. Extracts are then centrifuged again at 100,000 g, and the supernatant extensively dialyzed against PBS/antiproteases (octylglucoside is dialyzable). After dialysis, the extract is cleared again by ultracentrifugation, and sterilized by filtration (on 0.45 or better 0.22 μm). Protein content is determined by a suitable assay, and the extracts conveniently aliquoted and stored frozen at $-70°C$. The effects of both the soluble and the particulate organ fractions can be now assayed on the growth of cells in vitro.

Alternatively, whole organ extracts can be prepared by direct homogenization in extraction buffer (Rusciano et al., 1991).

Organ perfusates. Organs such as lung and liver can also be extracted in situ. It is, however, very much difficult to use more than one organ per animal. Organ retrograde perfusion is obtained through the use of a peristaltic pump connected to a blunted cannula (a 25-gauge blunted needle will work) which has to be introduced into either the portal vein or the inferior vena cava (from the right atrium of the heart) if liver is the target, or the pulmonary vein (from the right ventriculus of the heart) if lung is the target. Remember to cut open another large blood vessel from the target organ to allow flow and recovery from the perfusion. After rinsing the organ with PBS/heparin until most of the blood is removed (and the color of the organ has turned to be very pale), the organ is then extracted in situ by recirculating (at a very slow flow rate) at least three times

15 ml (liver) or 6 ml (lung) of extraction buffer (either high salt, or detergent, with antiproteases). This is done by fishing the buffer from a tube, and once the tube is almost empty, refilling it with the perfusate which has been accumulated in another container. Be careful to prevent air to enter the organ circulatory system. The perfusate is cleared by low speed centrifugation (∼1500 rpm) to remove cells and debris, dialyzed against PBS and concentrated. After sterilization by filtration, proteins are determined, and the perfusate is aliquoted and stored frozen (Rusciano et al., 1998a).

Organ conditioned medium. Organs are taken from animals as described above, however taking care to use sterile conditions, and cut in small pieces (≤ 1 mm^3) in a sterile dish. The organ fragments are then rinsed with sterile culture medium, their weight estimated in a preweighed sterile container, and finally placed in a 50 ml tube with 5 ml Hepes-buffered tissue culture medium, antibiotics and antimicotics, and (optional) 1% FCS. The air in the tube is equilibrated with an atmosphere of 95% air-5% CO_2, and the tubes are placed on a roller bottle apparatus for 4 h at 37°C. This medium is discarded, and replaced with SF-medium (DMEM:F12 in equal proportions can be used) containing 0.25 mg/ml BSA, at a ratio of 4:1 (vol/wt: 1 ml of medium per 0.25 g of tissue), so that different density of tissues is taken into account. In this way c.m. from different organs can be directly compared. The tubes are again equilibrated with 95% air-5% CO_2, and returned to the roller bottle apparatus at 37°C for a period of 12–24 h. The c.m. is collected, and cleaned from cells and debris by centrifugation at 900 g for 15 min. FCS to the same concentration as used in the growth assay (e.g. 0.2%) can be added now. The samples can then be aliquoted and stored frozen. To test organ c.m. growth effects, dilutions are prepared in fresh medium (as described before for autocrine growth stimulation), and added to cells in dishes (Horak et al., 1986; Nicolson and Dulski, 1986; Nicolson, 1987; Valle et al., 1992).

5.1.2.3. Tissue-specific cell co-culture

Whole organ extracts, or conditioned medium may suggest the presence of specific growth factors, however do not give any indications about their cellular source. For this purpose, it is possible to establish in vitro primary cell cultures from different organs, which then can be tested for their growth effects in a variety of assays. For instance, organ-specific fibroblasts can be obtained from almost every organ, and have shown specific growth effects (Mukaida et al., 1991; Cornil et al., 1991; Okumura et al., 1992). Hepatocytes (Sargent et al., 1988; Rusciano et al., 1993), osteoblasts (Goren et al., 1997), organ-specific endothelial cells (Hamada et al., 1992) have been shown to produce paracrine growth factors giving a specific growth advantage to malignant cells specifically metastatic to the organ from which the cell cultures were originated.

Conditioned medium can be prepared from organ-specific cells cultured in vitro, essentially as described under autocrine growth stimulation. Organ-specific cells can also be extracted in vitro, with the same procedure described for organ extracts. Conditioned medium and/or organ extracts are then tested in a growth assay on tumor cells.

Alternatively, co-culture of tumor and host cells can be made so that the two cell types are either in the same, or in separate compartments, with or without direct cell contact. Effects on tumor cell growth in the absence of direct cellular contact suggests the presence of soluble, secreted factor(s), whereas when the growth effects require direct cell contact it is a hint that the responsible growth factor(s) are insolubilized either on the cell surface, or in the BM.

Co-culture without direct cell-to-cell contact. This type of co-culture can be achieved in two different ways. In the first way, host cells are plated onto the bottom well of a transwell plate system, while tumor cells are plated in the transwell, on a tissue culture-treated membrane, with pores smaller than 3 μm, so to prevent migration of the cells across the membrane. The two cell types can be grown together until the end of the experiment, or the transwell can be moved to freshly plated host cells, in case the growth rate of the latter is higher than that of the former, or if tumor cells have been plated at very low density. Moreover, growth can be separately evaluated in both host and tumor cells, so to assess (with respect to appropriate controls) the reciprocal influence on growth of the two cell types. The second method is a slight modification of the soft agar colony formation assay (see above). The assay is prepared and carried out as before, with the tumor cells carefully dispersed as a single cell suspension in the upper layer of 0.3% agar in complete culture medium. The difference is that now the bottom layer (of 0.5% agar) contains a single cell suspension of host cells, which serve as source of diffusible growth factors. Therefore, the amount of colonies (defined either by size, or by cell number if countable) formed by tumor cells in the absence, or presence of host cells in the underlayer gives a measure of the growth effects exerted by host cells, which can be either stimulatory (Strobel et al., 1989), inhibitory (Kooistra et al., 1997), or both (Mukaida et al., 1991). Also, the effects measured in the double-layer soft agar assay are not necessarily identical to those tested in an assay where both cell types are grown as monolayers, and the two assays can in fact give different results (Mukaida et al., 1991). Moreover, host cells may produce factors that are either labile, or that are induced in response to the presence of tumor cells; in either case c.m. will show no effect, while the co-culture will (Mukaida et al., 1991; Rusciano et al., 1998b).

Co-culture with direct cell-to-cell contact. In a very elegant study published in 1988 by Sargent and collaborators, the lab of Max Burger has provided strong evidence that, not only paracrine growth stimulation can be critical for the development of site-specific metastasis (Burger and Madnick, 1983), but also that paracrine growth interactions may require direct cell-to-cell contact (juxtacrine interactions).

To achieve co-culture conditions that allow to address the presence of juxtacrine growth interactions, keeping the two interacting populations separated, cells are plated on both sides of a porous membrane. A pore size between 2.0 and 0.8 μm allows cell contact, but no migration through the filter, while pores smaller than 0.2 μm allow diffusion of soluble molecules, but no cell contact. Nucleopore polycarbonate membrane filters (Corning-Costar) offer a wide range of uniform pore size. To make the filters (35 mm diameter) suitable for cell plating, they are first prewet in a solution of 10 N NaOH, then extensively rinsed in deionized water (check the pH of rinsing water, until neutral), and finally coated with polylysine (0.1% in water, 15 min at room temperature). After a final rinse to remove the excess of polylysine, the filters are mounted on specially home-made metal frames, which create two chambers (one per each side of the filter) in which cells are plated (Fig. 5.1). The assembly is then sterilized by autoclaving, and posed into a 10 cm petri dish. Cells are plated first on one side of the filter, in 3–5 ml, and 24 h later, after inversion of the apparatus, onto the other side of the filter. Cells on the bottom side are kept wet by immersion into ~25–30 ml of medium in the petri dish (care has to be taken not to make air bubbles). At the desired times, cells from each side can be separately removed by trypsinization, and counted, or host cells can be wiped off the filter with a sterile cotton swab, and tumor cells on the other side quantitated either by colorimetric, or radioactive means (Sargent et al., 1988; Goren et al., 1997). Controls have to be prepared, in which host cells or tumor cells are plated alone on one side of the filter. Regular transwells can also be used to make co-cultures with cell-to-cell contact. However, transwells are not constructed to this purpose, and so plating cells to the bottom side of the filter is more tricky. Cells have to be plated in a small volume that can stay (just like a big bubble) on top of the filter, without running down the walls of the transwell, and so they will not be very homogeneously distributed over the filter surface.

Alternatively, if host cells grow at a much lesser growth rate as compared to tumor cells, then the two cell types can be plated

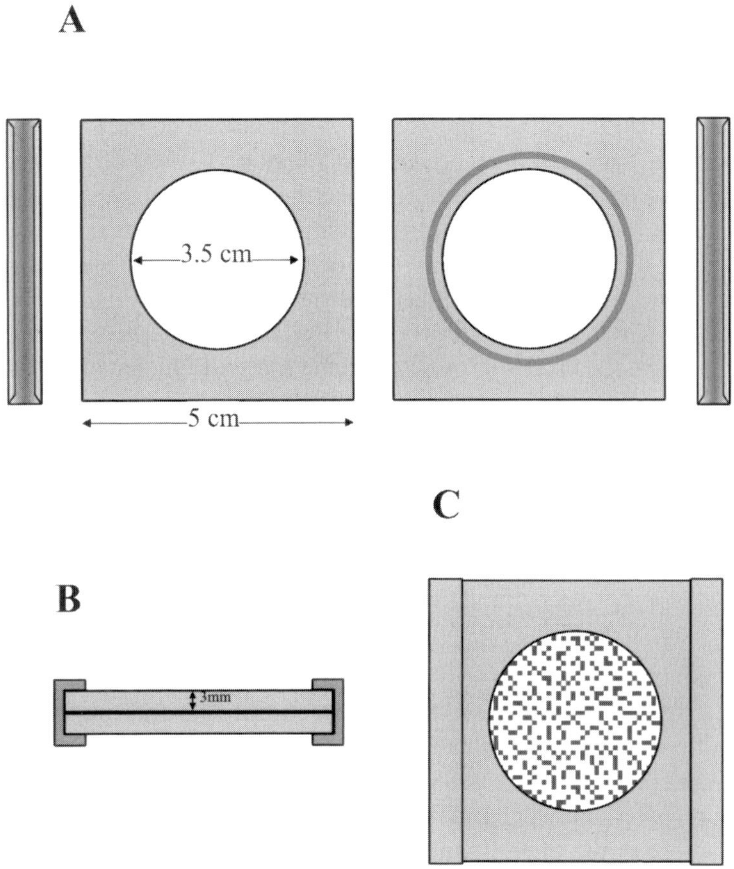

Fig. 5.1. Filter holder for co-culture assay. It is made of two holed aluminum plates, one of which has a fissure for an O-ring, and two clamps also made of aluminum (**A**). Once the membrane has been treated for tissue culture (see text), it is sandwiched between the two plates, which are kept in place by the two clamps (**B** and **C**), and the whole cluster is then sterilized by autoclaving. Once put together, the assemble contains an upper and a lower chamber, in which cells can be plated and, depending on the pore size of the filter, establish cell-to-cell contacts.

together in the same dish. Reference controls will contain host or tumor cells alone, under the same culture conditions. Then, growth of tumor cells in co-culture can be estimated, e.g. by ^{3}H-TdR labeling, after subtraction of background incorporation by host cells (which should not be more than 10% of the total amount incorporated in the co-culture system). Therefore

$$Rge = (cpmCo\text{-culture} - cpmHC)/cpmTC$$

where Rge is relative growth effects, HC are host cells, and TC are tumor cells (Sargent et al., 1988; Rusciano et al., 1993). However, this system does not allow to distinguish whether, or not, the tumor cells themselves may exert a growth effect on the host cells. A possibility is to treat the host cell monolayer with either γ-rays (4000 rads), or mitomicin C (4 μg/ml, 2 h at 37°C), to prevent incorporation of radioactivity by these cells (Cornil et al., 1991).

A way to distinguish either cell type in co-culture in the same dish, is to add a label to only one cell type. For instance, GFP stable expression can be used to label either the normal or the tumor cell type, whereas 5-chloromethylfluorescein-diacetate (CMFDA), which is taken up by cells in culture, is better used to label slowly growing cells, so that the dye is not diluted because of a high cell division rate. In this latter case, fibroblasts at confluence are treated with 5 μM CMFDA for 45 min in the incubator, then washed and incubated with complete medium for 1 h, and finally washed again to remove any trace of CMFDA. Tumor cells are then plated on top of the labeled monolayer, and the co-culture evaluated up to 4 days (CMFDA labeling is stable for even some longer time). Cells are detached by trypsinization, counted and pelletted by centrifugation. They are then resuspended in 1.0 ml, labeled with propidium iodide (25 μg/ml) and Hoechst 33342 (5 μg/ml), and subjected to FACS analysis in order to sort fibroblasts (CMFDA$^+$) from tumor cells (CMFDA$^-$), and each population evaluated for apoptosis (PI exclusion) and proliferation (amount of cells in S phase of the cell cycle) (Belloc et al., 1994). By using this method, it has recently been reported that direct co-culture with normal human fibroblasts decreases the apoptotic rate of human prostatic tumor cells in vitro, and enhances tumor growth rate in vivo (Olumi et al., 1998).

5.1.2.4. Growth on organ-specific ECM
The growth influence of organ-specific biomatrices can also be evaluated. A BM coating can be left on a dish after removal of the cells by nonenzymatic methods, as described for the preparation of the subendothelial matrix (see Chapter 2, Section 2.3.4.1). Alternatively, a whole organ-specific biomatrix can be prepared from organ homogenates, as detailed in Chapter 2, Section 2.3.3.4. Tumor cells are then plated, usually at clonal density, on the matrix, and growth of colonies is detected after a week or two (Doerr et al., 1989).

5.2. Growth interactions in vivo

The ability of tumor cells to respond to autocrine and/or paracrine growth factors may influence their growth ability also in vivo. When the cells activate an autocrine growth stimulation circuit, it might be expected that they have enhanced growth kinetics at any implantation site, as compared to cells of the same type, however unable to sustain autocrine growth stimulation. In contrast, when growth in vivo depends on paracrine growth interactions with a specific type of host cells, tumor growth is expected to be enhanced only at that organ site where such interactions can take place, as in the case of liver-specific B16 melanoma cells, which clearly show preferential growth with hepatocytes in vitro, and in the liver in vivo (Rusciano et al., 1993). Therefore, the effects of phenotypic changes of tumor cells, either by selection or genetic manipulation, after having been characterized in vitro, also need to be confirmed in vivo to establish their physiologic relevance.

5.2.1. Quantitation of tumor growth in vivo

There are basically two ways of detecting tumor growth in vivo. Tumor size and/or weight can be measured, and growth expressed in function of actual tumor size; or animals can be treated with a precursor of DNA, which will be preferentially incorporated by fast growing tumor cells, and growth then expressed in function of label uptake.

5.2.1.1. Size measurement

After subcutaneous injection of tumor cells, the resulting tumors, localized within the layers of the skin, are gently retracted from the body of the animal, and length (l), width (w) and height (h) measured to the nearest 0.5 mm with a caliper. Tumor volume (V) can be calculated thereafter using the formula for the volume of a hemiellipsoid (Rockwell et al., 1972; Janik et al., 1975):

$$V = 1/2 \ (4\pi/3) \ (l/2) \ (w/2) \ (h) = 0.5236 \ lwh$$

Tumor weight (wt) can also be extrapolated from the following formula (Haranaka et al., 1984; Giavazzi et al., 1986; Kurachi et al., 1991):

$$wt(mg) = (l \times w^2)/2$$

where l and w are expressed in mm.

From either value (volume or weight), the tumor doubling time (T_D) can be calculated as follows (Haranaka et al., 1984):

$$T_D = [(LN \ wt_i - LN \ wt_0)/LN2] \times 1/i$$

where wt_i is the mean tumor weight (but also the volume can be used) of a group at a given time, and wt_0 is the initial mean tumor weight for that group, 'i' being the number of days elapsed between the two measurements.

Measuring the size of tumors in this way, allows to keep the animals alive throughout the experiment, so that values are always referred to the same individuals, thus minimizing the variability of the experimental data. However, only growth of exterior primary tumors (such as subcutaneous or intrafoot-pad) can be followed by this way. If organ metastases have to be evaluated, groups of animals have to be sacrificed at certain time points, and the most suitable direct approach to evaluate metastatic tumor growth is weighing the organs with metastases, and the same organs from a control mice group matched by age, and subtract the latter from the former (Rusciano et al., 1993). Of course, cachectic effects of tumor growth on the organ size and weight are difficult to take into account with this method, which may then lead to an underestimation of tumor weight.

5.2.1.2. Incorporation of a labeled DNA precursor
Tumor cell proliferation in various organs can be evaluated by measuring DNA synthesis after that animals have been treated with DNA precursors which can be differently labeled.

Uptake of radioactive 5-iododeoxyuridine (^{125}IUdR) has been shown to nicely correlate with tumor growth at different organ sites (Bonmassar et al., 1975). Tumor-bearing animals are first injected i.p. with 5-fluorodeoxyuridine (25 μg/mouse) to decrease the endogenous availability of thymidine (Bennett et al., 1968). One hour later a second i.p. injection with ^{125}IUdR (0.5–2.0 μCi, specific activity 8.4-8.7 μCi/mg) is given to the same animals. After 24 h the organs are removed, rinsed well in PBS/heparin (1 U/ml), and soaked for 48 h in 70% ethanol (changing the ethanol solution every 12 h) to remove soluble radioactivity (Rusciano et al., 1992). Each organ is then transferred to a new clean vial, and the radioactivity measured in a gamma counter. Normal control mice, matched by age and sex, should also be injected to establish the background level of radioactive incorporation per each organ. This value has then to be subtracted from the value obtained for tumor-bearing mice, to obtain the actual amount of radioactivity incorporated by tumor cells. A direct correlation between tumor size (or amount of metastatic colonies) and radioactive counts has been demonstrated (Bonmassar et al., 1975).

Alternatively, BrdU labeling can also be used in vivo. BrdU is initially dissolved by vigorous stirring at pH 9.0, and once in solution, the pH is brought to 8.0. Thirty minutes, up to 24 h before sacrifice, animals are given an i.p. injection of BrdU at a dose between 25 and 100 mg/Kg, in 100 μl of PBS (Saito et al., 1992; Marengo and Chung, 1994; Kobayashi et al., 1996). Organs of prelabeled animals are then fixed in 70% cold ethanol, embedded in paraffin, and sections cut at different levels throughout the organ, so to allow a panoramic of metastatic deposits. At least 1000 tumor cells have to be considered per each organ, and the BrdU labeling index (LI) is then defined as the number of BrdU-labeled cells divided by the total number of cells counted expressed as percentage (Christov et al., 1994). A higher LI will correspond to a higher growth rate of the metastatic deposit.

5.2.2. Paracrine growth interactions in vivo

Differential tumor growth at different organ sites suggests that paracrine growth interactions occur, that may facilitate metastatic outgrowth in a site-specific way. However, the possible source of these factors remains elusive. Testing the effects of different cell types in vitro may give some evidence towards the role of certain cells, but this is not a final proof that the same effects will also be relevant in vivo. In order to study the growth interactions between tumor cells and a specific cell population, the two cell types can be co-injected in the animal, and tumor development observed in presence, or absence, of the paracrine partner.

Co-injection can be done by simply mixing the two cell populations in suspension in an adequate injection vehicle (e.g. HBSS), usually in a 1 : 1 ratio. By using this method, the growth promoting effect of bone stromal cells on prostatic cancer has been demonstrated, after subcutaneous injection of a mixture containing 10^6

bone fibroblasts, and 10^6 cancer cells (Gleave et al., 1991; Wu et al., 1994). Tumor cells and fibroblasts alone, also at higher cell number, have also to be injected as controls, and histochemical analysis of the tumors developing after co-injection has to show that only tumor cells are contributing to tumor size, and not fibroblasts, or host infiltrated inflammatory cells. If soluble factors produced by host stromal cells are responsible for the paracrine growth effects, a possible way to test their effects in isolation is to absorb them on a binding matrix, which is then co-injected with tumor cells. Gelfoam (Pharmacia-Upjohn), has been successfully used to this purpose (Gleave et al., 1992). Under sterile conditions, Gelfoam, a solid gelatin sponge, is presoaked with 100 μg/ml collagen IV for 12 h at 4°C, followed by either collagen-binding growth factors at 1 μg/ml, or a tenfold concentrated conditioned medium, for 1 h. The Gelfoam is then finely minced with a polytron homogenizer to allow s.c. inoculation with an 18-gauge needle.

Alternatively, cell populations to be co-injected can be mixed together within a natural extracellular matrix, such as Matrigel^{TM}. However, since Matrigel^{TM} by itself can have an enhancing effect on tumor growth (Fridman et al., 1991; Topley et al., 1993), also because of its content of growth factors (Vukicevic et al., 1992), controls have to be made injecting either tumor or host cells plus Matrigel^{TM} alone, and a mix of cell populations with growth factor-depleted Matrigel^{TM}.

Tumor and host cells to be injected are resuspended at the desired concentration in 0.25 ml of cold serum-free medium (or HBSS), mixed with the same amount of cold liquid Matrigel^{TM} (10 mg/ml), and injected subcutaneously in laboratory animals (Noel et al., 1993, 1998).

Growth factor-free Matrigel^{TM} can either be bought as such, or prepared by ammonium sulphate precipitation of Matrigel^{TM}. Ice-cold Matrigel^{TM} is precipitated with 20% ammonium sulphate, centrifuged, and the pellet resuspended in TBS, pH 7.4. A second precipitation with 20% ammonium sulphate can be repeated, for better cleaning. Thus treated Matrigel^{TM} is dialyzed for 2 h against

TBS containing 0.5% chloroform, then for 2 h against PBS alone, and finally against serum-free medium (or HBSS) (Taub et al., 1990). Protein concentration should be adjusted to ~10 mg/ml, as in the original Matrigel^{TM}.

It is interesting to note that from co-injection experiments with Matrigel^{TM}, it has emerged that the cooperative effects of host fibroblasts is exerted not only through direct secretion of growth factors, but also by release of proteases that can liberate growth factors adsorbed to the extracellular matrix, making them available to the cancer cells (Noel et al., 1998).

5.2.3. Autocrine growth stimulation in vivo

To measure the influence of autocrine growth stimulation on tumor growth in vivo, there are basically two ways. A first approach is to artificially induce, e.g. by transfection, an autocrine loop in a given cell line, preferentially with the transgene under the control of an inducible promoter (so that the same transfected cell line serves as an internal control), and then evaluate the rate of tumor growth in the presence or the absence of an activated autocrine loop (Jiang et al., 1998). However, if the relevance of an ascertained autocrine loop, naturally present and activated in certain tumors, has to be evaluated, the approach has to be different. In this case, a host whose paracrine (or endocrine) contribution to the loop is negligible should be chosen or created, and then tumor growth in such a host evaluated when the autocrine loop is functioning, or when there is an interference (e.g. specific antibody treatment) with the loop (Kurachi et al., 1991). Interestingly, interfering with autocrine growth stimulation with specific monoclonal antibodies, can have also some therapeutic value, as shown by preclinical trials with small cell lung cancer (Avis et al., 1991), or clinical trials with breast cancer (Baselga et al., 1996).

CHAPTER 6

Selection of metastatic variants

A striking property of malignant tumors is their heterogeneity, so that at the time of diagnosis, most animal and human neoplasms are populated by cells with different biological properties (Heppner, 1984). Cells derived from the same tumor have been shown to differ with respect to their antigenic or immunogenic profile, growth rate, karyotype, pigment production, cell surface receptors for lectins, hormone receptors and response to cytotoxic drugs (reviewed in Hart and Fidler, 1981). Although in certain cases a tumor can be of multicellular or multifocal origin (Muto et al., 1995; Heim et al., 1997), and thus its heterogeneity can be initially explained by the diversity of its original cells, most tumors are monoclonal in their origin (Fialkow, 1972, 1976; Nowell, 1976), and so diversification and progression should result from acquired genetic variability within the original clone. Indeed, genomic instability appears to be a hallmark of most cancers (Cheng and Loeb, 1993), either of monoclonal or polyclonal origin, and mutations of genes involved in DNA repair or in cell cycle checkpoint controls have been implicated in the raising of a mutator phenotype in cancer cells (Lindahl, 1994; Perucho, 1996; Imai et al., 1998; Gualberto et al., 1998).

Also the metastatic phenotype is not homogeneous within a tumor population, which is rather composed of a variety of cell subpopulations with widely differing invasive and metastatic capabilities (Hart and Fidler, 1981). By using a modification of the classical fluctuation test of Luria and Delbrück to distinguish between selection and adaptation in the origin of bacterial mutants, metastatic variants have been shown to pre-exist within a tumor

(Fidler and Kripke, 1977). That metastasis is a selective, rather than a simple adaptive phenomenon is also indicated by experiments showing that, while active selection of truly metastatic variants (derived from the spreading of a primary tumor) results in an increase of the metastatic ability of cells recovered from metastatic colonies (Raz et al., 1981), the mere lodging and growth of cells in an organ (for instance, after injection of cells on microcarrier beads, which favors mechanical trapping in the first capillary bed) will not necessarily yield a more metastatic cell population (Nicolson and Custead, 1982). Beside being sequential and selective, the process of cancer metastasis also contains stochastic (involving chance or probability) elements (Price et al., 1986). In fact, if the process of metastasis depended solely on selective tumor cell properties, cell lines derived from metastases or by clonal isolation of pre-existing variants, should display a much greater metastatic efficiency than the results indicate. Moreover, individual metastases also tend to become heterogeneous, and while early metastatic colonies of clonal origin contain cells with similar metastatic abilities, late colonies contain a heterogeneously metastatic cell population (Poste et al., 1982). Accordingly, it has been suggested that the higher metastatic ability of a selected cell line derived from the B16 murine melanoma may also depend on its higher mutation rate, which could continuously favor the generation of more metastatic cells (Hill et al., 1984).

In conclusion, it is on the one hand the presence of a heterogeneous population within a tumor, and on the other hand the fact that intrinsic properties of tumor cells, and not adaptive mechanisms, are important for metastatic ability, that renders a tumor (either primary, or a metastasis) amenable to selection of metastatic variants. We will describe next different selection methods for such metastatic variants, which can be performed either in vivo, or in vitro, depending on the type of metastatic ability that is being looked at.

6.1. Selection of organ-specific metastatic variants

Clinical data obtained from patients with histologically defined primary tumors indicated a tendency for metastases to occur in particular distant organs (Sugarbaker, 1981). This nonrandom pattern of metastasis was already evident to the British surgeon Stephen Paget who, in a seminal paper published in 1889, proposed the now famous 'seed and soil' theory, stating that metastases resulted from the proliferation of a few tumor cells (the seed) in the favorable milieu provided by some organs (the soil). In 1928 Ewing challenged this theory, arguing that the pattern of metastases distribution is influenced by purely mechanical factors, mainly attributable to the vascular connections of the primary tumor. Ewing's proposal, however, does not explain the observation that some organs (such as brain, bone and adrenals) are served by a very small fraction of the circulatory system, yet they are often involved by metastatic deposits of certain cancers. Moreover, other organs, such as heart, muscle, skin, kidney and spleen, each receiving a considerable supply of blood, are only sporadically colonized by cancers. The current opinion today is that both Paget and Ewing were partially correct, and that the two theories need not to be mutually exclusive. Some metastatic tumors can colonize a wide variety of tissues, their lack of preference suggesting that the first site encountered by this kind of metastatic cells can be the most common site of metastasis formation. Other tumors can be far more selective, bypassing proximal organs and selectively colonizing specific distal organs (Willis, 1973; Sugarbaker, 1981). This selectivity has been elegantly demonstrated by Tarin and collaborators in 1984. In the reported study, patients with incurable abdominal ascitic cancer were treated with peritoneovenous shunting in order to alleviate abdominal pain and distension. In this procedure, the abdominal effusion is returned to the circulation via an anastomosis, containing a one-way valve, between the peritoneal cavity and the lungs. Therefore, a large number of tumor cells are directly infused into the circulation. Despite this huge tumor load in the circulatory system, many patients did not develop evident metastases, and among those who did, the distribution of secondary deposits was unexpected, in that metastases did not form in the organ containing the first capillary bed encountered (i.e. the lungs). Organ-specific colonization by metastatic tumor cells has been observed in many rodent model systems (Rusciano and Burger, 1992). Already in 1952, Sugarbaker injected tumor cell suspensions from different types of tumors into the same site in rats, and observed that each type established its own pattern of metastases. Organ selectivity was also maintained towards ectopic implants of the preferred target tissue: following injection of lung-colonizing tumor cells, metastatic foci developed both in the in situ lungs and the grafted lung, but not in other grafted control organs (Kinsey, 1960; Sugarbaker et al., 1971; Hart and Fidler, 1980).

Murine cell lines with different organ-specific metastatic ability have been described. B16 melanoma cells have been selected in vivo to yield cell lines specific for lung- (Fidler, 1973a; Fidler and Nicolson, 1976), brain- (Brunson et al., 1978), liver- (Tao et al., 1979; Rusciano et al., 1993) and ovary- (Brunson and Nicolson, 1979) specific colonization. RAW117 lymphosarcoma cells have been selected for liver- (Brunson and Nicolson, 1978) or liver and lung- (Nicolson et al., 1982) specific colonization. Metastatic variants of the Lewis lung carcinoma (3LL) with either lung- or liver-specific metastatic ability have been reported (Brodt, 1986). The T-cell hybridoma cell line BW-14 preferentially colonizes the kidney, whereas its counterpart BW-19 metastasizes mainly to the spleen and the liver (Schmidt et al., 1994). F9 teratocarcinoma cells have been shown to colonize the liver with high selectivity; however, on retinoic acid treatment, they showed a remarkable change in their organ preference, shifting toward lung colonization (Terrana et al., 1987). For a comprehensive list of organ-specific tumor lines (see Chapter 8, Table 8.1).

The selection for organ-specific colonizing cells has been carried out mostly on in vivo animal model systems. Tumor cells are directly injected into the blood stream. In certain cases it seems advantageous that the target organ for which selection is desired

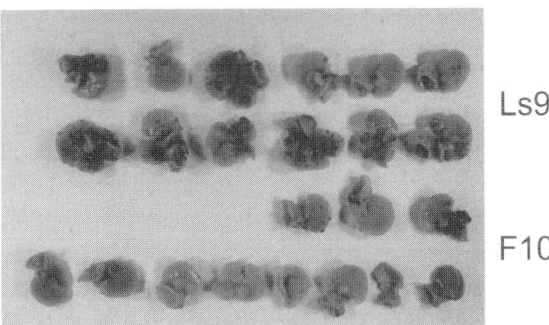

Fig. 6.1. Selection of organ-specific variants with the B16 melanoma cell line. (**A**) B16-F10 were selected after repeated injections in the tail vein, and subcultivation of lung colonies. B16-LS9 have been selected by means of serial injections through the mesenteric veins, and subcultivation of liver colonies. (**B**) Livers of C57 Bl/6 mice taken 2 weeks after intrasplenic injection of B16-LS9 (two upper rows), or B16-F10 (two lower rows) cells. The superior liver colonization ability of B16-LS9 cells is evident also when the liver is the first capillary bed along the route of intravenously injected cells.

is the first capillary bed encountered. Therefore, selection of lung metastatic lines (B16 and RAW117) has been achieved after tail vein inoculation (Fidler, 1973a; Nicolson et al., 1982); selection for liver metastatic B16 melanoma cells was obtained after mesenteric vein (Tao et al., 1979), or intrasplenic (Rusciano et al., 1993) injection (Fig. 6.1), whereas tail vein selection with B16 cells failed to give an increase in liver colonization ability (Rusciano, unpublished). Similarly, Lewis lung (LL) tumor cells could be selected for increased liver metastatic ability after intrasplenic injection (Pàl et al., 1983). However, the liver preference of selected LL tumor cells was evident after intracardiac injection (Paku et al., 1989), and that of selected B16 cells after either intracardiac or tail vein inoculation (Tao et al., 1979). Intrasplenic injection in nude mice was also used to select a liver metastatic variant of the human melanoma cell line A2058 (Ladanyi et al., 1990). Selection of brain colonizing B16 melanoma cells initially required intracardiac injections, and was completed after tail vein inoculation (Brunson et al., 1978), whereas ovary colonizing B16 cells were selected after tail vein injections (Brunson and Nicolson, 1979). In one reported case (Bresalier et al., 1987), selection for liver-metastatic variants was done starting from primary tumors developing after orthotopic implantation of colon cancer cells in the cecal wall of the syngenic mouse, and subcultivation of the liver metastases thus produced.

In each of the above studies, the procedure adopted is basically the same as originally described by Fidler (1973a, b): a single cell suspension, containing more than 90% viable cells as judged by trypan blue exclusion, is injected directly into the blood stream. The injection volume varies depending on the injection site, and up to 0.25 ml can be injected in the tail vein, while 0.1 ml are inoculated in the spleen or the mesenteric veins. Two to three weeks after the injection, depending on the growth rate of the injected cells at the desired site (it can take longer if human cells are injected in nude mice), animals are sacrificed and dissected under sterile conditions (they are usually dipped in 80% ethanol, and opened in two layers: skin first, peritoneal and pleural membranes next). The desired target organ containing the metastatic colonies is carefully removed, and placed in a tube or a dish containing sterile PBS. After further rinsing with PBS, metastatic colonies are excised from the organ parenchyma as clean as possible from host tissue, placed in another dish of appropriate size, and finely minced with the help of forceps and a scalpel, in order to set free the tumor cells. Then complete medium is added, or thus treated pieces can also be subjected to enzymatic treatment (trypsin, collagenase or dispase) to increase the amount of cells released from the tumor mass. Enzymes are neutralized by the addition of FCS-containing medium, and cells pelleted by centrifugation and replated in complete medium, containing antibiotics and an antimycotic (e.g. amphotericin). Tumor cells are grown in the dish until confluent, and split at least once, in order to eliminate all possible host contaminating cells. At this point some of the cells can be frozen to establish a freezing lot. When those left in culture are close to confluence, they can be prepared again for injection, and the whole procedure is repeated once more. Usually, 8–10 rounds of selection are required before an organ-specific cell line can be generated. This most likely reflects the stochastic nature of cancer metastasis (Price et al., 1986), so that at each round of selection a mixture of specific and nonspecific cells is recovered from metastatic colonies, and only after a certain number of rounds, the balance leans in favor of the specificity. Moreover, organ specificity can be enhanced, apparently when it is based on preferential adhesion to target organ tissue, by rounds of in vitro selection, which work by increasing the proportion of metastatic cells highly adhesive to the kind of tissue used as substrate. In this way, B16 melanoma cells selected for enhanced adhesion to lung cryostat sections also showed higher lung colonization ability (Netland and Zetter, 1985), and increased liver colonization was observed after selection of RAW117 large-cell lymphoma cells for adhesion to liver microvessel endothelial cells (LaBiche et al., 1993), or of Lewis lung 3LL H59 cells for adhesion to hepatocyte monolayers (Brodt, 1989). However, when organ-specificity appears to rely more on paracrine growth interactions, such as in the case of brain-colonizing (Nicolson et al., 1990), or liver-colonizing melanoma cells, selection in vitro by co-culture with the target tissue fails to give an increase in the organ-specific colonization ability (Netland and Zetter, 1985; Rusciano, unpublished). Similarly, co-culture of B16 melanoma cells with organ fragments in vitro did not result in a clear increase of metastatic ability (Price et al., 1988). If the relevant paracrine growth interaction that contributes to the organ specificity is known, though, it can be used as a tool for selecting cells with increased metastatic ability. Human colon cancer cells metastatic to the liver of nude mice after intracecal injection expressed higher amounts of the EGF receptor, and cell sorting on the basis of their EGF receptor expression resulted in increased liver metastatic ability for those cells expressing the highest amount of the receptor (Radinsky et al., 1995). Similarly, selection of rat mammary carcinoma cells for increased expression of transferrin receptor, yielded a cell population with enhanced metastatic ability to the regional lymph nodes (Cavanaugh and Nicolson, 1998).

6.2. Selection of metastatic variants with enhanced or decreased metastatic abilities

Surface properties are critical determinants of the metastatic behavior of malignant tumor cells. Adhesion molecules, proteolytic enzymes, antigenic determinants and hystocompatibility molecules, cytokine and growth factor receptors are all displayed on the cell surface, and known to play a specific role in the metastatic cascade. Therefore, selection of cell populations based on each and any of these properties can yield tumor cell sub-lines with varying metastatic capabilities.

6.2.1. Lectins, carbohydrates and other adhesion molecules

The first demonstration that changes associated with cell surface molecules could influence the metastatic behavior of cells came from a seminal study by Tao and Burger (1977), following the observation that lectins (which bind specifically to surface carbohydrates: Elgavish and Shaanan, 1997) could be used as selective agents to obtain cells with membrane carbohydrate alterations (Gottlieb et al., 1975; Stanley et al., 1975). Therefore, B16 melanoma cells were selected in vitro for their ability to grow in presence of a toxic concentration of the lectin wheat germ agglutinin (WGA). After 4 rounds of selective growth, the resulting cell line expressed less WGA receptors, and showed a dramatic decrease of metastatic capability through both lymphatic and vascular channels (Tao and Burger, 1982). The profile of cell surface glycoproteins was indeed found to be changed in the selected metastatic variants (Finne et al., 1980), and the change to be due to an increase in fucosyltransferase activity in WGA resistant cells (Finne et al., 1982). Accordingly, when WGAr cells were subjected to a further selection with a fucose-binding lectins, the resulting cell lines had lost expression of the fucosylated antigen SSEA1, and reacquired a high metastatic potential (Finne et al., 1989). Many other reports followed the original discovery of Tao and Burger (1977), and lectin resistance has been used to generate more metastatic variants (Kerbel and Man, 1984; Dennis et al., 1984; Ishikawa and Kerbel, 1989). Moreover, the altered balance between fucosyltransferase and sialyltransferase that emerged from the analysis of the original WGAr mutants (Finne et al., 1982) came to corroborate the idea that cell surface sialylation had a positive role in the invasive and metastatic potential of tumor cells (Yogeeswaran and Salk, 1981), a role that has been confirmed in further studies (Dennis and Laferté, 1987; Passaniti and Hart, 1988). More generally, an influence of cell surface glycoconjugates on the metastatic phenotype has been established (Humphreys and Olden, 1989; Hakomori, 1996), that can guide cell adhesion (Kannagi, 1997), or even tumor cell interactions with cells of the immunosurveillance system such as natural killer (NK) cells (Dennis and Laferté, 1985).

A special class of lectins are the 'selectins', carbohydrate-binding cells surface molecules transiently expressed by endothelial cells, that facilitate leukocyte arrest and extravasation at sites of injury. E-selectin has been proposed to facilitate attachment of tumor cells to the endothelium in a similar fashion (Bevilacqua and Nelson, 1993). Remarkably, tail vein injection of B16-F10 melanoma cells, transfected with fucosyltransferase to increase their expression of cell surface E-selectin ligand, produced lung metastases in normal syngenic mice, but gave predominantly liver metastases in transgenic mice stably expressing E-selectin in liver sinusoidal cells (Biancone et al., 1996), thus showing the importance of lectin-carbohydrate interactions also in directing metastatic colonization to specific organs (Kieda, 1998). More generally, transfection of glycosyltransferase (Gorelik et al., 1995, 1997; Yoshimura et al., 1995), or even glycosidase (Tokuyama et al., 1997) genes has been used to change the metastatic phenotype of several cell lines.

Lectin-based cell selection can be basically carried out in two main different ways: either by sequential subcultivation of cells resistant to toxic doses of the chosen lectin (Tao and Burger, 1977), or by panning cells on a nonadherent surface coated with the desired lectin, so that only those cells possessing the adequate carbohydrate receptors may attach to the dish.

6.2.1.1. Lectin resistance
In order to start the selection with an appropriate amount of selecting agent, a dose-response curve has to be determined to establish the LD50. To this purpose, 1 to 2×10^4 cells are plated per each well of a 96 multiwell plate in complete medium, and, once adhered, treated (at least in triplicate) with a wide range of lectin doses. Twenty-four hours later, cells are pulse-labeled with radioactive thymidine, and processed as described in Chapter 5, in order to detect the amount of proliferating cells. LD50 is defined as the lectin concentration that gives 50% inhibition of DNA incorporation. This concentration (or a little big higher, not over a 75% inhibitory effect) is chosen for the selection. Cells are plated in 10 cm dishes at 2×10^6, in complete medium, and treated with the chosen dose(s) of lectin. The first round of selection is terminated when only few

surviving cells are left in the dish, which should happen in a window of time between 2 and 10 days. In case the selection time extends over 2 days, a change of medium with freshly added lectin is advised every 48 h. Surviving cells are kept in the dish with fresh medium without selecting agent until they form colonies. At this point, further selection can proceed either on single colonies (which are cloned into separate dishes), or on the colony pool (in which case all the colonies are passed together in a new dish). One has to be aware that cloning may select by chance other phenotypic changes, not necessarily dependent on glycoconjugate biosynthesis. Then, the selection with the lectin is repeated several times as above, until most of the cells survive the toxic treatment with the lectin. Populations coming from each selective round can be compared by their LD50. It might happen that cell survival to a certain lectin is too low, so not to allow selection. In this case, treatment with a mutagen of the cell line before the start of selection, may result in a better yield of resistant colonies (Dennis, 1985; Dennis et al., 1981).

Panning. This method aims at separating cell populations on the basis of their ability to attach to an immobilized ligand (lectins and antibodies have been used, but probably other kind of ligands could be effective as well) (Smith et al., 1990; Wysocki and Sato, 1978; Ernst et al., 1989; Koch et al., 1992). Lectin panning dishes are prepared by coating bacteriological culture dishes (60 mm) with 4 ml of a 10 μg/ml solution of the lectin in PBS (certain lectins, such as concanavalin A, lentil and pea require the presence of Ca^{++} and Mg^{++} for binding, so 0.1 mM of each ion is added to PBS), for at least 2 h at room temperature. Panning with antibodies is done with an antibody solution diluted at 10 μg/ml in 50 mM Tris, pH 9.5, overnight at 4°C. After thorough rinsing with PBS, the dishes can be used immediately, or stored indefinitely in the cold. Primary antibodies against a cell surface determinant can be used for direct panning, or secondary antibodies against a primary. In the latter case, cells to be panned have to be incubated first with a dilution of the primary antibody (made at 10 μg/ml in tissue culture medium buffered with 10 mM Hepes, pH 7.4, 2% FCS), 30–60 min at 4°C. Before panning, cells are washed free of the primary antibody and the FCS by centrifugation, and resuspended in an appropriate medium for panning (see below) (Ernst et al., 1989). The choice of bacteriological culture dishes is made to prevent unspecific attachment to plastic. However, tissue culture plates can also be used, provided that after lectin coating they are saturated with 1 mg/ml BSA. Usually, also after BSA saturation, they tend to give a higher background of nonspecifically adherent cells.

Cells to be panned are prepared and resuspended in ice-cold serum-free (SF) medium, at a concentration of 10^6/ml. Cells are then kept at 4°C throughout subsequent steps in the panning procedure. 3 ml aliquots of the above suspension, containing 3×10^5 cells, are plated per each 60 mm panning dish, and incubated on ice for 1 h. Nonadherent cells are next removed by gently rinsing the dishes 5 times with 5 ml each time of ice-cold SF-medium. Finally, serum-containing medium is added to the dishes, which are then returned to the incubator for 48 h. Then, viable adherent cells are detached by enzymatic treatment, and replated onto standard tissue culture dishes. Specific displacement of the cells from the lectin ligand can also be obtained by incubating the dish with a large excess of the specific lectin-binding sugar or oligosaccharide. Once grown close to confluence, cells can be tested for their lectin binding ability, and either used as such in experiments, or further selected by a second panning procedure (Smith et al., 1990). Although a consistent enrichment can be obtained already after one panning selection, repeated panning selection ensure a more homogeneous population of cells. If the phenotype is not stable, a panning selection can be carried out routinely every month or so, in order to keep a selective pressure on the cell population. Panning selection can also work in the reverse order, and cells can be selected that DO NOT adhere to a specific ligand, or a lectin. In this case the procedure remains basically identical, only after the 1-h incubation at 4°C on the dish coated with the ligand, the nonadherent cells, that remain in suspension and come off with the gentle rinsing, are collected all together by centrifugation pelleting, and replated on a tissue culture dish. Negative selection for adhesion can also be repeated more than once, to enhance the homogeneity of the selected population.

Adhesion to specific substrates. Adhesion of malignant tumor cells to cellular or acellular components of the target organ is a necessary step for further invasion and growth as metastatic colonies. Hence, adhesion to ECM components or to parenchymal cells have been used to select cell populations with altered adhesion properties, and a changed metastatic phenotype.

Laminin is a major constituent of BM, which can interact with tumor cells and modulate their metastatic behavior. For instance, laminin can increase the production of collagenase IV from malignant cells, which can in turn enhance their invasive ability (Turpeenniemi-Hujanen et al., 1986); long-term culturing of B16 melanoma cells in presence of exogenously added laminin resulted in increased lung colonization after tail vein injection (Terranova et al., 1984). Two different bioactive peptides have been identified in the laminin molecule (YIGSR in the B1 chain, and SIKVAV in the A chain), both of which can promote tumor cell adhesion and migration (Graf et al., 1987a, b; Tashiro et al., 1989). However, while after co-injection of B16-F10 cells with YIGSR-containing peptides there was a reduced number of lung colonies (Iwamoto et al., 1987; Saiki et al., 1989), co-injection with SIKVAV-containing synthetic peptides resulted in an enhanced lung colonization ability (Kanemoto et al., 1990). Interpretation of these results, though, are complicated by the fact that these peptides have been found to also act on vascular endothelial cells of the host, with YIGSR inhibiting (Iwamoto et al., 1996), and SIKVAV promoting (Kibbey et al., 1992) angiogenesis and vascularization of the tumors. Nonetheless, human colon cancer cell populations selected in vitro for differential adhesion to laminin acquired very different tumorigenic properties, with the laminin-adherent subclone forming large tumors in nude mice, and the laminin-nonadherent subclone failing to form sizeable tumors (Kim et al., 1994). Moreover, selection of B16 melanoma cells by adherence or nonadherence to either one of the above laminin peptides yielded three metastatic variants with enhanced (selection by adhesion to YIGSR), or decreased (selection of nonadherent cells to either YIGSR or SIKVAV) metastatic abilities (Yamamura et al., 1993).

Fibronectin is another prominent constituent of BMs and extracellular matrix, and malignant transformation is usually associated with a failure to deposit extracellular matrix components such as fibronectin (Hynes, 1976). Co-injection of a synthetic peptide (GRGDS) containing the cell binding domain of fibronectin with B16 melanoma cells dramatically inhibited the formation of lung metastatic colonies in the syngenic mouse (Humphries et al., 1986b), suggesting that adhesion to fibronectin can be relevant to metastasis formation. On the other hand, transfection experiments leading to the overexpression of the classical fibronectin receptor ($\alpha_4\beta_1$) in CHO cells resulted in loss of the tumorigenic phenotype (Giancotti and Ruoslahti, 1991). Similarly, selection of human adenocarcinoma cells for increased adhesion to fibronectin coated dishes, yielded clones with decreased tumorigenic ability (Stallmach et al., 1994), indicating that strong adhesion to fibronectin may keep the cells in check for growth, and even reverse their tumorigenicity. However, to our knowledge, no metastatic variants have been derived by direct selection on fibronectin.

Adhesion to organ-specific structures can also be used to select metastatic variants, that will usually show increased homing to the tissue used for selection. In this way, B16 melanoma cells have been selected on lung cryostat sections (Netland and Zetter, 1981), Lewis lung carcinoma cells on hepatocyte monolayers (Brodt, 1989), and RAW117 large-cell lymphoma cells on hepatic sinusoidal cells (LaBiche et al., 1993).

Selection procedure. The first step is to appropriately coat a tissue culture dish with the substrate chosen for the selection. ECM proteins (commercially available) are diluted in PBS, and the coating done with an amount ranging between 1 and 10 μg/cm^2, by incubating the dish 1 h at 37°C, 6 h at room temperature, or overnight at 4°C. If the solution is not sterile, coating can be done under UV light in a laminar flow hood, until the solution is dried. In any case, after coating the dishes are rinsed three times with PBS, and either used immediately, or stored at 4°C for a few weeks. In case of cells, these have to be used when they form a confluent monolayer. Primary hepatocytes and endothelial cells do not proliferate very fast in vitro, and so they are unlikely to contaminate the selected cell population. However, if other cells, with a higher proliferation rate are used as substrate, care has to be taken not to carry them along during the selection procedure, and killing of the monolayer by, for instance, colchicin treatment might be considered (thorough rinsing of the dish has to be carried out, to remove all traces of the killing agent before the selection procedure). Next, saturation of free plastic sites with BSA at 1 mg/ml in PBS is advised, to prevent unspecific attachment. Cells to be selected are resuspended as single cells in serum free medium, and plated onto the coated dish at concentrations ranging between 5×10^4 and 5×10^5 per cm^2. After 15–30 min at 37°C, with or without gentle shaking, the dish is rinsed three times with serum free medium to remove nonadherent cells. Adherent cells are detached by a short incubation in PBS/EDTA, and gentle pipetting, and transferred to a new dish with complete culture medium, to let them grow. The selection procedure, similar to the panning technique described above, is repeated several times, until a cell population is obtained, that shows enhanced adhesion to the chosen substrate as compared to the initial population. The selected line can be kept under selective pressure, and be subjected to a new panning once every five or ten passages.

6.2.2. Resistance to immunity

During the metastatic process malignant tumor cells come in contact with many components of the immune system, and only if able to survive to the combined attacks of immune cells, they will manage to establish metastatic colonies. Two different main strategies can be adopted by tumor cells to escape the immune surveillance of the host: develop resistance to toxic cytokines, or hide their diversity, so to deceive immune cells. Loss of major histocompatibility complex (MHC) class I expression is a common finding on tumors of many origins, and may represent a way for tumor cells to escape T cells immune surveillance. In fact, T lymphocytes recognize foreign cell surface antigens almost exclusively in association with molecules encoded by the MHC. Specifically, in the murine system the H2 class I antigens appear to direct the recognition of neoplastic and virus-infected cells by cytotoxic T lymphocytes (CTL). Therefore, at least for those tumor cells that express high levels of tumor-specific antigens (TSA), it might be convenient to hide their diversity by an unbalanced expression of the H2 antigens. Indeed, many mouse tumors, irrespective of their mode of induction, showed a preferential loss of H2K and enhanced expression of H2D (Gopas et al., 1989). Furthermore, transfection of the H2K^b gene into metastatic H2K^b-deficient cells resulted in a reversal of the metastatic phenotype (Plaksin et al., 1988; de Giovanni et al., 1991). Finally, MHC antigens on tumor cells can also be masked, and made inaccessible to immune cells, through increased production of a glycocalix covering the cell surface, as happens in the case of epiglycanin secretion by TA$_3$ murine mammary carcinoma cells (Codington et al., 1979).

So-called 'natural, or nonadaptive immunity' also appears to have a prominent role in controlling the malignant phenotype of tumor cells. Both mice and humans with a low natural killer (NK) activity are more susceptible to develop metastases, whereas adoptive transfer of purified large granular lymphocytes (LGL: a subset of cells closely associated with NK activity) to immunosuppressed rodents restored their resistance to metastasis (reviewed in Whiteside et al., 1998). LGL and Kupffer cells are involved in the defense system of the liver under various physiological and pathological conditions. In a study with nude mice it was reported that only after depletion of NK activity was a murine colon carcinoma cell line able to produce liver metastasis on intraspleen injection, indicating that liver-resident NK cells can play an important role in preventing liver metastasis (Arisawa et al., 1990). Similarly, B16-F1 or B16-F10 melanoma cells dramatically increased their liver colonization ability after tail vein injection in triple immunodeficient mice, which, among other deficiencies, also lack NK activity (Calorini et al., 1992). Suppression of either NK cells or macrophage activity in nude mice was also enough to boost the liver colonizing ability of a human lung carcinoma cell line (Yano et al., 1997). The susceptibility, however, of tumor cells to macrophage killing does not yet have clearly defined mechanisms. On the one hand, Fogler and Fidler (1985) failed to select in vitro tumor cells that were resistant to macrophage-mediated lysis, concluding that tumor cell destruction by activated macrophages was nonselective and, therefore, could not be related to the metastatic potential. Based on this findings, Fidler proposes macrophage activation as a possible modality of cancer therapy (Killion and Fidler, 1998). On the other hand, Yamamura and colleagues (1984), working with a mammary murine adenocarcinoma that did not produce spontaneous metastases, found that an occasional metastatic subline was less susceptible to both macrophage-mediated cytolysis in vitro, and in vivo host antitumor mechanisms. Moreover, other authors have reported selection in vitro of macrophage-resistant 3LL cells of increased tumorigenic and metastatic potential (Remels and de Baetselier, 1987), which resulted to be correlated to their decreased expression of TNFα receptors (Remels et al., 1989). Interferon (IFN) treatment of tumor cells also gave somewhat intriguing results. The cellular response to either IFNα or IFNγ was an upregulation of the expression of H2 class I antigens, which resulted in decreased tumorigenicity of cell lines transfected with either one IFN gene (Lollini et al., 1993; Kaido et al., 1995). The decreased tumorigenicity is conserved also in T and NK deficient mice, and therefore might depend on an enhanced sensitivity to macrophage attack (Lollini et al., 1993). The lung colonization ability, though, of IFNγ expressing cells was dramatically increased, most likely because of an increased resistance to NK cells-mediated killing (Lollini et al., 1993). It is in fact known that NK cells are activated by the absence of H2 class I antigens, the expression of which can, conversely, protect tumor cells from lysis by NK cells (Moretta et al., 1998). LaBiche and Nicolson (1993) selected in vitro, by repeated subculturing in presence of 10 U/ml of IFNγ, and 1 μg/ml of lipopolysaccharide (LPS), cell populations that were highly resistant to the growth-retarding effect of these agents. After 7–13 sequential selections, the resulting variant lines were completely refractory to the growth-inhibitory effects of IFNγ and LPS, and had completely lost their tumorigenicity on s.c. or i.m. injection, and metastatic potential after tail vein inoculation. The selected variants conserved the full sensitivity to NK cytolysis, but were less adhesive to microvascular endothelial cells, suggesting that IFNγ selection in this model system resulted in lines that have altered properties, such as growth and adhesion, that are important in both the tumorigenic and metastatic process.

6.2.3. Complement-mediated cell killing

This method of selection is based on differential expression of cell surface antigens by a cell population. So, if the relevance of a specific cell surface component to the malignant phenotype of a cell line has to be tested, a possibility is to select a cell population with a strongly reduced amount of that component by selectively killing all the cells expressing high levels of that antigen, by using specific antibodies in presence of serum-derived complement.

This technique has been mainly used to purge bone marrow from tumor cells in case of autologous transplantation (Nimgaonkar et al., 1996; Duerst et al., 1991), but it can be nicely applied also to in vitro selection of tumor cell phenotypic variants (Starkey et al., 1982; Tsuruoka et al., 1993).

Selection procedure. This is basically done by exposing the cell population to an antibody specific for an exposed cell membrane component (e.g. a specific protein, or an oligosaccharide determinant), and then adding complement to lyse all the cells that have bound enough antibody to trigger a cytotoxic effect. A selection procedure carried out on cells in suspension is described next. The same selection can also be applied to cell monolayers.

A cell suspension in PBS or culture medium, containing 1 or 2% heath-inactivated (56°C, 30 min) FCS is prepared, at a cell density of 1 to 3×10^6/ml, and incubated with an appropriate dilution of the antibody for 30 min at 37°C. If presence of complement within the antibody preparation is suspected, its inactivation by incubation at 50°C for 30 min (before it is used with cells) is advised. Cells are then rinsed in the centrifuge three times with PBS or medium alone, resuspended in 1 ml of PBS or medium, and fresh complement (commercially available) added, usually at a dilution of 1 : 10, for 30 min at 37°C. One or two more additions of fresh complement at 30 min intervals have been shown to increase the rate of cell killing (Mabry, 1985; Duerst, 1991). Release of DNA from lysed cells can result in cell clumping, which can be avoided by adding Dnase (50 μg/ml) to the incubation mix (Scollay and Shortman, 1985). Thus treated cells are then rinsed free of complement by centrifugation (which can be also performed in a Ficoll or Metrizamide gradient, to remove dead cells), and plated onto a tissue culture dish in complete growth medium, to allow selected cells to grow. The selection procedure is then repeated a few more times (3–5), until a population with a stable survival index (see below) is obtained.

Optimal conditions for both antibody and complement dilutions have to be determined empirically, by a cytotoxicity test (see below). The complement-mediated lysis reaction proceeds between antibody-saturated target cells as one reactant, and complement as a second reactant. Therefore, when target cell surface antigens are saturated with the corresponding antibody, lysis will not increase with increasing antibody concentration. On the other hand, antibody aggregates, which can form at high concentrations (usually >5 µg/ml), can significantly inhibit killing, if not clarified by centrifugation immediately before use (Duerst et al., 1991).

The cytotoxic effects of antibody and complement on target cells can be evaluated by any of the assays already described in Chapter 5. Briefly, after cells have been treated with the selective mixture, they are rinsed and plated in multiwell dishes, at least in triplicates. Untreated cells are used as control. Viable cells are then stained with either crystal violet, or with MTT, and the dye solubilized in the appropriate buffer, to obtain optical density readings at the spectrophotometer. The ratio of treated vs. untreated cells o.d. readings gives a survival index, which should increase through successive selections. Alternatively, a more specific cytotoxicity test can be adopted, such as release of lactate dehydrogenase (LDH) from dying cells. LDH is a stable cytosolic enzyme, and is released from dying cells in a similar way to ^{51}Cr release from prelabeled target cells (Korzeniewsky and Callewaert, 1983). An LDH kit (Cyto-Tox 96^{TM}) is available from Promega, and has been used to assay cytotoxic lymphocyte killing (Burke et al., 1995).

A valid alternative to this method is now under development, and makes use of antibodies coupled to magnetic particles, so that cells expressing a specific antigen can be sorted by a magnet.

6.2.4. Deformability, migratory and invasive ability

In the process of cancer metastasis, tumor cells have to migrate from their primary location to distant sites. Consequently, those cells endowed with higher motility and invasive abilities, and that can resist the shear stress of being transported through narrow capillaries, will have higher chances to complete the metastatic cascade, and develop into metastatic colonies. Hence, selection in vitro for cells in which any of those characteristics is expressed at the highest level could result in some increase of the metastatic ability.

6.2.4.1. Active deformability

Tumor cells surviving in the microvasculature must be resistant to the shear stresses arising in the vascular bed (Brooks, 1984), the frictional forces originating between their peripheries and vessel walls (Weiss and Dimitrov, 1984) as well as have the ability to traverse capillaries which generally are rigid and smaller in diameter than tumor cells (Sato and Suzuki, 1976; Sato et al., 1977; Weiss and Dimitrov, 1984). Moreover, they have to be able to actively migrate, squeezing their way between endothelial capillary cells, into the organ parenchyma (Roos and Dingemans, 1979). Therefore, depending on the type of tumor, the degree of deformability of tumor cells may play a critical role in their ability to form metastatic foci. In fact, Tullberg and Burger (1985) managed to isolate tumor cell variants able to penetrate filter pores ten times smaller than their cell diameter, and found that they showed an increased spontaneous metastatic capacity when compared to unselected cells.

Selection procedure. Originally, cells were selected through polycarbonate Nucleopore filters, which had to be pretreated to give them a positive surface charge to allow cell attachment. Nowadays there are several suppliers of transwell chambers for tissue culture, with electronically controlled pore sized filters, that can be used. B16 mouse melanoma cells, plated at a density of about $1.5 \times 10^5/cm^2$, penetrate in 24 h with increasing efficiency pores of 3, 5 and 8 μm, as visualized at the microscope after fixation and staining of the filters (for details see Chapter 3 on motility assays). Therefore, selection was carried out on 2 μm pore filters, since this size was close to the penetration limit. For different cell types the penetration limit has to be empirically determined in a similar way. Cells are plated at a density between 1 and $2 \times 10^5/cm^2$, taking care to keep the density well beyond confluence, otherwise, specially with cells of epithelial origin, increased cell cohesion can prevent migration. After 48 h cells that have migrated to the underside of the filter are harvested by incubation with trypsin/EDTA which is placed in the bottom well, so to wet the filter underneath, and after

3–5 min at 37°C, growth medium is added, and used to rinse out cells from the underside of the filter with either a sterile pasteur pipette, or a pipette tip. Cells migrated from replicate filters are pooled together, and replated in a tissue culture dish, until they reach a sufficient number to perform a second selection procedure. Parallel filters during each selection are used to fix, stain and count the migrated cells, to evaluate the increase in migratory ability. Two rounds of selection were already enough to reach a plateau with B16 cells (Tullberg and Burger, 1985). Thus selected cells can be either used as a population, or further cloned, and the chosen clones tested for their migratory ability to confirm the change in phenotype. The increased penetration ability through filter pores appeared to be, in the case of B16 cells, a stable property, at least for 30 passages in culture (Tullberg and Burger, 1985). Cell deformability of the selected clones was then measured as the percentage of cells traversing 10 μm diameter pores at constant pressure as a function of time (see Chapter 3 on deformability), and found to correlate with both increased migration and metastatic ability (Ochalek et al., 1988).

6.2.4.2. Invasive ability
A simple variation on the method described above was used by Tullberg et al. (1989) to select B16 melanoma cells of increased invasive (and metastatic) ability. Filters with pores of 10 μm are employed to avoid a selection for pore penetration, and coated with a thick layer (approximately 1 μm) of Matrigel^{TM}, a reconstituted BM derived from EHS tumors (see Chapter 4 for more details). Cells are plated on the coated filters at a density that is lower than that used for the migration selection, since in this case the time it takes for the cells to migrate can be considerably higher, and excessive growth on the upper side of the filter has to be prevented. Optimal conditions have to be empirically determined for each cell line. Cells that have invaded and passed to the underside of the filter are collected and expanded as described above, and further selections can be made to obtain a fairly homogenous cell population of increased invasive potential, from which clones can be derived.

Basically, harvest of invasive cells after any invasion assay (see Chapter 4) can be used to select a cell population with increased invasive ability. Poste et al. (1980) have used the chorioallantoic membrane (CAM) assay, a canine blood vessel perfusion-invasion chamber, or retrograde injection of tumor cells in the mouse bladder, to select tumor cell variants with respect to their invasive ability. B16-BL6 melanoma cells, widely used as a prototype of highly invasive metastatic cells have been subjected to six rounds of selection with the bladder system. The application of these three systems for selection is described in detail in the original paper (Poste et al., 1980), and will not be treated further here.

CHAPTER 7

Genetic tagging as a means of studying tumor progression or metastasis-related genes

7.1. Clonal dominance in tumor progression

As already mentioned at the beginning of Chapter 6, any given tumor, despite its clonal origin, will eventually evolve to become composed of a highly genotypically and phenotypically heterogeneous population of neoplastic cells. This dynamic (i.e. continuously changing over time) heterogeneity results in a nonuniform pattern of biological parameters (such as karyotype, gene expression, drug sensitivity, etc.) within a tumor. Tumor progression can thus be seen as the consequence of stepwise emergence of new subclones, which temporarily take over the primary tumor, to be eventually replaced by the progeny of even more aggressive, derivative subclones (Nowell, 1976). A variety of genetic and phenotypic markers have been used to support this hypothesis, including cytogenetic markers, enzyme polymorphism, immunoglobulin and drug resistance markers (reviewed in Woodruff, 1988), beside molecular approaches such as RFLP (Dracopoli et al., 1985, 1987). However, the usefulness of those approaches to study malignant tumor progression is limited by the weariness and difficulty of the analyses involved, and/or by the small numbers of uniquely marked cell populations that can be isolated and examined. A way around these problems has been found through a methodology developed to study the destiny of single cells during embryonal development (Price, 1987), and consists in the artificial genetic tagging of tumor cells with a foreign marker gene (Kerbel et al., 1989). In fact, the integration of transferred DNA sequences after plasmid transfection (Southern and Berg, 1982) or retrovirus vector infection (Price, 1987) occurs randomly in the genome of recipient cells, so that it can be used to generate very large numbers of uniquely marked cell clones, which can be detected by Southern analysis using a nucleic acid probe which specifically hybridizes with the chosen DNA marker. Thus, for instance, a vector containing the gene that confers resistance to neomycin can be used as genetic marker, allowing isolation of transduced cells in selective medium. A genetic probe within the neomycin sequence can then be used in the southern blot analysis of genomic DNA after cutting with restriction enzymes (RE). The choice of RE will dictate the kind of pattern resulting from southern analysis. An enzyme cutting once within the probed sequence will generate two bands in case of a single integration, and multiples of two in case of further integration events. An enzyme cutting outside the DNA marker sequence will generate a band per each integration event. The bands will be all of the same size if the restriction site is chosen at each end of the marker sequence, whilst they will be of different sizes if a restriction site is chosen that is present within the host DNA and not in the marker DNA. In this case, each integration event, either in the same cell, or in different cells, will generate a unique band, depending on the site of integration (Fig. 7.1). Therefore, all transfectants would be distinguishable from each other because of the random nature of the integration events.

When a bacterial plasmid containing the neomycin selectable marker is used for cell transfection, most of the isolated clones contain each a unique integration site band (Talmadge and Zbar, 1987; Waghorne et al., 1988). When the clones are pooled and grown together, and analyzed on a southern blot, a smear of bands

GENETIC TAGGING

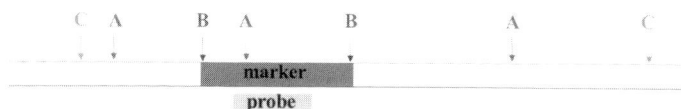

RE "A": Cuts once within the probed sequence, and at variable sites within the host DNA sequence, depending on the integration site, therefore generating two bands of variable sizes per each integration site, either in the same cell, or in different cells.

RE "B": Cuts at the ends of the marker DNA sequence, thus generating a band of that size, independent of the integration site.

RE "C": Cuts at variable sites within the host DNA sequence, depending on the integration site, and produces one band, the size of which depends on the specific integration site within the host DNA.

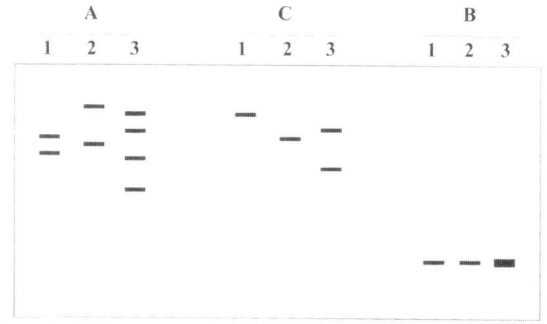

Clones 1 and 2 show one integration site, while clone 3 shows 2 integration sites

Fig. 7.1. The figure illustrates the hypothetical results obtained from three different cell clones after random integration of a marker gene, following either plasmid transfection or retroviral infection. Genomic DNA digestion with the restriction enzymes A, B or C will result in a unique band pattern, which is suggestive of the integration events in each cell clone.

is seen, indicating that no single clone predominates on the others. However, when the pool is injected subcutaneously in a host animal, to produce a primary tumor and metatases, a single clone appears to predominate in the primary tumor, which is the same that gives rise to the metastatic colonies (Waghorne et al., 1988). Indeed, among a population comprising both metastatic and nonmetastatic cells, a metastatic clone is the one that eventually predominates within the primary tumor, giving rise to metastatic colonies (Theodorescu et al., 1991b). However, when two different metastatic clones of similar growth and metastatic abilities are challenged against each other in a primary tumor, one will nonetheless predominate, thus suggesting that clonal dominance is independent of metastatic ability (Samiei and Waghorne, 1991).

When a retroviral vector is used for infection of cells with a selectable marker, a much larger number of resistant cells is obtained, due to the higher efficiency of transduction. Specifically engineered retroviral vectors, that are capable of only one round of infection, and are unable to direct expression of genes adjacent to the integration site, have to be used to try and avoid phenotypic effects that depend on the integration site (Yu et al., 1986). Upon subcutaneous injection of the resistant population, advanced primary tumors again show dominance of only few clones, while metastases contain the progeny of one or at most two clones, which not necessarily are the most evident in the primary tumor (Korczak et al., 1988). This suggests that, within a metastatically homogeneous population of cells, dominant growth properties do not necessarily correlate with metastatic potential.

The fact that clonal dominance depends on local factors has been shown by a study in which renal carcinoma cells were genetically tagged by retroviral infection with a neomycin-resistance gene, and implanted in vivo at either the orthotopic site (under the renal capsule), or at other ectopic sites (subcutaneously and intracecally) (Staroselsky et al., 1992). Results showed that clonal dominance occurred in every tumor, but that different clones became dominant in different organ environments. Metastases retained the dominant pattern expressed by the parent tumor from which they were derived, with the exception of subcutaneous growing tumors, which showed a random pattern of clonal dominance in both their primary and metastatic sites. It is therefore concluded that local organ factors produce a selection pressure on tumors resulting in the outgrowth of different clones with a varied growth potential. Most importantly, cytogenetic analysis of tumors performed in this study also supported previous observations (Bell et al., 1991) showing that, despite clonal dominance as assessed by an added genetic marker, tumors remain heterogeneous. In other words, clonal dominance does not imply tumor homogeneity.

Clonal dominance has been reported to occur also in mammary tumors and metastases from human patients (Symmans et al., 1995), although a similar study performed on squamous cell carcinomas of the larynx found a more marked heterogeneity between primary tumors and their metastases (Sun et al., 1995), suggesting either subclone heterogeneity within the primary tumor at the time of establishment of metastasis, or further clonal evolution of both tumors and metastasis.

The most serious criticism that can be raised against the experimental data showing the emergence of dominant metastatic clones is that the very same act of tagging (i.e. the random insertion of a piece of DNA somewhere in the host genome) could be the cause of phenotypic changes of the cell, resulting in its altered malignant properties. The use of a crippled, deficient retrovirus should prevent activation or enhancement of transcription of oncogenes, and the fact that each primary tumor analyzed consisted of a different set of clones argues against selection for a provirus integration site (Korczak et al., 1988). However, the possibility cannot completely be ruled out that inactivation of tumor or metastasis suppressor genes by the insertion of a foreign DNA sequence is at least contributing to the selective advantage of some of the clones. Indeed, as we will see later, a similar tagging strategy has been employed to identify putative metastasis related genes.

7.2. Visualization of cancer metastasis

The use of a selectable marker as a genetic tag in malignant cells allows a fine genetic analyses of the progeny developing as tumors and metastases. However, the early stages of tumor progression and micrometastasis formation have been arduous to analyze, due to the difficulty in identifying small numbers of tumor cells against a background of many host cells. To facilitate this kind of analysis, genetic tags conferring visual cues to the transfected cells have been exploited. Both histochemical and fluorescent tags have been used to detect single tumor cells or small collections of tumor cells at ectopic or orthotopic injection sites, and at multiple target organs for metastasis (Lin et al., 1990a, b; Chishima et al., 1997a, b). Each technique presents advantages and drawbacks, which have to be carefully weighed against the type of analysis that has to be performed.

7.2.1. Histochemical marker genes

Histochemical marker genes that have been used to tag tumor cells include the bacterial lacZ gene, human placental alkaline phosphatase (ALP) and the Drosophila alcohol dehydrogenase genes (Lin and Culp, 1991). Histochemical staining allows detection of tumor cells by light microscopy in thin sections of fixed tissue, thus consenting a very good evaluation of the topological relationships of tumor cells with respect to neighboring normal tissue cells and blood vessels. Fluorescence is very much decreased by tissue fixation and staining, making thus these relationships more difficult to resolve. Histochemical gene products are remarkably strong enzymes, whose typical colored products are easily identified, and resistant to degradation with appropriate storage of the samples. Moreover, the dye generated by the histochemical reaction is stably retained within the borders of the tumor cell, and does not readily disperse into the microenvironment (Culp et al., 1998). Finally, a large array of histochemical substrates for β-galactosidase, alkaline phosphatase, and alcohol dehydrogenase, yielding differently colored products are commercially available (e.g. from Research

Organics, Cleveland, OH, USA), thus allowing injection and later identification of differently tagged tumor cell populations.

Here follow protocols to detect histochemical marker genes in cells or tissues.

In case of tissue cultured cells, these are rinsed with PBS, and next incubated in fix solution (2% v/v formaldehyde in PBS) for 5 min at 4°C. In case of animal tissues, immediately on excision they are rinsed well in PBS, and then incubated in fix solution for 60 min at 4°C. These samples are then rinsed three times with PBS, and stained as described below, depending on the reporter gene.

7.2.1.1. Escherichia coli lacZ reporter gene staining

Fixed cultured cells or whole organs containing tumor cells and expressing the *E. coli lacZ* gene can be stained with X-gal (blue) or Red-gal (red). The *lacZ* gene encodes for the pH-neutral enzyme β-galactosidase which breaks down X-gal to yield a bright blue, or Red-gal to yield a bright red color (Lin et al., 1992).

Materials

Stain solution: 5 mM potassium ferricyanide, 5 mM potassium ferrocyanide, 2 mM magnesium chloride in PBS;
X-gal (or Red-gal) stock solution: 40 mg/ml in dimethylsulfoxide (DMSO);
nonidet P-40;
sodium deoxycholate;
3% (vol/vol) dimethylsulfoxide (DMSO) in PBS;
storage solution: 0.02% (wt/vol) sodium azide in PBS.

Procedure

After fixation (see above), cultured cells are incubated in stain solution containing X-gal (1 mg/ml, final) overnight at 37°C. Animal tissues are incubated in stain solution with X-gal (1 mg/ml, final), 0.02% (vol/vol) nonidet P-40, and 0.01% (wt/vol) sodium deoxycholate overnight at room temperature. Following staining, cultured cells are rinsed three times with PBS, while animal tissues are rinsed twice with 3% (vol/vol) DMSO. Cells and tissues are maintained stably in storage solution at 4°C; however, tissue staining results should be recorded within the same day to avoid nonspecific substrate breakdown and increased background staining (Lin et al., 1992).

7.2.1.2. Placental alkaline phosphatase reporter gene staining
Cells expressing the human placental alkaline phosphatase gene (ALP) may be specifically stained various colors by incubating fixed cells or tissues with colorless substrates which yield colored precipitates on cleavage by ALP. These cells stain red using fast red TR and naphthol AS-MX phosphate (or naphthol AS-BI phosphate), black using nitroblue tetrazolium (NBT) and 5-bromo-4-chloro-3-indolyl phosphate (X- phosphate), red-brown using iodonitrotetrazolium (INT) and X-phosphate, or blue using X-phosphate alone (Lin et al., 1992).

Materials
Red stain solution: 2 mg/ml fast red TR salt and 1 mg/ml naphthol AS-MX phosphate in 0.1 M Tris pH 8.5 (must be prepared immediately before use for best results);
black stain solution: 1 mg/ml X-phosphate and 1 mg/ml NBT in 0.1 M Tris pH 10.0;
red-brown stain solution: 1 mg/ml X-phosphate and 1 mg/ml IBT in 0.1 M Tris pH 10.0;
blue stain solution: 1 mg/ml X-phosphate in 0.1 M Tris pH 10.0;
nonidet P-40;
sodium deoxycholate;
3% (vol/vol) dimethylsulfoxide (DMSO) in PBS pH 7.4;
Storage solution: 0.02% (wt/vol) sodium azide in PBS pH 7.4.

Procedure
Animal tissues are heated to 65°C in PBS for 30 min following fixation and rinsed three times in PBS. Heat treatment of animal

tissues reduces background staining by inactivating endogenous phosphatases found in blood vessels and certain other tissues. Heat treatment of tissue-cultured cells is unnecessary. After fixation and subsequent PBS washes, tissue-cultured cells are incubated in the desired staining solution for 30–60 min at room temperature. Animal tissues are stained overnight at 4°C in the desired staining solution supplemented with 0.02% (vol/vol) Nonidet P-40 and 0.01% (wt/vol) sodium deoxycholate. In both cases enough staining solution is added to completely cover the cells or tissues. When the staining reaction is complete, tissue-cultured cells are rinsed with PBS. Animal tissues are rinsed first with 3% (vol/vol) DMSO in PBS and then with PBS. Cells and tissues submerged in storage solution may be stored for 1–2 months at 4°C (Lin et al., 1992). Due to the high background staining and low sensitivity of the fast red TR/naphthol AS-MX staining solution, this staining solution is only suitable for staining tissue-cultured cells. To reduce brownish background staining of tissue-cultured cells, the fast red TR/naphthol AS-MX staining solution should be prepared less than 10 min before use and the staining reaction should be stopped once cells expressing ALP have turned pink (often within 10 min of adding staining solution) (Lin et al., 1992).

7.2.1.3. Double staining: lacZ and ALP

Double-histochemical staining makes it possible to distinguish lacZ- from ALP-expressing cells in tissue culture- or animal tissue-derived cell populations containing two genetically-different tumor cell populations. Many combinations of lacZ and ALP substrates may be used, allowing investigators to identify cells carrying different reporter genes using colors that give the best contrast in their particular system. The only limitation is that lacZ staining must be done before ALP staining because the diazonium salts used in ALP staining inhibit $E.\ coli\ \beta$-galactosidase. Also, the 65°C heat treatment of animal tissues prior to ALP staining inactivates the $E.\ coli\ \beta$-galactosidase, in addition to the endogenous phosphatases that it is meant to inactivate (Lin et al., 1992).

Procedure
Tissue-cultured cells or animal tissues are prepared, fixed, and stained for lacZ expression as described in preceding sections. Instead of storing the cells in 0.02% sodium azide in PBS, the cells are rinsed again with PBS. ALP staining should be performed as described in the previous section, starting with the 65°C heat treatment and reducing the number of PBS washes that follow from three to one (Lin et al., 1992).

7.2.1.4. Embedding and staining of tissue sections
Previously stained animal tissues may be embedded and sectioned for microscopic examination of single tumor cells in 4–10 μm thin sections. Alternatively, tissues can be embedded first such that β-galactosidase and alkaline phosphatase activities are preserved, and subsequently stained for expression of the lacZ, ALP, or both reporter genes (see above) (Lin et al., 1990a, b).

Materials
Paraformaldehyde;
0.1 M phosphate buffer, pH 7.4;
Sucrose;
50, 95 and 100% acetone;
Embedding medium: 20 ml JB-4 A, 0.5 ml JB-4 B and 0.09 g catalyst.

Procedure
Previously stained tissues are fixed at 4°C for 2 h in 2% (vol/vol) paraformaldehyde; nonstained tissues are fixed at 4°C for 3 h in 4% (vol/vol) paraformaldehyde. These samples are then rinsed with phosphate buffer. Previously stained tissues are incubated for 3 h in 3% (wt/vol) sucrose in phosphate buffer, while nonstained tissues are incubated for 3 h in 7% (wt/vol) sucrose followed by 15% (wt/vol) sucrose in phosphate buffer for 3 h. Under vacuum, tissues are dehydrated in 50% (vol/vol) and 95% (vol/vol) cold acetone for 10 min each, and for 20 min in 100% cold acetone. Tissues are

incubated under vacuum in glycol methacrylate monomer (JB-4A) overnight at 4°C for previously stained tissues and at −20°C for nonstained tissues. For polymerization, previously stained tissues are kept at 4°C under vacuum and nonstained tissues are kept at −20°C for 24–48 h in vacuum desiccator. Serial sections (4–10 μm) are cut with a glass knife microtome (Lin et al., 1990a, b).

For lacZ staining, dried sections are incubated at 37°C overnight in X-gal stain solution (see above). Sections are then rinsed with water and mounted. For lacZ and ALP double-staining, sections are first stained for lacZ, then rinsed twice with distilled water and incubated for 1 h at 37°C with the ALP stain solution yielding the desired color (see above). Sections are rinsed with water and mounted (Lin et al., 1990a, b).

7.2.2. Fluorescent tags

In contrast to histochemical markers, fluorescent tags do not require any specific tissue preparation, and can be seen in fresh living tissues, with no interference from endogenous fluorescence. The green fluorescent protein (GFP) gene from the jellyfish *Aequorea victoria* is the most popular and effective tag, used in a variety of cell types (Chalfie et al., 1994; Cheng et al., 1996). Recently, among other gain-of-function mutants of GFP with enhanced fluorescence, a humanized clone was isolated, the brighter fluorescence of which allows for easy detection of GFP expression in transfected cells (Zolotukhin, 1996). In order to use GFP as a marker for in vivo experiments, it is of paramount importance to have stable transfectants that can express GFP for many generations under nonselective conditions. Retroviral transfer of GFP has been shown to result in stable transfection of human cancer cells in vitro (Levy et al., 1996). However, in this case use of crippled variants unable of further infection rounds (Yu et al., 1986) is advised, to avoid the risk of interference with an unstable genotype such as the one of cancer cells. Alternatively, a dicistronic GFP plasmid based on the incor-

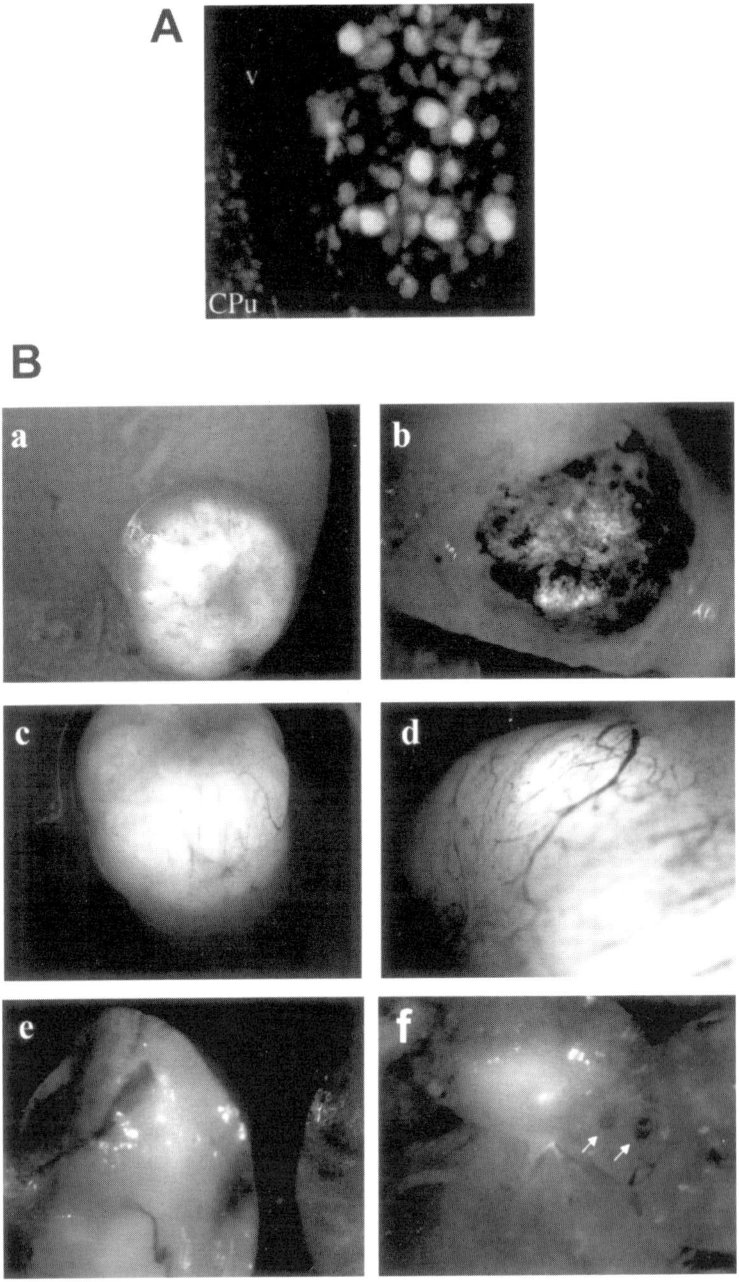

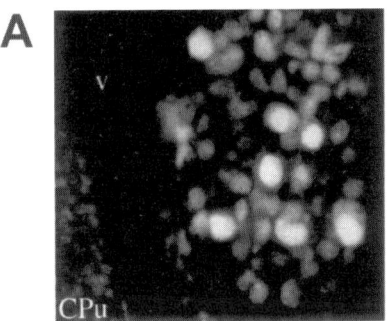

poration of an internal ribosomal entry (Kaufman et al., 1991) has been used as tumor cells tag (Chishima et al., 1997a). In this construct a DHFR gene conferring resistance to methotrexate (MTX), and a GFP gene are inserted respectively downstream and upstream of the internal ribosomal entry site, to produce a dicistronic mRNA in which translation of the DHFR gene is not reduced; rather, transfectants must amplify both the GFP and DHFR genes during MTX selection, because of the particular dicistronic structure (Chishima et al., 1997a). GFP-tagged tumor cells have been detected in vivo and ex vivo in histoculture of tumor-bearing organs, both at the single cell level, and at colony level (Chishima et al., 1997a-e; Yang et al., 1998). Usually, 1 mm sections are cut from fresh tissue samples, and directly observed under a fluorescence microscope equipped with a GFP filter set. However, whole metastatic colonies can also be nicely visible under u.v. illumination. In our laboratory we have produced a GFP-tagged derivative of the B16-LS9 cell line, which conserves the liver specificity of the parental line, and allows visualization of liver colonies and lung micrometastasis (Fig. 7.2, panel B).

Below, we describe a method for transfecting neuroblastoma cells with GFP using Lipofectamine reagent, selecting stably GFP-

Fig. 7.2. (**A**) N2a murine neuroblastoma cells transfected with GFP, injected into syngenic mouse brain lateral ventriculum (v). The difference in brightness and color between GFP-transfected cells and those of the nucleus Caudatus-Putamen (CPu) is clearly visible in the micrograph, taken after one week. (Photograph courtesy of Prof Laura Calzà, Modena, Italy.) (**B**) Liver and lung metastases visualized by GFP tagging of B16-LS9 melanoma cells. (a) and (c) show two different colonies on the surface of the liver, brilliantly glowing under u.v. illumination. The outline of the liver lobes is also visible as a dark shaded area: (b) in a dark melanotic colony, GFP fluorescence is heavily quenched by the melanin production of the cells, and only a faint glowing is evident in areas with less pigment: (d) A close-up of the colony presented in (c), to show how nicely the dark (since GFP negative) blood vessels vascularizing the tumor are delineated on the surface of the glowing colony. (e) and (f) show micrometastases on the surface of lung lobes. Also in this case, melanotic colonies are less visible under u.v. light (arrows in (f)).

transfected clones for injection in syngenic or nude mice, and then detecting GFP-fluorescent micrometastases.

Materials

Reduced serum medium, Opti-MEM (Gibco-BRL);
plasmid pEGFP (Clontech);
Lipofectamine™: plus reagent and Lipofectamine (Gibco-BRL);
fluorescence microscope, with 480 exc.-505dichr.-535 em. filters.

Procedure

Cells are grown in 10-cm culture dishes with 10% FCS to 40–50% confluency, and rinsed twice with Opti-MEM. For each culture dish, 4 μg of pEGFP plasmid are diluted with 30 μl of Plus reagent, then mixed with 750 μl of Opti-MEM and incubated 15 min. In a separate vial 30 μl of Lipofectamine reagent are diluted in 750 μl of Opti-MEM. Plasmid solution and lipofection mixture are then combined and allowed to stand for 15 min at room temperature (polystyrene tubes only). The plasmid/liposome preparation is finally poured into the culture dish and cells incubated at 37°C for 4 h, after which 3–5 ml of growth medium supplemented with 10% FCS is added to each dish, and incubation proceeds for 4 days.

Selective medium (500 mg/l of neomycin G418) is then added, and changed every 3 days. Positive fluorescent clones appear after two weeks. Isolate positive clones by aspiration with a thin polystyrene pipette, and seed in a 24 multiwell for propagation. Where necessary for observation under microscope (see above), fix the cells for 30 min with 3% paraformaldehyde in 0.1 mM $CaCl_2$, 0.1 mM $MgCl_2$, 250 mM HEPES, pH 7.4; quench free aldehyde groups with 50 mM NH_4Cl in HBS (150 mM NaCl, 1 mM $CaCl_2$, 20 mM HEPES, pH 7.4). Harvest the cells and resuspend in PBS w/o Ca^{++} and Mg^{++}. Depending on cell type and experimental design, the cells may be injected (2 to 5×10^4 cells in 200–500 ml) into syngenic or nude mice s.c., i.m., i.v., i.p., into brain ventricles, etc. At suitable time intervals, animal tissue slices can be analyzed in a dark-room by fluorescent microscope with 480 exc.-505dichr.-535 em. filters. The results may be quantitated using a fluorescent image analyzer.

Thus tagged GFP-neuroblastoma cells injected into the brain ventricles of syngenic mice were easily visualized up to one week, but then failed to grow and metastasize, and traces of fluorescence were barely detectable after 2 weeks (Fig. 7.2, panel A) (Garbisa et al., in preparation).

7.3. Genes controlling the metastatic phenotype: use of gene tags to identify metastasis-related genes

Somatic cell hybridization studies have shown that the metastatic phenotype of solid tumors is under a genetic control distinct from that causing unrestrained growth of primary tumor cells, and that in most cases it behaves as a recessive trait (Ramshaw et al., 1983; Sidebottom and Clark, 1983; Layton and Franks, 1986). The complexity of the metastatic process also suggests that its genetic regulation relies on the activation and/or deactivation of several specific genes (Table 7.1). Depending on the apparent relationship between a specific gene product and the related function in the metastatic cascade, we have arbitrarily classified these genes in three categories. There are gene products whose activity make them obvious regulators of the metastatic ability of tumor cells: cell surface adhesion molecules (e.g. cadherins or integrins) can on the one hand restrain the movements of tumor cells, thus preventing their invasive ability (Behrens, 1994–95; Ruoslahti, 1994–95), while on the other hand the expression of a particular integrin subset might confer a certain degree of specificity for the colonization of organs containing a higher amount of the related adhesive substrate (Sung et al., 1998a). Similarly, an increased production and/or secretion of proteolytic enzymes will contribute towards a higher invasive ability, which is necessary for the formation of metastatic colonies

TABLE 7.1.
Genes involved in the metastatic phenotype

Genes whose product has a direct relevance to the metastatic process
Adhesion receptors and mediators
Proteases and their inhibitors
Immunohistocompatibility antigens

Oncogenes
Inducers of the metastatic phenotype (e.g. met, ras, src, mos, raf, fes, fms)
Enhancers of the metastatic phenotype (e.g. myc, p53, E1a)

Metastogenes
Downregulated in metastatic cells (e.g. nm23, WDNM1, WDNM2)
Upregulated in metastatic cells (e.g. pGM21, CD44, p16/mts1, mdm2, mta1)

(Duffy, 1996), and might also influence motility and growth of tumor cells by the release of bioactive proteolytic fragments (Noel et al., 1997). Finally, escape from recognition by the immune system enhances the survival probabilities of tumor cells, thus increasing the likelihood that some will manage to complete the metastatic cascade, giving rise to distant organ colonies (Garrido et al., 1991).

Transfection experiments have shown that some oncogenes such as ras, or the kinase family of oncogenes, including *src, mos, raf, fes*, and *fms*, are potent inducers of metastasis when transfected in certain recipient cells (Wright et al., 1990). In many cases, however, no special care was taken to distinguish between the possibility that such cells have become generally more tumorigenic, or whether, or not, they have become specifically more metastatic, since the assays used did not allow a clear distinction to be made. Among these genes, a specific mention is deserved for the proto-oncogene *c-met*, the cellular receptor for hepatocyte growth factor/scatter factor (Bottaro et al., 1991). Ligand binding to this tyrosine kinase receptor triggers a series of events, such as motility, invasion, growth regulation, angiogenesis (Jiang and Hiscox, 1997), which are all relevant to the metastatic phenotype (Vande Woude et al., 1997). Indeed, c-met overexpression and/or constitutive activation has been correlated to the metastatic potential of several human tumors (Jeffers et al., 1996). Most intriguingly, it might be the first pair of receptor/ligand also involved in organ-specific colonization, as suggested by data obtained in experimental animal model systems (Rusciano et al., 1995, 1998; Otsuka et al., 1998), and in a human clinical survey of liver metastasis derived from colon cancer (DiRenzo et al., 1995).

Finally, differential screening of cDNA libraries derived from metastatic and nonmetastatic variants of the same tumor type, coupled with subtractive hybridization, has resulted in the cloning of specific genes associated with the metastatic phenotype. The *pGM21* gene is characterized by strong overexpression in a highly metastatic rat mammary adenocarcinoma cell line (Phillips et al., 1990). The protein product of *pGM21* has homology to the human elongation factor-1α (EF-1α), which is known to have a relevant role in the motility of metastatic cells by its ability to bind to actin filaments (Yang et al., 1990). A good correlation was also found between the levels of EF-1α gene expression and metastatic ability in fos-transfected rat fibroblasts and in a mouse melanoma cell line (Taniguchi et al., 1992). The *mta1* gene was isolated from the 13762NF rat mammary adenocarcinoma metastatic system, and its expression was found to be four times higher in metastatic cells than in related nonmetastatic cells (Toh et al., 1994). Overexpression of the *mta1* gene also correlated with tumor invasion and metastatic potential of human colorectal and gastric carcinomas (Toh et al., 1997). Its function seems to be related to cell motility (Nicolson and Moustafa, 1998), and recently a similarity emerged between *mta1* and EGL-27, a protein involved in cell migration and polarity in *C. elegans* (Herman et al., 1999). Another metastogene with a putative role in motility and invasion is *mts1*, originally cloned from a mouse mammary carcinoma cell line (Ebralidze et al., 1989). The *mts1* gene encodes a calcium-binding protein of the S100 subfamily, which is frequently, but not always, overexpressed in metastatic tumors and not in their nonmetastatic counterparts. In normal cells, *mts1* is usually expressed in those with invasive behavior. The gene product of *mts1* seems to interact with tropomyosin and myosin heavy chain, thus suggesting that it may have an influence on cell motility (reviewed in Lukanidin and Georgiev, 1996).

In addition to genes inducing the metastatic phenotype, the same approach of subtractive hybridization cloning has also yielded genes that may suppress the metastatic phenotype. The first member of this group, *nm23*, was isolated from murine melanoma cell variants (Steeg et al., 1988). Reduced nm23 expression, either at the RNA or protein level, was correlated with high metastatic potential in several rodent tumor cell lines, whereas in other systems no such correlation was evident (reviewed in Freije et al., 1996). In the human clinical situation, a significant correlation of low *nm23* expression and high metastatic potential has been observed for some tumor types, but not others (de la Rosa et al., 1995), suggesting that either *nm23* expression could be relevant to the metastatic phenotype in a limited number of cell lines and tumor types, or that the activity of the gene product could be deregulated by means other than its simple reduced expression. Homologouses of the *nm23* gene have been independently cloned, and named awd (abnormal wing disk) in Drosophila, and NDPK (nucleoside diphosphate kinase) in other species, suggesting a role for the gene product in differentiation and/or proliferation (reviewed in Freije et al., 1996). The subtractive hybridization method made it possible to isolate two other metastasis suppressor genes, *WDNM1* and *WDNM2*, whose transcription was down-regulated in rat metastatic mammary adenocarcinoma cells (Dear et al., 1988, 1989). Sequence analysis showed no homology to known genes. However, further studies have implicated *WDNM1* in the involution of the mammary gland after lactation (Baik et al., 1998).

The use of an immunological approach has revealed another gene, whose activity is tightly related to the metastatic ability of tumor cells. In this case, a monoclonal antibody specific for a metastatic variant of a rat pancreatic carcinoma cell line was used to identify the related gene product, that turned out to be a splicing variant of the *CD44* gene (Günthert et al., 1991). The transfection of this splicing variant into nonmetastatic cell lines converted these cells into rapidly spreading metastatic tumor cells (Günthert et al., 1991). The same variant was also found to be expressed by different metastatic tumors and cell lines (Hofmann et al., 1991; Heider et al., 1993a, b). Several splicing variants of CD44 have been reported so far, encoding different backbones onto which a number of functionally important carbohydrate moieties may be added (Cooper and Dougherty, 1995). Consequently, the metastatic behavior of tumor cells may depend on both the type of splice variant expressed, and their ability to appropriately modify the polypeptide core generated. Some recent evidence suggests that heparan sulfate moieties present on exon v3 may act as low affinity receptors for FGFs, and that this growth factor presentation mechanism may account for the function of this CD44 variant in the metastatic phenotype (Herrlich et al., 1998).

7.3.1. Proviral tagging

Proviral tagging has been widely used, mainly in vivo, to identify genes involved in oncogenesis and development in various animal model systems after infection with slow transforming retroviruses (van Lohuizen et al., 1991; Lee et al., 1995; Gaiano et al., 1996; Kapoun and Shackleford, 1997). Application of the same method in vitro should allow the identification of genes involved in conferring to cells any selectable phenotype, the rationale being that proviruses integrate randomly in the genome, and consequently a gene involved in the acquisition of a specific phenotype may be mutated or activated by insertion of the proviral DNA within the gene or in its adjacency (Habets et al., 1994; Lu et al., 1995).

Proviral activation of putative proto-oncogenes can occur by different mechanisms, all governed by the unique properties of the viral long terminal repeat (LTR) sequences present at both ends of the provirus. LTR sequences contain both promoter and enhancer elements as well as polyadenylation signals, capable of influencing flanking cellular transcription. Activation by promoter insertion occurs when the provirus integrates upstream or within the 5′ end of a gene, allowing transcription from either the 5′ or the 3′ LTR-promoter to drive transcription of the downstream gene. Activation by enhancement occurs when the provirus integrates either at the 3′ end of a gene in the same transcriptional orientation, or at the 5′ end, usually in the opposite transcriptional orientation. In the former situation, integration may take place within 3′ untranslated sequences of the target gene, thereby removing potential RNA-destabilizing sequences. Finally, proviral insertion can result in message truncation, thus leading to the expression of either inactive proteins (however, the presence of a second, wild type allele will usually mask the effects of gene inactivation), or proteins with an altered function. (For a comprehensive review, see van Lohuizen and Berns, 1990.) Therefore, slow transforming retroviruses can be used as tools to induce phenotypic changes by altered gene expression, and subsequently identify the responsible target gene(s). After molecular cloning of provirus-host junction fragments from transformed cells DNA, and further isolation of unique host specific probes, it is possible to screen genomic DNA from independently transformed cell clones showing the same phenotype (e.g. after selection for a desired property) for rearrangements in the same region of DNA. Once a common proviral integration cluster is recognized, the relevant gene(s) affected by the proviral insertion(s) can be identified.

Infection with Moloney murine leukemia virus of a lymphoma cell line in vitro resulted in the identification of a gene—*Tiam1*—that affects the invasive ability of tumor cells (Habets et al., 1994), and thus has the potential to affect also their metastatic behavior. *Tiam1* acts as an activator of *rac1*, thereby influencing cytoskeletal architecture and cell motility (Collard et al., 1996). Although no clinical evidence has been reported to date of a role of *Tiam1* in human malignancies, transfection of truncated *Tiam1* constructs in NIH 3T3 cells resulted in tumorigenic transformation of the selected population (van Leeuwen et al., 1995). In a separate study, infection of a melanoma cell line, coupled with selection for drug resistance, allowed the isolation of a putative gene associated with resistance to a specific chemotherapeutic drug (Lu et al., 1995).

We will briefly delineate the essentials of this technique, which has the potential of permitting the identification of other likely metastogenes, given the availability of an appropriate selection mechanism.

The choice of the infecting provirus has to be compatible with the target cell line. The Moloney murine leukemia virus (M-MLV) can be used with most murine and rat cell lines, whereas choice of an amphotropic virus is recommended for infection of cell lines derived from other organisms (Goff, 1987). Viruses can be replication competent, but usually replication defective viruses are used, so that the number of insertions per cell can be better controlled by the infection conditions: a multiplicity of infection (MOI) of 10 should (at least theoretically) result in 10 insertions per cell. If the virus contains a selectable marker (such as neomycin resistance), this can be used for both selection of infected cells, and as a specific probe to detect viral insertions. For infection, recipient cells should be at subconfluent density ($\sim 2 \times 10^4/cm^2$), and are infected for a period up to 24 h. After removal of the infection medium, further growth of infected cells is allowed (at least 48 h) before any selection is attempted. Several independent dishes are usually infected, and selection carried out independently on each population. Any selectable property can be used, such as growth under restrictive conditions, invasion, motility, resistance to exogenous agents, and so forth. After several rounds of selection (usually more than 5) applied independently on each infected population, the resulting cells are cloned, and at least 5 clones from each selected population are analyzed by southern blot with a specific retroviral probe. In fact, there are usually multiple insertion sites per cell, and many of the cells resulting from the selection screening can be spontaneous mutants, carrying proviruses at irrelevant loci. If several independently selected clones show a common cluster on the southern blot, which is absent from unselected control clones, they may then contain a specific target gene related to the selected phenotype. The next step is then cloning of the viral flanking sequences. To this purpose a genomic library is constructed with the DNA of one of the clones containing the putative gene of interest, and screened using a provirus specific probe. Positive clones are then isolated, and characterized for the presence of 3′ and/or 5′ flanking regions. Upon subcloning of a suitable fragment containing the flanking regions only (after removal of the retroviral cassette), this new probe is challenged against a southern blot of selected and unselected clones. It is expected that unselected clones, and selected clones whose phenotype does not depend on proviral integration show only one hybridization band, corresponding to the two alleles of the involved gene. However, those selected clones whose phenotype results from a mutation caused by proviral insertion should show two bands: one corresponding to the normal allele, and the other corresponding to a rearranged cluster containing the mutated allele. After the identification of the rearranged locus, more genomic clones and probes can be isolated from this region, to screen for additional proviral insertions, and to search for the affected gene. To this purpose, the chosen genomic clones can be used to screen a relevant cDNA library, likely to contain the putative gene of interest. Sequencing and search through a database will give final information about the nature of the gene.

CHAPTER 8

In vivo cancer metastasis assays

Metastasis is defined as the formation of tumors that are discontinuous from the primary tumor. These secondary tumors can be at nearby or distant sites and can form following dissemination of cells via lymphatic, hematogenous, coelomic cavities or epithelial cavities (Willis, 1973). The most common routes for metastatic spread are lymphatic and hematogenous metastasis; so, secondary tumor formation via those routes will be the focus in this chapter. However, it must be noted that these routes are not necessarily the common ones for the spread of some tumor types (e.g. ovary (Cannistra, 1993)).

For a cell to successfully colonize a secondary site, it must complete *every* step of a complex, multistep cascade (Fidler and Radinsky, 1990; Radinsky and Fidler, 1992). Malignant cells invade adjacent tissues and penetrate into the lymphatic and/or circulatory systems. Then tumor cells detach from the primary tumor and disseminate. During transport, cells travel individually or as emboli composed of tumor cells (homotypic) or tumor cells and host cells (heterotypic). At a secondary site, cells or emboli either arrest because of physical limitations (e.g. too large to traverse a capillary lumen) or by binding to specific molecules in particular organs or tissues. Once there, tumor cells then proliferate either in the vasculature or extravasate into surrounding tissue (Chambers et al., 1995; Koop et al., 1996; Luzzi et al., 1998). To form macroscopic metastases, cells must then recruit a vascular supply (Ellis and Fidler, 1995; Folkman, 1995b; Kohn and Liotta, 1995; Rak and Kerbel, 1996; Weinstat-Saslow and Steeg, 1994; Weinstat-Saslow et al., 1994) and respond appropriately to the tissue's environmental milieu by proliferating (Nicolson, 1994; Radinsky, 1995a, b). Less than 0.1% of cells that intravasate survive to form clinically detectable, macroscopic metastases (Fidler, 1970; Tarin et al., 1984a). At which step(s) of the metastatic cascade circulating tumor cells commonly succumb is debatable (Chambers et al., 1995; Koop et al., 1995, 1996).

The very nature of the metastatic process dictates that the organization of this chapter differ from other chapters dealing with in vitro studies. At the outset, it is important to acknowledge that there are as many variations for how to study metastasis in vivo as there are people doing them. While the chapter will conclude with specific technical recommendations, most of the chapter will outline the theoretical considerations and rationale for experimental design and interpretation of metastasis assays.

8.1. Why study metastasis in vivo?

Metastasis is not equivalent to invasion, adhesion, growth rate, susceptibility to immune cell killing, or any of the many steps in the metastatic cascade. These processes are *necessary* for successful colonization of secondary sites, but they are *not sufficient* for a cell to be metastatic. Failure to distinguish between individual steps from the complete process of metastasis has contributed greatly to confusion and misinterpretation in the scientific literature. Just because a cell line is highly invasive or adheres strongly to extracellular matrices does not necessarily translate to its having the ability to metastasize. Yet, this type of faulty extrapolation is common.

Despite the wealth of useful information than can be gleaned from in vitro assays measuring a step(s) in the metastatic cascade, the *only* method for assessing metastatic potential involves the use of in vivo models. In vitro models are simply not of sufficient complexity to recapitulate the multitude of steps of the metastatic process. This is not to demean the usefulness of in vitro models.

On the contrary, they are extremely useful because in vitro studies minimize or eliminate variables that complicate interpretation, contribute to interexperimental variability, and cloud model development. Other chapters in this volume will focus on the proper use of and interpretation of in vitro assays of adhesion, invasion, etc. The focus of this chapter will be on the methodologies used to study metastasis in vivo. Parallel studies utilizing both in vivo and in vitro approaches are powerful strategies for dissecting the biochemical and molecular basis for cancer metastasis.

Another common issue relates to the relationship between tumorigenicity and metastasis. They are distinct phenotypes. Tumorigenicity is a prerequisite to metastasis, but metastasis is a property of only a small portion of cells within a neoplasm (reviewed in Chambers, et al. 1995; Fidler and Nicolson 1987; Luzzi et al., 1998; Weiss, 1990). Metastatic cells represent a subpopulation having additional capabilities to those required for uncontrolled growth. So, in order to study metastasis, models designed specifically for this purpose are necessary. Moreover, the appropriate use of these models is required.

8.2. What defines an appropriate model of metastasis?

As long as a model allows testing of a defined hypothesis, its use is appropriate. However, the quality of the model for emulating the pathobiology of human disease determines whether, or not, the model is relevant. Relevance is difficult to define as all models are inadequate for studying some aspects of metastasis. It is beyond the scope of this chapter to evaluate the validity of individual models. Readers are referred to other reviews which more adequately cover this area (Welch et al., 1986; Welch, 1986, 1997).

Two criteria must be met if a model is to be considered useful for studying metastasis. First, one must use metastatic cells. Unfortunately, it is still commonplace that many studies utilize nonmetastatic cells for studies of metastasis. The primary reason for this is the rationalization that if cells are derived from metastases, they are de facto metastatic. This conclusion does not necessarily follow. For example, virtually all human breast carcinoma cell lines were isolated from metastases or pleural effusions; however, few (e.g. MDA-MB-435 and MDA-MB-231) reproducibly form macroscopic metastases in a majority of immunocompromised mice (Sung et al. 1998b). Additionally, some investigators started their studies with metastatic cells, but the conditions under which the cells were maintained rendered them nonmetastatic. This does not necessarily mean that cell culture maintenance was flawed. Rather it could simply reflect the inherent phenotypic instability of tumor cells (Nowell, 1976; Cheng and Loeb, 1993). Nonetheless, phenotypic instability and inconsistent culture conditions have resulted in several variants of cells with the same name, but profoundly different biologic properties. Therefore, it is incumbent on every investigator to replicate previously published experiments in his/her laboratory prior to initiating further studies. This point cannot be overemphasized.

The second criterion is that tumor cells must be compatible with the animal model. This is analogous to Paget's (1889) 'Seed and Soil' hypothesis. This parameter may explain the difficulties in obtaining relevant models for certain tumor types and/or sites of metastatic colonization.

8.3. Cell lines

The most critical component for the study of metastasis is metastasizing cells. Since most people obtain their cells from cell culture, it is important to outline some of the criteria for the proper preparation and maintenance of those cultures. Details are omitted regarding verification of species of origin (karyotype, isozyme expression, etc.), tissue of origin (surface markers, enzyme expression patterns) and absence of opportunistic infections. It is presumed that such characterization will be done by readers of this volume before any studies are undertaken. Another assumption is that the conditions used to culture the cells have been optimized. That is, the cells are not allowed to undergo stress due to neglect. If this happens, unintended selection will have occurred and stability of cellular behavior cannot be ensured. The third assumption is that the culture conditions used are standardized. That is, cells will be cultured under *identical* conditions (e.g. same medium, culture plates, serum type and concentration, detachment procedures, etc.) to those utilized in previously published papers. Although the latter is intellectually apparent, it is the one least adhered to by investigators new to the metastasis field. The following section outlines the reasons why this parameter deserves focused attention of an investigator.

Cells should never be used unless they are *at least* 90% viable. Dead cells can modify the behavior of viable cells. Fidler, using B16 melanoma cells, showed that presence of lethally irradiated cells within an inoculum greatly changed metastatic potential (Fidler, 1973b). This situation is reminiscent of the so-called feeder layer or Revesz effect seen in survival curve data (Puck et al., 1956; Revesz, 1956, 1958). Yet the mechanisms are unknown.

Even when culture conditions are optimized and great care is taken to minimize iatrogenic effects, tumor cell populations display greater genetic instability than their normal counterparts. Metastatic cells are no exception. At diagnosis, tumors are already complex mixtures of cells despite the fact that the vast majority of tumors are clonal in origin. Among the earliest detectable changes in transformed cells (anchorage-independent growth, not contact-inhibited, and immortal) is genetic instability (reviewed in Greller et al., 1996; Heppner and Miller, 1997; Tlsty, 1997). Even before they become tumorigenic, transformed cells display genomic instability that is apparently the driving force for further progression. Genomic plasticity is crucial for the generation of intratumoral heterogeneity (Cheng and Loeb, 1993; Tlsty, 1997; Welch and Tomasovic, 1985). Heterogenous populations are subjected to selection pressures that drive evolution toward increasingly malignant characteristics (i.e. invasion and metastasis). Besides the inherent differences between cells at the genetic level, cells also respond to signals from the host and other tumor cells. Thus, it is common to observe phenotypic drift over time (reviewed in Nowell, 1976; Welch and Tomasovic, 1985). This change is often gradual, presumably as proportions of different clones change within the population, but the rate and direction of drift are clone-dependent. To minimize the impact of phenotypic drift (since there is currently no known way to eliminate it), behavior must be periodically compared to a baseline. When behaviors change (i.e. when metastatic potential increases or decreases, or when distribution of metastasis is altered), frozen aliquots from a lower passage should be retrieved and used. As long as metastatic potential does not change within that interval, use of cells is acceptable.

Further complicating the situation is a series of experiments which show that 'trivial' culture conditions can profoundly affect metastatic potential of cells (Welch, 1997 and references therein). Examples of culture conditions that affect metastatic potential or metastasis-associated phenotypes include pH of the culture medium (Martinez-Zaguilan et al., 1998), the type of medium in which the cells are grown (Prezioso et al., 1993) and confluence (see Welch, 1997 and references therein).

The number of metastases in patients and in animal models is proportional to the number of tumor cells present (i.e. primary tumor size). This correlation is imperfect, but still a reasonably good rule of thumb (Glaves 1983; Hejna et al., 1999; Liotta et al., 1974). Recall that Tarin and colleagues demonstrated that the mere presence of tumor cells in the circulation does not always portend development of metastases. Perhaps a more critical parameter is the number or proportion of tumor cell-containing emboli in the blood (Fisher and Fisher, 1967a, b; Lane et al., 1989; Zeidman and Buss, 1952). This has been shown in experimental systems. Specifically, the number and size of the emboli determined the frequency and efficiency of metastasis (Fidler, 1973b; Liotta et al., 1976; Updyke and Nicolson, 1986). These results demonstrate the need to control the inoculum; specifically, single cell suspensions should be used.

To achieve predominantly single cells, the majority of clumps can be dissociated by gentle pipetting. This is maximized by the use of smaller bore pipets. However, it is important that the bore not be too small since cell killing can occur if the diameter is too small. Another condition that maximizes single cells is the use of *ice-cold* media or saline throughout the cell preparation steps of the procedure. The number and size of emboli increases as temperature rises. And finally, cell clumping is a time-dependent phenomenon, necessitating that inocula-containing syringes be prepared *immediately* before injection (Updyke and Nicolson, 1986). An estimate of cell clumping can easily be determined by processing cells exactly as would be done for injection, except that they are delivered into a small dish and examined under a microscope. Single cells and clumps of various sizes can be counted directly.

Cells are loaded into the syringe when no needle is in place. The negative pressure combined with the relatively small bore of a needle causes damage and death to a relatively large number of cells. Of course, the sensitivity of cells to this particular manipulation is cell line-dependent. Until proven otherwise, it is better to be safe than sorry. As soon as the inoculum is loaded into the syringe, This makes the subsequent process of injection easier since it is easier to see the amount injected.

To obtain suspensions of mostly viable single cells, the methods used to obtain the cells from a culture are crucial. For cells growing in suspension, the technique is simple washing and dilution in an appropriate inoculation fluid (i.e. isotonic, nonallergenic). However, for adherent cultures, other methods must be employed. Scraping cells followed by gentle pipetting, filtration and/or sedimentation has been used but the yields can be inconsistent. Viability and proportion of single cells can be suboptimal using this approach (D. R. Welch, personal communication). For these reasons, enzymatic or chemical detachment are more commonly used.

The most common variations involved solutions containing the proteolytic enzyme trypsin (0.05–0.25%) or the chelating agent ethylenediaminetetraacetate (0.5–5 mM EDTA). Detachment times vary from less than 1 min to more than 1 h and this must be determined empirically. However, it is important to emphasize that exposure to any detachment agent be minimized. Prolonged treatment affects survival and metastatic potential. In the same vein, failure to remove all cells from a plate imposes a selection for weakly adherent cells which have different survival, growth and metastatic potentials than strongly adherent cells (Akiyama et al., 1995; Albelda, 1993; Behrens, 1993; Hart et al., 1991; Roos, 1991; Tang and Honn, 1994; Weiss, 1994). And while the method of detachment may seem trivial, it is not. Metastatic potentials are profoundly affected by the conditions chosen and the magnitude and direction of the change are cell line-dependent (Welch, 1997). So, until otherwise determined, conditions for cell lines received from another laboratory should not be altered. And, when a cell line is being developed and characterized, systematic evaluation of these conditions would be recommended. This notion is illustrated by a recent experience in Dan Welch's laboratory. Following transfection of human breast carcinoma cells with a candidate metastasis-suppressor gene, the cells, which are routinely subcultured using a mixture of trypsin and EDTA, became exquisitely sensitive to the presence of trypsin. In fact, even minuscule amounts of trypsin proved toxic (Md. J. Seraj and D. R. Welch, unpublished observations).

Once a properly diluted and a single cell suspension of viable cells at the 'correct' level of confluence is obtained, there is yet another parameter that should be controlled. Suspensions should be maintained in polypropylene containers rather than polystyrene. Typically, tumor cells adhere better to the latter and the number of cells being injected can change as cells adhere to the walls.

8.4. Considerations regarding animals

Ultimately, the decision of which host to use is determined by the metastatic cells to be evaluated. Initially, immunologic considerations predominate. A few rules of thumb apply.

Whenever possible, syngeneic animals should be used. It is intuitively obvious that tumor development and progression are most closely recapitulated in syngeneic mice, more so in autochthonous models (Potter et al., 1983; Price et al., 1984). Syngeneic models are those in which tumor cell lines are derived from the same inbred strain. Autochthonous models are those in which the tumor arises within a host and the experiment is carried out in that host. The latter can be spontaneous tumors, carcinogen-induced tumors or tumors that arise in knockout, or knock-in animals. The major limitation to the use of autochthonous models relates to the large number of animals required to achieve statistically valid interpretation. This issue mostly relates to the incidence of tumors developing and the proportion of those which develop metastases. The use of genetically engineered animals may overcome these limitations (Webster and Muller, 1994); however, the number of transgenic models that metastasize is still relatively limited. Table 8.1 lists the currently available transgenic and knockout mouse cell lines which reportedly metastasize. Readers are cautioned that this list is neither exhaustive nor does it imply a recommendation. While the use of transgene and specific gene knockouts holds great promise, a great deal more research is required to determine the impact on our understanding of the metastatic process. As well, more studies are required to demonstrate that the models mimic the pathobiology of human disease. Of some concern is whether, or not, more subtle aspects of the metastatic process will be discernible in these systems.

Analysis of tumor growth and metastasis of human cancers requires the use of immunodeficient animals. The most common xenograft host is the athymic mouse (*nu/nu*, T-cell deficient), also known as the 'nude' mouse. Xenografts generally retain morphologic and biochemical characteristics following transplantation. Unfortunately, not all tumor types appear amenable to growth in athymic mice, making this host suboptimal. To some extent, this is alleviated with the availability of other immunodeficient strains such as the SCID (*xid*, T- and B-cell deficient), beige (*bg*, NK cell

TABLE 8.1

Transgenic and knockout mice which develop metastases

Transgene or knockout	Site of primary tumor	Reference
Review of transgenic metastasis models	Mammary	(Dankort and Muller, 1996)
Nf2$^{+/-}$	Multiple	(McClatchey et al., 1998)
p27KIP1$^{-/-}$	Prostate	(Corlon-Cardo et al., 1998)
p53$^{-/-}$ + TGF-β1	Skin	(Akhurst and Balmain, 1999)
tg: SV40T	Prostate	(Foster et al., 1997; Gingrich et al., 1997; Gingrich et al., 1996)
tg: × SCD, beige (immune-deficient mice)	Pancreas	(Gallo-Hendrikx et al., 1994)
tg: cryptdin-2-SV40T	Prostate	(Garabedian et al., 1998)
tg: Metallothionein—Ret	Melanoma	(Asai et al., 1999)
tg: MMTV-LTR-mtsl × GRS/A	Mammary	(Ambartsumian et al., 1996)
tg: polyomavirus middle T	Mammary	(Ritland et al., 1997)
tg: MMTV-Fgf8b	Mammary, salivary gland	(Daphna-Iken et al., 1998)
tg: Tyr-SV40E + u.v. irradiation	Melanoma	(Kelsall and Mintz, 1998)
tg: Tyr-SV40T	Retinal pigment epithelium	(Penna et al., 1998)
tg: fetal globin-SV40T	Prostate	(Perez-Stable et al., 1997)
tg: Rip1Tag2 × E-cadherin dominant negative	Pancreatic	(Perl et al., 1998)
tg: RET/PTC3	Papillary thyroid	(Powell, Jr. et al., 1998)
tg: MMTV-neu	Mammary	(Ritland et al., 1997)
tg: MMTV-polyoma middle T	Mammary	(Ritland et al., 1997)
tg: MMTV-polyoma middle T	Mammary	(Lifsted et al., 1998)
tg: keratin-p53^{172H}	Keratinocyte	(Wang et al., 1998)
tg: Probasin-SV40T	Prostate	(Kasper et al., 1998)
tg: C3(1)-SV40T × p534$^{-/-}$	Prostate	(Maroulakou et al., 1997)
tg: HGF/SF	Melanoma	(Otsuka et al., 1998)

Note: Several papers utilize tumor cell implants into knockout or transgenic mice with associated changes in metastatic potential (Araki et al., 1997; Biancone et al., 1996; Bourguignon et al., 1998; Davies et al., 1996; De Vries et al., 1995; Driessens et al., 1995; Eitzman et al., 1996; Goldfarb et al., 1998; Hall and Thompson 1997; Kruger et al., 1998; Lloyd et al., 1998; Marvin et al., 1998). The citations listed above are those in which metastases are observed in the genetically engineered mice without inoculation of tumor cells.

TABLE 8.2

Commonly used metastasis models

Cell line	Route(s) of injection	Organ tropism	Species	Reference
B16 melanoma				
B16-F10	i.v., s.c., i.m.	Lung, brain, ovaries, intestine	mouse, C57BL/6	(Fidler, 1973a)
B16-F1	i.v., s.c., i.m.	Lung	mouse, C57BL/6	(Fidler, 1973a)
B16-L8/LS9	i.v., i.a.	Liver	mouse, C57BL/6	(Tao et al., 1978; Rusciano et al., 1993)
B16-O10	i.v.	Ovary, lung, other	mouse, C57BL/6	(Brunson and Nicolson, 1979)
B16-B15b	i.v. (carotid)	Meninges, lung, other	mouse, C57BL/6	(Kawaguchi et al., 1983; Miner et al., 1982)
B16-BL6	i.m. (with leg amputation)	Lung	mouse, C57BL/6	(Poste et al., 1980)
13762A mammary	i.p.	Lymph node, ascites	rat, F344	(Osbakken et al., 1986)
13762NF mammary	m.f.p., i.v.	Lung, lymph node	rat, F344	(Neri et al., 1982; Welch et al., 1983)
C8161 melanoma	i.v., i.d.	Lung, brain, liver, ovaries	human (nu/nu, SCID)	(Welch et al., 1991)
Co-3 colon	colon	Liver	human (nu/nu)	(An et al., 1997)
Dunning R3327 prostate	s.c.	Lung, lymph node	rat, Copenhagen	(Isaacs et al., 1978)
Esb/Eb lymphoma	s.c.	Liver, lung, spleen	mouse, DB A/2	(Altevogt et al., 1985)
F9 teratocarcinoma	i.v.	Liver, lung	mouse, Sv 129/Ter	(Cotte et al., 1982; Terrana et al., 1987)
HRCC	kidney	Lung	human (nu/nu)	(Naito et al., 1986)
K1735 melanoma	i.v.	Lung	mouse	(Volk et al., 1984)
Lewis lung carcinoma (3LL)	s.c., i.v.	Lung	mouse, C57BL/6	(Giraldi et al., 1977; Gorelik et al., 1980; Gorelik et al., 1982; Hilgard et al., 1976; Young et al., 1990)
LOX melanoma	i.v., s.c., i.d.	Lung	human (nu/nu)	(Shoemaker et al., 1991)
M24met melanoma	i.v., s.c.	Lung	human (nu/nu, SCID)	(Mueller et al., 1991)
M4Be melanoma	s.c.	Lung	rat (nu/nu)	(Bailly and Dore, 1991)
MDA-MB-231	i.c.	Bone	human (nu/nu)	(Sasaki et al., 1995)
MDA-MB-231	i.v.	Lung	human (nu/nu)	(Price et al., 1990)
MDA-MB-435 breast	m.f.p.	Lung, lymph node	human (nu/nu)	(Price et al., 1990)
MDAY-D2 lymphoma	s.c., i.v.	Liver, lungs, spleen, kidney	mouse, DBA/2	(Kerbel et al., 1978)
MelJuSo melanoma	i.d., i.v.	Lung	human (nu/nu)	(Miele et al., 1996)
MeWo melanoma	i.v., s.c.	Lung	human (nu/nu)	(Ishikawa and Kerhel, 1989; Ishikawa et al., 1988)
MMTV mammary tumors	s.c., i.v., m.f.p.	Lung, liver, lymph node	mouse, BALB/c	(Heppner et al., 1978; Miller, 1981)
P574 mammary	spontaneous	Lung, adrenal, ovary, kidney	mouse, C3H	(Price et al., 1984; Tarin and Price, 1981)

TABLE 8.2

(continued)

Cell line	Route(s) of injection	Organ tropism	Species	Reference
PAN-12 pancreas	pancreas	Liver, kidney, lymph node	human (nu/nu)	(An et al., 1996)
ras-transfected NIH-3T3 fibrosarcoma	i.v.	Lung	mouse (nu/nu)	(Chambers and Tuck, 1988)
RAW 117 lymphosarcoma	i.v., s.c.	Liver, spleen	mouse, BALB/c	(Brunson and Nicolson, 1978)
SPl mammary	s.c.	Lung	mouse, CBA	(Frost et al., 1987)
UV2237 fibrosarcoma	i.v.	Lung, skin, mesentary	mouse, C3H/HeNCr	(Kripke et al., 1978)

The cell lines listed in this table are incomplete, particularly with regard to variants selected from the parental. Since the B16 melanoma is the most commonly used metastatic cell line used, the most popular variants are listed with associated references.

Abbreviations used: i.v., intravenous; s.c.: subcutaneous; i.c.: intracardiac; m.f.p.: mammary fat pad; i.d.: intradermal; i.p.: intraperitoneal.

Note: This list of tumors is not exhaustive and is provided only as a resource to identify some commonly used metastatic tumor cell lines.

deficient) or mice with a combination of immune deficiencies. Intuitively, one would predict that metastatic potential would inversely correlate with relative immunodeficiency, though this is not always the case (Clarke, 1996; Garafalo et al., 1993; Mueller et al., 1991; Phillips et al., 1989; Xie et al., 1992). To date, there is no certain method to predict behavior in each strain.

Regardless of strain, it is crucial that all animals are tested and found to be free of infections with endoparasites (pinworm, tapeworm, ...), ectoparasites (lice, mites, ...), viruses (minute mouse virus, mouse hepatitis virus, hepatitis, pneumonia virus, ...), bacteria (*Pseudomonas, Staphylococcus*, ...) and *Mycoplasma*. Infections can profoundly affect experimental outcome. Therefore, sentinel animals should be tested frequently (monthly or bimonthly) for infestation by opportunistic pathogens using sentinel animals from every animal room throughout the facility. 'Routine' animal maintenance conditions (caging, light/dark cycles, diet, water chlorination, etc.) are also important and it is incumbent on each investigator to monitor animal conditions throughout the course of the experiment.

Another consideration is the natural killer (NK) cell activation. In short, metastasis has been shown to correlate inversely with NK activity (Hanna, 1982, 1985; Hanna and Schneider, 1983; Hanna and Fidler, 1980; Urdal et al., 1982). Since young mice (3 weeks) have lower NK activity than older mice (6–8 weeks) (Hanna, 1982; Hanna et al., 1982; Pollack and Fidler, 1982), there is greater likelihood of observing metastases in younger, rather than older, mice. Generally, the use of young mice is recommended if metastasis is a desired endpoint (Fidler, 1986).

8.5. Site of injection

Animal models for metastasis typically involve two approaches. The first involves inoculation of tumor cells into tissue sites (i.e. subcutaneously (s.c.), intradermally (i.d.), intramuscularly (i.m.), or into specific organs or tissues (e.g. mammary fat pad (m.f.p.)) which results in the formation of a local tumor from which spontaneous metastases eventually form. The second approach bypasses local tumor growth and intravasation, by introducing tumor cells directly into the vasculature (usually intravenously (i.v.), but also intra-arterially (i.a.) or intracardially (i.c.)). This results in formation of experimental metastases. Both methods have contributed to our understanding of the multigenic, multistep metastatic phenotype; however, the experimental metastasis assay has been maligned by some. While there are valid reasons for questioning the direct vascular injection of tumor cells (reviewed in Welch et al., 1983; Welch, 1997), there are equally cogent assertions that this model is appropriate. It is crucial, however, to establish whether, or not, results from both assays are equivalent. An example where distribution of metastases is significantly different depending on route of injection involves use of the MDA-MB-231 human breast carcinoma cell line. Intravenous injection produces lung metastases with occasional extrapulmonary metastases. However, inoculation directly into the left ventricle results in formation of osteolytic metastases in the long bones, a condition rarely seen in mice (Guise and Mundy, 1998; Rabbani et al., 1999; Yang et al., 1999; Yin et al., 1999; Yoneda et al., 1994), but common in human breast carcinomas.

Some argue that bolus inoculation of thousands of tumor cells directly into the vasculature does not reflect the situation in humans; however, quantification of tumor cells in patient blood indicates that numbers between 10^4 and 10^7 are not unreasonable (Tarin et al., 1984a; Willis, 1973). Indeed, mere presence of large numbers of tumor cells in the blood does not necessarily mean that macroscopic metastases will develop (Luzzi et al., 1998; Tarin et al., 1984a; Weiss, 1990). Perhaps the more relevant concern is the condition of the tumor cells at the time of injection and the site at which cells enter the vasculature.

Although injection of tumor cells into the vasculature results in wide distribution of tumor cells throughout the body (Chan et al., 1988; Juacaba et al., 1989; Potter et al., 1983), the most cells interact with and are arrested in the first capillary bed encountered. It follows, then, that the site of injection can be used to enhance development of metastases at a particular organ. This strategy has been taken advantage of during the selection of subpopulations with increased propensity to colonize a particular organ (Chambers et al., 1982; Chambers and Wilson 1988; Fidler 1973a; Giavazzi et al., 1986; Kawaguchi et al., 1983; Miner et al., 1982; Nicolson et al., 1989; Sargent et al., 1988). Intravenous inoculation into the lateral tail vein is the most common route of injection for the experimental metastasis assay. As expected, the typical site colonized is at the first capillary bed encountered—lung.

A great deal of attention has been recently afforded the orthotopic injection of cancer cells. Even a cursory review of the literature shows that most investigators have injected tumor cells subcutaneously. The reasons are simple—injections are convenient, tumor monitoring and measuring are convenient and skill level required is minimal. For a substantial number of tumorigenicity studies, more complicated protocols are unnecessary. However, most tumors fail to metastasize from this site (Liotta, 1986), despite maintaining morphological and biochemical characteristics. It is logical to contend that injection into an orthotopic site is more relevant and some investigators argue that orthotopic implantation is essential (Fidler, 1990, 1991; Hoffman, 1994; Kerbel et al., 1991; Kubota, 1994; Meyvisch, 1983). Orthotopic implantation (colorectal carcinoma—caecum (Bresalier et al., 1987; Dong et al., 1994; Fidler, 1991; Fu et al., 1991; Morikawa et al., 1988; Singh et al., 1997), renal carcinoma—kidney or subrenal capsule (Clayman et al., 1985; Naito et al., 1982, 1986; Singh et al., 1994), cutaneous melanoma—intradermal (Juhasz et al., 1993; Miele et al., 1996; Welch et al., 1991), ocular melanomas—intraocular or choroidal (Albert et al., 1980; Niederkorn et al., 1981), bladder carcinomas—bladder wall (Ibrahiem et al., 1983; Kawamata et al., 1995a, b; Kerbel et al., 1991; Theodorescu et al., 1990, 1991a), breast carcinoma—mammary fat pad (Bao et al., 1994; Kaufmann et al., 1996; Levy et al., 1982; Miller and McInerney, 1988; Phillips et al., 1996; Price, 1996; Price and Zhang, 1990; Price et al., 1990), prostatic carcinoma—prostate (Dong et al., 1999; Knox et al., 1993; Rembrink et al., 1997; Stephenson et al., 1992; Yang et al., 1999), pancreatic carcinoma—pancreas (An et al., 1996; Marincola et al., 1989; Reyes et al., 1996; Tan and Chu, 1985) , osteosarcomas—bone (Berlin et al., 1993; Crnalic et al., 1997; Simon et al., 1998), gastric adenocarcinoma—stomach wall (Fujihara et al., 1998; Togo et al., 1995), lung tumors—intrabronchial or intrapleural (Howard et al., 1991; McLemore et al., 1987; Nagamachi et al., 1998)) often results in greater metastatic efficiency and more relevant colonization patterns (i.e. similar to human cancer) than ectopic implantations into age- and sex-matched mice. Hoffman and colleagues further contend that surgical implantation of tumor fragments increases metastatic potential compared to inoculation of single cells into an orthotopic site (Hoffman, 1994). The mechanisms for the orthotopic effect remain largely unknown, but insights are forthcoming (Hoffman, 1994; Kerbel et al., 1991; Singh et al., 1997).

Another common occurrence is injection of tumor cells into multiple sites of the same animal. This practice may complicate interpretation since tumor cells communicate with each other and profoundly influence the biological behavior of distant cells. Several examples are described in the clinical and experimental literature (Brunschwig et al., 1965; Clark et al., 1989; Fidler and Lieber, 1972; Fisher et al., 1989; Howard, 1963; Koike et al., 1963; O'Reilly et al., 1994; Southam and Brunschwig, 1961; Sugarbaker et al., 1977; Warren et al., 1977; Woodruff, 1980, 1990) or inhibit (Gorelik et al., 1982; Isoai et al., 1990; Torosian and Bartlett, 1993). So, unless cell-cell communication is being tested, in vivo studies should not employ inocula into multiple sites of the same animal.

8.6. Materials needed

All of the materials needed for these studies are available through a variety of scientific product distributors. Unless specifically warranted, specific brand names are not recommended since they vary considerably according to availability. Rather, key manufacturing criteria (e.g. construct material or components) are provided.

It is advised that all the cell inoculum be prepared with 25–50% more volume than the amount calculated to be necessary. During the course of injections, volume is lost. Rather than cutting it close, it is advisable to prepare extra. Also, during the injection process, trituration of the cell suspension (with the needle absent) is encouraged to minimize cell clumping or sedimentation.

8.7. Spontaneous metastasis assays

There are so many variations of the spontaneous metastasis assay that it is impossible to describe all of them here. Therefore, two approaches are provided as examples from which the reader is to extrapolate to his/her situation. Once cells are prepared according to the criteria outlined above, the next decision is to determine the site of injection. Unless a bona fide scientifically-based reason is presented, orthotopic injection is recommended. The two examples provided will be injection of melanoma cells intradermally and breast carcinoma cells into the mammary fat pad. In both cases, the experiments outlined involve mice, but other animals can be used as warranted by the model.

There are two common variations of the spontaneous metastasis assay: (1) animals in which the primary tumor remains throughout the experiment; and (2) animals in which the primary tumor is removed (to allow time for metastases to grow to detectable size). With regard to the former, it is becoming increasingly common that animal facilities will not allow tumors to achieve sizes >1 cm. Under these conditions, the likelihood of developing metastases, especially in xenograft models, is low. The honorable intention of avoiding suffering has an undesirable byproduct for the metastasis researcher—studies of late-stage tumor biology becomes increasingly difficult, perhaps impossible. Therefore, it is important that two things take place. First, institutional animal care and use committees should be educated regarding this issue. Emotional arguments must be countered with rational, persuasive, yet still compassionate polemic. Second, appropriate safeguards should be introduced so that euthanasia or treatment is initiated at the first sign of distress.

Some of these issues are alleviated when the primary tumors are surgically removed. Timing of this operation is critical. Since tumor size is proportional to the likelihood of developing metastases (Price et al., 1990; Safarians et al., 1996), one must balance probability that metastases will have developed with the complexity of the surgery. Generally, a mean tumor diameter (square root of the product of orthogonal measurements) or geometric tumor diameter (cubed root of the product of three orthogonal measurements, distinct from mean tumor diameter because depth is measured) of 1.0–1.5 cm achieves the balance. If metastases are going to develop, they will likely have done so once a tumor has reached this size. And, even in small mice, a tumor can be removed without difficulty or post-surgical complications.

The choice to focus on intradermal and mammary fat pad injection also allows a discussion of other anatomic considerations. While obvious for skin, it is less well known that the mammary fat pads are extensive and span nearly the entire length of the ventral surface of mature rats and mice. There are even vestiges of mammary tissue on the flanks and backs of rats and mice. Orthotopic introduction of breast tumor cell lines into these animal could, therefore, be done over a wide area. Nonetheless, assurance that one is actually injecting into the mammary fat pad is easiest near the teats. Which teat? Kyriazis and Kyriazis (1980) indicate that there is a so-called 'cranial-caudal' gradient that influences tumor behavior. MDA-MB-435 human breast carcinomas metastasize more often from the thoracic mammary fat pad than from the inguinal mammary fat pad (Meschter et al., 1992). The pattern of metastases also changed—tumors in the inguinal region produced extensive intra-abdominal lymph node metastases; whereas, thoracic tumors developed more blood-borne metastases. Similar findings with melanoma cells have been described (Bani et al., 1996; Welch, 1997). Clearly, the frequency and location of metastasis is altered based on implantation distance from the head (Price, 1996). Placement also will impact the ability to remove the locally growing tumor if needed.

Inoculation volumes vary according to the site of injection. Maximum inoculum volume should not exceed 100 μl for mammary fat pad injections, 50–100 μl for intrasplenic injections, 25–50 μl for intra-adrenal injections or 50–100 μl for intradermal injections. Subcutaneous injections can utilize volumes as high as 0.5 ml. Keeping the volume below these levels minimizes leakage into surrounding tissues and stromal and epithelial damage. On the other hand, if the volume is too small, accuracy of injection becomes difficult to control. For volumes less than 100 μl, accuracy of delivery volume is best accomplished with a sterile Hamilton syringe or tuberculin syringes. The viscosity of solution being inoculated is also important. It is therefore recommended that cell concentrations no greater than 1×10^8 cells/ml be used. In addition, it is important that the inoculum not leak from the needle tract. While this is not always easy to insure, some precautions should be instituted to minimize the possibility.

Injections into the subcutaneous or intradermal sites do not require that the animals are anesthetized; however, intradermal injections are significantly easier when the animals are unconscious or sedated. Metastatic potential does not seem to be affected by anesthetics at this step. Methoxyflurane (Metofane®, Pitman-Moore) inhalation anesthetic is relatively inexpensive and effective for this purpose. A homemade anesthesia jar will suffice. Place a small volume of Metofane® under a wire screen suspended above the floor of a container with a lid. Mice and rats will be unconscious within 3–5 min and the effects will last less than 5 min. It is crucial that the animals never come into contact with the fluid. Transdermal absorption can be lethal. Also, anyone using this inhalant should also be cognisant of the content within the Material Safety Data Sheets (MSDS) and the chances of pathologies if overexposed. Hint: construct the anesthesia jar with a wire mesh that allows feces to drop as this maintains clean conditions. This provides for easier and more complete cleaning of the jar.

For injection into other sites (mammary fat pad, intrasplenic, other orthotopic sites), a mixture of Ketaset-Rompun (Ketamine-HCl, xylazine) injected intramuscularly provides excellent results. This anesthetic is also useful for simple surgical procedures like tumor removal. A stock solution of 10 ml Ketamine (100 mg/ml) containing 1.6 ml xylazine (20 mg/ml) works well for rats. A female Fisher 344 rat weighing 150–180 gm inoculated i.m. with 0.1–0.15 ml will remain unconscious for 1–4 h. For mice, the stock solution should be diluted 1 : 10 in saline. For most mice, we have found that 0.1 ml/10 g bodyweight is sufficient to anesthetize for 30 min to 1 h. However, nude mice require a higher dose (0.15 ml/10 g bodyweight). The reasons for this difference in dosage are not completely understood.

Figures 8.1 and 8.2 show intradermal and mammary fat pad injections, respectively. For photographic purposes, the animals were also sedated for intradermal injections. Both types of injection used 27 gauge needles attached to tuberculin 1-cc syringes. The single-cell suspension was prepared in ice-cold Hank's Balanced Salt Solution (however, any isotonic liquid will work as long as it does not contain serum). Note that the bevel faces the syringe markings, making it easy to see injection volume. Also note that for all of the injections the bevel is oriented so that the bore is visible from the top.

For the mammary fat pad injection, a small incision is made toward the midline of the teat using a scissors (a scalpel also works well) (Fig. 8.2A, B). The mammary fat pad is exposed in the incision by inverting the tissue using a finger and sliding the needle into the fat pad immediately under the teat (C). A 'blister' forms at the site of injection that is translucent and relatively fragile. For this type of injection, slower injection rates are recommended. When the inoculum has been injected, the needle is withdrawn and the incision is closed with sterile wound clips (D). Removal of the wound clips is usually not necessary since they usually fall out with 1–2 weeks. If, however, the clips remain, they pose no adverse threat to the animal or the tumor.

Intradermal injections are similar to mammary fat pad injections in many respects (Fig. 8.1). Photographically, the intradermal injection is indistinguishable from a subcutaneous injection. The needle is inserted into a skin fold and then reintroduced into the skin from the internal surface. For a subcutaneous injection, the needle is simply placed into the skin fold. A hallmark of intradermal injection is greater resistance during the process. Subcutaneous injections seldom require much pressure to be placed on the plunger.

If the local tumor (sometimes called the 'primary' tumor) is to be removed, the animals are anesthetized using the same Ketamine-Rompun solution described above. Adequate sedation is easily

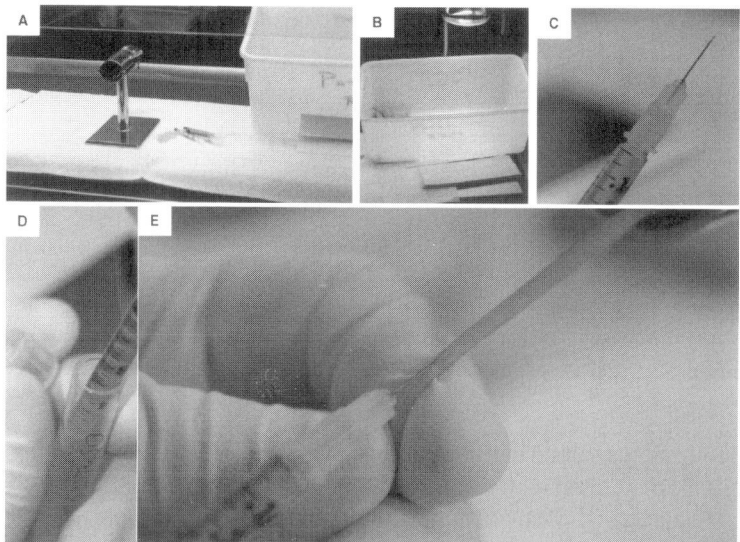

Fig. 8.1. Depiction of the set-up and intravenous injection of tumor cells for the experimental metastasis assay. Panel A shows a restrainer sitting atop a sterile surgical pad. Sufficient needles and syringes are opened and immediately available to complete all of the injections. To the right of the restrainer is a cage which sits underneath an infrared light used to cause dilation of the tail veins (Panel B). The syringe is loaded with cell suspension when there is no needle attached to the syringe (Panel D). Following syringe loading, the needle is placed with the bevel opening aligned with the syringe markings (Panel C). The lateral tail vein is located, visualized and the needle is inserted under the skin and 'tracked' up the vein (not the dark vein). Tail coloration in this photograph is the result of room lighting conditions and proximity to the heat lamp. Tail reddening should never occur.

determined by pinching a foot between ones fingers. If the animal responds, it is not adequately sedated. Tumors are removed with a wide margin (highly variable) using a scissors or scalpel. At this stage premeditation is key in order to produce a cosmetically and medically acceptable result. Deliberate maneuvers utilizing the fewest possible cut angles is best. This leaves smooth edges which are less prone to infection and which heal faster. The wounds can be

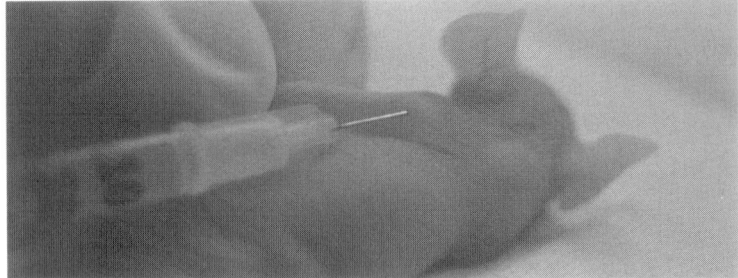

Fig. 8.2. Intradermal injection of cells into the athymic mouse. The needle is inserted subcutaneously and then tracked to injected into the skin from the internal side. A bleb indicates that the inoculate is being injected.

sutured (consult a veterinarian for advise regarding the best composition) or stapled with sterile wound clips. Depending on the facility and the site, topical antibiotics can be applied to further decrease the chances for infection. Standard aseptic techniques are typically all that is necessary for infection not to be a problem. Dissolving sutures is recommended in order to minimize follow-up anesthesia. This is particularly desirable if visceral tumors are removed. Wound clips are not recommended for internal wound closure. It is not necessary to remove them as they tend to fall out shortly after healing is complete anyway.

8.8. Experimental metastasis assays

The mechanics for experimental metastasis assays are similar to those for the spontaneous metastasis assay. As above, volume and viscosity of the inoculum are important parameters. Volumes should never exceed *0.2 ml* because plasma blood volume exceeds the normal range ($\pm 10\%$) such that the distribution pattern of cells is altered. Tumor density is even more critical for intravascular injections since introduction of emboli can cause death because of vascular obstruction. Therefore, the same concentration maximums reported above are also applicable for this type of assay.

Ch. 8　　　　　　　IN VIVO CANCER METASTASIS ASSAYS　　　　　　　229

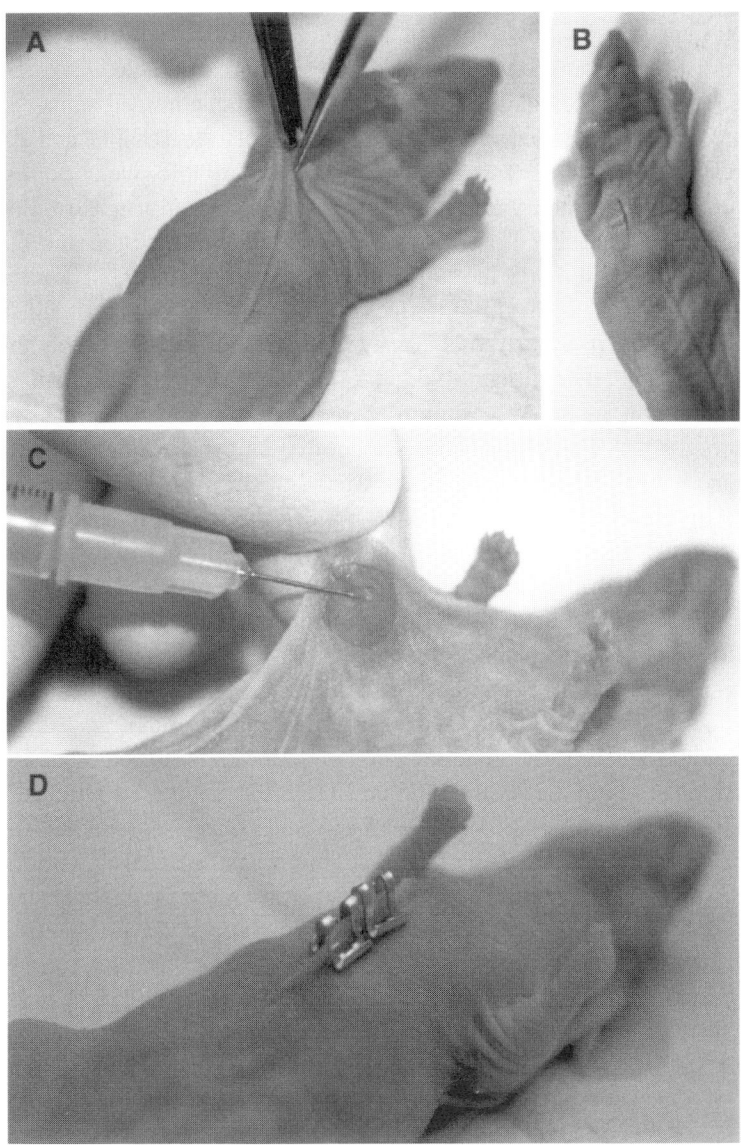

The key difference for intravenous injection is the use of a restrainer which allows the tail to extend outside of the enclosure for i.v. injections. Several restrainer designs are available. The most common are plexiglass. The preferred design is a hinged stainless-steel tube suspended on a weighted pedestal (Please note that the restrainer shown (Fig. 8.3) is custom made for D. R. Welch.). This unit is autoclavable; however, rinsing with a dilute bleach solution suffices for sterilization. A mouse is placed into the tube and enclosed while holding the tail (Fig. 8.3, panels A and E). The dark environment has a calming effect (a similar effect can be obtained in the plexiglass restrainers by wrapping with electrical tape and attaching a piece of lead to the bottom of the restrainer to increase the weight in order to minimize movement during the injection process.). A lateral tail vein is identified and the needle is gently inserted to just below the skin. After 'tracking' for some distance to minimize leakage and backwash when the needle is withdrawn, the cells are injected using a 27 gauge needle fitted onto a 1 cc tuberculin syringe (some prefer a 26-g needle) (Fig. 8.3, panel E). During the process, a slow, steady rate of injection is the objective. For i.v. injections, successful inoculation is evident by the lack of resistance during the process. If any resistance is felt, the process should be restarted. It is best to begin injections at the most distal part of the tail. If the injection is missed, one can proceed cranially. If one starts at the base of the tail, the effort cannot be redeemed without subsequent inocula leaking from the hole(s) generated by prior injections. Although i.v. inoculations can be done without the aid of procedures to dilate tail veins, the process is facilitated when

Fig. 8.3. Key steps involved in orthotopic mammary fat pad injections. An incision is made medially to the teat under which the fat pad will be exposed (Panels A and B). The fat pad is externalized by inverting the skin using a finger (Panel C). The needle is inserted into the fatty tissue below the teat and inoculum injected. Panel D shows the mouse following closure of the incision with wound clips. As noted in Fig. 1, coloration is distorted due to room lighting, flash and sterile hood lighting conditions.

tail veins are enlarged. This can be accomplished by dipping the tails into hot water, swabbing with irritants (such as xylenes) or brief warming under a heat lamp (preferred). A high-walled container is useful in order to decrease the chances for mice escaping (Fig. 8.3, panel B). This is a particular concern for 'hyperactive' mice such as C57BL/6. Use of a heat (infrared) lamp requires close monitoring to assure that the animals do not get overheated and that personnel are careful not to touch the bulb since second and third degree burns can result.

8.9. Enumeration of metastases

After injecting the tumor cells. The hardest part of the experiment follows—waiting. The interval can be a few days for particularly aggressive cells to several months for others. When a new model is being developed, it is important to allow adequate time for metastases to develop. Therefore, periodic euthanasia of a subset is advised. If, at any time, animal health appears to diminish, the experiment should be terminated for humane reasons.

For some studies, the mere presence or absence of metastases is sufficient; however, quantification of metastases is often desired. To assess metastasis, animals are killed, organs are examined then removed and rinsed in cold water to remove excess blood.

Euthanasia methodology affects the ease by which metastases can be detected and quantified. Three methods have been used by my laboratory—carbon dioxide asphyxiation, cervical dislocation and overdose using anesthetics. All methods for euthanizing animals must be approved beforehand by the Institutional Animal Care and Use Committees (IACUC). Guidelines for euthanasia are evolving; hence, regular consultation with veterinarians concerning procedures is advised.

Carbon dioxide is not recommended because of a substantial number of petechia in the lungs. They complicate counting of tumors, particularly for less inexperienced lab personnel. Cervical dislocation works well but sometimes can result in the presence of clots in the lungs. Nonetheless, cervical dislocation is still used as the primary method for mouse euthanasia (see below). For rats and as an alternative for mice, Metofane® inhalation works extremely well. The animals fall asleep and then die. Lungs are clear and quantification of metastasis is unimpeded . However, this is more expensive.

Identification of macroscopic metastases is easier if coloration is different from parenchyma. This was a major advantage when studying melanoma; however, this characteristic was not available for other tumor types. It is possible to identify metastases in most tissues. They appear as clear or white raised gelatinous surface structures. Examination is facilitated by the use of a dissecting or stereomicroscope. However, if there are imperfections in the tissue or if the tumor is small, errors can occur. Therefore, different approaches have been employed to enhance visual contrast between tumor and parenchyma.

For lung metastases, the trachea can be injected with 1.5–2.5 ml of a 15% solution of India ink in neutral buffered formalin (37% formaldehyde (100 ml), tap water (900 ml), sodium phosphate monohydrate (4 g), disodium phosphate (6.5 g), pH 7.0). Following sealing with surgical suture to prevent leakage, the lungs are then suspended in a beaker of tap water. Tumor colonies are then bleached with Fekete's solution (37% formaldehyde (10 ml), glacial acetate (5 ml) and 70% ethanol (100 ml)). Metastases appear as white colonies against a black background (Fig. 8.4, panel B). While this approach works well, it can be impractical when lots of animals are being necropsied simultaneously. Also, it can be done more easily with assistance. Therefore, the preferred method is fixation of organs and tissues in Bouin's solution (saturated picric acid (300 ml), neutral buffered formalin (100 ml) and glacial acetate (20 ml)). After fixation, the tumors appear as white or pale yellow spots against a darker yellow background (Fig. 8.4, panel A). The use of Bouin's is not without problems, however. Tissues become brittle making subsequent confirmatory histology difficult. To partially al-

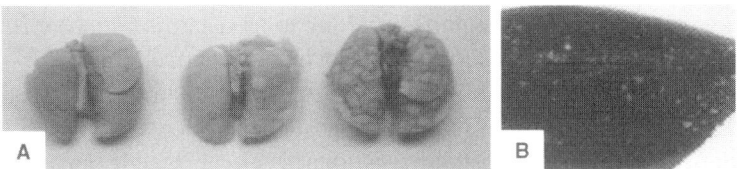

Fig. 8.4. Visualization of lung metastases following staining with Bouin's fixative (lighter colored nodules on the surface, Panel A, two lungs to the right of the normal lung) or following injection of India ink solution into the trachea followed by destaining with Fekete's solution (Panel B, depicting white nodules against the background.

leviate this problem, we use a mixture of one part Bouin's fixative in five parts neutral-buffered formalin.

One important consideration when removing tissues is to avoid touching the surface with forceps. Striations caused by the teeth complicate visualization of small metastases. Most lung (Polissar and Shimkin, 1954; Welch et al., 1983; Wexler, 1966; Wood et al., 1954) and liver (Lafreniere and Rosenberg, 1986) metastases develop near the surface; therefore, anything which compromises visualization will affect quantification. Similar precautions should be taken during the preparation of other tissues.

Random samples of tissues should be submitted for histologic confirmation of presence/absence of metastases. Ideally, one could quantify metastasis by serially sectioning tissues, measuring the surface area of the lesions, calculating total tumor volume and comparing that to the organ/tissue volume (Boeryd et al., 1966). Unfortunately, this is impractical. The use of histologic preparations allows determination of total tumor burden. Most investigators merely count the *number* of metastases. Small lesions are not equivalent to large lesions, however. Assessment of metastatic tumor burden should not only include assessment of number, but also size/*volume* of lesions. In the lung, the majority of metastases are spherical (Welch et al., 1983) making this calculation straightforward. Side-by-side comparison of the number and volume of metastases can give important information regarding mechanism responsible for developing metastasis by different cell lines.

Although well established, counting and measuring metastasis can be tedious, especially for large-scale experiments such as drug screening for antimetastatic compounds. Organ weight can offer an option. However, care must be taken to properly control for experimental read-out.

First, organ weight is not a valid measure alone since this is proportionate to animal weight/mass. Therefore, a ratio of organ weight to animal weight should be calculated. Second, measurements must accompany verification that ancillary variables have not caused changes. For example, some treatments can cause edema which would increase organ weight thereby mimicking tumor burden. Organ weight and metastatic tumor burden correlates well in some tumor models (e.g. Lewis lung carcinoma (Gorelik et al., 1980)), but not in others (Welch, 1997). Therefore, use of organ weights as a measure of metastasis must be confirmed on a case-by-case basis.

Welch has considered development of image analysis software for quantification of metastasis (D. R. Welch, personal communication). While this would greatly aid the metastasis researcher, three limitations remain—variable 'staining' of the metastases; difficulty distinguishing organ surface imperfections from bona fide small metastases; and three-dimensional nature of organ surfaces. Staining conditions could be worked out with systematic evaluation. The imperfection issue is more difficult because such imperfections are not predictable or consistent, making programming very difficult. Finally, image analysis is dependent on two-dimensional photographic or imaging systems. Unless multiple images from different angles are prepared and assimilated using computer algorithms, this is not likely to be cost-effective in the short term.

Another method for estimating the number of metastatic cells successfully colonizing a tissue is accomplished using tumor cells tagged with a drug resistance, genetic or color marker (see also Chapter 7: genetic tagging). Visualization of metastases can be enhanced if the cells are tagged with the lacZ gene (Brunner et al.,

1992; Brunner et al., 1993; Fujimaki et al., 1993; Kurebayashi et al., 1993; Lin et al., 1990a, b; McLeskey et al., 1996) or fluorescein (Potter et al., 1983). The tumor cells appear blue or yellow-green, respectively. However, stability of the transfectants in vivo can vary considerably and a substantial proportion of macroscopic metastases will no longer be colored or the metastases will contain mixtures of colored and colorless cells. Therefore, if coloration were used as a criterion, the number and volume of metastasis would be underestimated (Fujimaki et al., 1993). If tumor cells are tagged with a drug resistance or genetic marker, cells colonizing different organs can be recovered from dissociated tissues (Miller et al., 1990) and the proportion compared to the inoculum. Similarly, cells labeled with ^{125}IUdR or BrdU can be detected in dissociated tissues using a gamma counter or ELISA, respectively (Fujimaki et al., 1993). To use these approaches, prior verification that the label does not affect biological behavior of cells must be obtained.

Cells transfected with enhanced green fluorescent protein (GFP: Cubitt et al., 1995) are exposed to blue light and the tumor cells fluoresce in the green range. This approach was first utilized by others to visualize metastases in vivo (Chishima et al., 1997a, b; Farina et al., 1998; Yang et al., 1998, 1999) and offers a significant increase in sensitivity since metastases are more visible. However, a caveat to these techniques deserves mention. *Presence of single cells in an organ should not be equated with metastasis formation.* Clinically relevant metastases are those which have grown to sufficient size to disrupt cellular or tissue function. This is usually not the case with single cells. As such, single cells do not qualify as having completed all steps in the metastatic cascade. That being said, single cells can remain as occult disease for several years until stimulated to divide (i.e. complete the metastatic cascade). Presence of single cells should not be ignored, but should be categorized differently than macroscopic metastases.

8.10. Statistical considerations

New investigators are often shocked by the relatively high variability observed when doing in vivo studies. While it should not be surprising, given the increased number of variables incorporated into the experimental design, the 'shock' that accompanies an inadequately designed experiment is real. It is critical, therefore, that each experiment include sufficient 'power' to provide a statistically valid result. For most studies, groups of a *minimum* of 8–10 animals are required, but sometimes more are needed. The total number of test animals needed can be determined using appropriate power calculations (Heitjan et al., 1993).

As with most statistical calculations, increasing the number of measurements increases the likelihood that one will obtain normally distributed data. If this is the case, then parametric assays can be used. However, for many studies, nonparametric statistics are necessary. The most common reasons for needing nonparametric tests are nonnormal distribution of low metastasizing cell lines (i.e. the numbers of metastases are mostly zero) or counts greater than a countable number (i.e. >250 metastases per lung). In both cases, the data are not normally distributed. In this situation, the statistic used is the Mann-Whitney U-test which evaluates differences between groups using ranks. Readers are encouraged to consult with a statistics textbook and/or a statistician prior to beginning a study and when interpreting the results.

8.11. The influence of stress

The concept that tumor progression is amenable to control by the immune system dates back to the end of eighteenth century; exciting developments progressively and continuously occurred leading to modern immunology, including the field of tumor immunology. For several decades, the immunologists pictured the immune system as an independent homeostatic system, engaged in defending the host from the invasion of foreign pathogenic organisms. Only recently a robust evidence has been provided, showing that the central nervous system can modulate the action of immune effectors via neurovegetative and neuroendocrine circuits (Ader et al., 1994). At the same time, an abundant experimental evidence has been collected, showing that stress paradigms display the capacity to cause distinct immunitary alterations (Bellinger et al., 1994), which may lead to a modification of tumor incidence and growth in laboratory animals (Justice, 1985).

It is interesting to note that, in spite of the large mass of accumulating experimental evidence indicated above, a remarkably limited attention has been given to two crucial factors. The first one is that almost the totality of the experiments performed to study the effects of stress paradigms on tumor progression in laboratory rodents considered primary tumor incidence and growth, and did not consider malignant metastatic dissemination of the tumor (Bammer, 1981). Second, considering of the accepted view that stress may affect the growth of tumors in laboratory animals, the possible influences of the stress resulting from the housing as well as from the experimental procedures have been largely neglected (Steplewski et al., 1987).

On the basis of these considerations, the data available in the literature on the effects of experimental stressors on tumor metastasis and response to chemotherapy in laboratory mice will be briefly described.

The uncontrolled conditions occurring when maintaining laboratory mice in a conventional animal house may result in a stressful environment, capable of inducing significant stress responses such as increased glucocorticoids levels and atrophy of lymphatic organs (Riley et al., 1981). Housing conditions were correspondingly shown to significantly affect tumor growth in mice (Riley et al., 1981), and specifically to increase the formation of spontaneous lung metastasis (Giraldi et al., 1989). The effects on metastasis were first observed in mice bearing Lewis lung carcinoma. This tumor is weakly immunogenic into syngeneic female BD2F1 (C57BL/6×DBA/2F1) mice, with a TD_{50} of 8.6×10^4

cells per mouse when animals are maintained in conventional housing (Perissin et al., 1991). If rotational stress is applied to mice implanted with 10^6 tumor cells, no difference on metastasis is observed. On the contrary, significant effects on metastasis were observed when the experiments were performed keeping the animals in a low stress environment (Giraldi et al., 1989). In order to control stressful variables, the animals were kept in the protected environment for 2 weeks before each experiment and throughout its duration. To avoid overcrowding or isolation effects on tumor progression, the animals were kept in groups of five in plastic cages measuring $27 \times 21 \times 14$ cm with a stainless steel grid cover. To minimize acoustic, olfactory and visual communication the cages were kept in cabinets allowing laminar air flow in between and placed in a room remote from the animals rooms; staff entrance was limited to the supply of water and food. The light-dark cycle in the room was 12–12 h, with an intensity in the cages of approximately 5 lux; temperature and relative humidity were constant at 20°C and 60%, respectively (Riley, 1981). When the animals were kept in these controlled protected conditions, the weight of spontaneous lung metastasis was remarkably lower, as compared with mice kept in conventional housing. Moreover, the application of rotational stress significantly increased the weight of lung metastasis in these animals; a similar increase was caused also by handling the animals to perform an intraperitoneal injection of a small volume of physiological saline (Giraldi et al., 1989). When BD2F1 mice maintained in the protected environment were implanted with 5×10^5 tumor cells, no tumor take was observed; tumor take occurred after the application of rotational stress, or when the experiment was performed keeping the animals in the conventional housing. After a tumor implant of 10^6 cell all of the mice developed tumors, and metastasis weight was significantly increased by rotational stress (Perissin et al., 1991; Giraldi et al., 1994a). The differences observed in metastasis as a function of housing conditions and application of rotational stress were accompanied by relevant modifications in the main murine glucocorticoid, corticosterone, as stress index (Perissin et al., 1989).

The effects of stress on Lewis lung carcinoma metastasis are not limited to housing and rotational stress. Other stress paradigms, including avoidable foot-shock, early maternal separation, and restraint stress, do modify metastasis number and weight (Giraldi et al., 1994b). These effects are accompanied in general by smaller or negligible modification of primary tumor growth, suggesting that the stress paradigms examined influence tumor cell replication in metastatic foci rather then the malignant process of tumor metastasis.

At the same time, the effects of stress paradigms on tumor progression do not appear to be limited to Lewis lung carcinoma. Indeed, the survival time of female CBA mice implanted with a different inoculum size of TLX5 lymphoma has been studied as a function of the application of rotational stress. The inoculum with 10^3 or 10^4 tumor cells resulted in survival times which were not affected by rotational stress. On the contrary, in mice implanted with 10^2 cells, the mean survival time was decreased by rotational stress from 14.8 to 12.9 days. In mice implanted with 10 cells, tumor take occurred in 6/20 mice, and this proportion was significantly increased to 16/20 by rotational stress; the mean survival time of the mice with tumors was 15.5 and 14.1, respectively (significant) (Perissin et al., 1997). The effects reported above for rotational and restraint stress are observed also in female CBA mice implanted with MCA mammary carcinoma (Giraldi et al., manuscript in preparation).

The relationships between immune functions of the host and metastasis as a function of stress have been examined in mice bearing Lewis lung carcinoma subjected to rotational stress. Since some immune parameters were shown to display rhythmic variations, the occurrence of seasonal factors has been also taken into consideration. Indeed, the increase in metastasis number and volume caused by rotational stress varied in magnitude with a highly significant circannual rhythm, the acrophase coinciding with summer solstice.

Rotational stress caused a significant reduction in the number of CD3+ and CD4+ T-lymphocyte subsets in summer, whereas in winter the number of CD3+ subset was significantly increased; the CD4+/CD8+ ratio and the number of NK 1.1 antigen positive cells were not significantly modified by rotational stress in both periods considered. The increase in metastasis formation by rotational stress thus appears to negatively correlate with the number of splenic CD3+ and CD4+ T-lymphocyte subsets (Perissin et al., 1998).

In the perspective outlined so far, it is conceivable that the magnitude of the effects of antitumor and antimetastatic agents might be affected by stress. Indeed, when the effects of drug treatment are determined at necroscopy after sacrifice, the magnitude of the cytotoxic action of cyclophosphamide and of the selective antimetastatic effects of razoxane was significantly reduced by rotational stress (Perissin et al., 1991). The therapeutic effects of cyclophosphamide are attenuated also by restraint stress, and the attenuation is likewise evidenced by the analysis of the animals' survival. The increase in survival time caused by cyclophosphamide is significantly reduced by restraint stress, and the fraction of mice cured by the drug is abolished. Also in this case, the results are consistent with a neuroimmunomodulation caused by restraint stress. Physical restraint reduced the number of CD3+ and particularly of CD4+ positive splenic lymphocytes, as well as of CD4+/CD8+ ratio; the reduction in the helper/suppressor ratio is additive with the effects of cyclophosphamide (Zorzet et al., 1998). Additionally, the increase in survival time of CBA mice with small implants of TLX5 lymphoma ($10–10^2$ cells per mouse) by CCNU is similarly attenuated (Perissin et al., 1997).

The conclusions that can be drawn from the experimental evidence reported above can be summarized as follows. First, laboratory mice maintained in a conventional animal house display, in comparison with mice kept in a protected low stress environment, display distinct humoral and immunity characteristics of exposure to stress that accompany significant differences in spontaneous tumor metastasis; handling the animals for experimental procedures may cause similar effects. Second, the sensitivity to stressful events may display a rhythmic seasonal pattern, evident on metastasis and on T-lymphocyte subsets. Third, the magnitude of the effects of cytotoxic and antimetastatic drugs may depend on the levels of stress to which the animals were exposed. The implications of these observations appear of interest for their experimental relevance, and may account for an unexpected lack of repeatability or for an excessively large variability of experiments lacking a careful control of these factors. Moreover, the relevant experimental settings appear to offer valuable experimental models, applicable to protocols with laboratory animals useful to investigate elements of interest for clinical oncology and psycho-oncology.

Investigations aiming to study the effects of stress on metastasis in laboratory mice, should focus on a careful control of the stress resulting from housing and handling the animals for the experimental procedures.

Special attention has to be given to the following issues, which were shown to influence tumor growth and/or metastasis in mice:

- The use of an animal house, or animal room(s), sufficiently detached from the conventional animal housing (Riley et al., 1981).
- Entrance of staff for food and water supply to the animals limited to the minimum (Riley et al., 1981).
- Maintenance of the cages containing the animals in cabinets limiting acoustic, olfactory and visual communication between the mice in different cages. Such cabinets were described by Riley, can be custom built, and are commercially available from suppliers of animal housing hardware (Riley et al., 1981).
- Maintenance of the mice in the protected environment for at least two weeks before any experimental procedure, in order to avoid the effects resulting from shipment (LaBarba et al., 1970; Riley et al., 1981).
- Constant, and appropriate animal density in the cages, since isolation or overcrowding may affect tumor growth and/or metastasis (LaBarba et al., 1970; Riley et al., 1981).

- Use of a constant light-dark cycle, in order to limit the influences deriving from the existence of neuroendocrine and immunitary rhythms (Perissin et al., 1998).
- Limit to the minimum handling the animals (Riley et al, 1981). For instance, the administration of drugs might be performed dissolving them in drinking water or admixed to the food (Giraldi et al., 1989; Zorzet et al., 1998).
- Control of the stress level of the animals by measuring their blood level of corticosterone. The assay can be performed fluorimetrically or by RIA, and values above 60–100 ng/ml are indicative of a significant stress response (Riley et al., 1981).

8.12. Concluding remarks

Metastasis assays are, by their very nature, complicated. This is largely because the process of metastasis is itself complex and variable. The key elements to successfully studying metastasis are careful consideration of the question(s) being asked, quality characterization of the cell lines being used for the study, and utilization of appropriate model(s). The technical components of the studies require due care, but attention to the details will enhance the likelihood of success.

CHAPTER 9

Angiogenesis and metastatis

Over the last decade the study of tumor-induced angiogenesis has exponentially grown. It is now clear that neovascularization of the tumor mass is a 'sine qua non' condition for neoplastic growth and dispersion into the vascular system; for this reason the production of angiogenic factors is a fundamental step of tumor progression (Folkman, 1984, 1992; Folkman et al., 1989; Weidner et al., 1993b). On the other hand, tumor cells are also able to release antiangiogenic factors. This ability could explain the apparent control exerted by primary tumors on the metastatic spread (Chen et al., 1995; Folkman, 1995a). These observations suggest that tumoral angiogenesis is linked to a switch of the equilibrium between positive and negative regulators (Hanahan et al., 1996). Recent studies of the antiangiogenic peptides angiostatin and endostatin (respectively the proteolytic fragment of plasminogen and type XVIII Collagen) have shown the ability of these molecules to cause the regression of primary tumors, thus representing a strong support to the antiangiogenic therapy (O'Reilly et al., 1996, 1997; Wu et al., 1997).

Numerous in vitro and in vivo tests have been developed to study specific steps of the neoangiogenic process. Several of these experimental protocols use the same techniques applied for assessment of tumor cell behavior (Albini, 1998). The activated endothelial cell acts similarly to a metastatic cell in the initial phases of angiogenesis, degrading the capillary basement membrane (BM), extravasating, digesting the extracellular matrix and moving toward the angiogenic stimulus (Liotta et al., 1991). For this reason adhesion, migration, protease production and invasion can be detected with the same tests as those used for cancer cells. The main difference between the endothelial and the tumoral cell is that angiogenesis is a strictly controlled phenomenon of normal endothelial cells, which ultimately differentiate into new vessels or undergo apoptosis (Risau, 1997). This difference can also be evaluated in vitro using morphology tests which assess the final differentiation of endothelial cells into capillary-like structures (Albini et al., 1995; Montesano et al., 1992).

The angiogenic process involves not only endothelial cells but also many other cell types and signals coming from other districts of the organism. Therefore in vivo studies are a necessary checkpoint of any experimentation of angiogenic stimulators or inhibitors.

In order to develop angiogenic and antiangiogenic strategies, there are concerted efforts to provide animal models for more quantitative analysis of in vivo angiogenesis. In vivo techniques include the cornea pocket and iris implant in the eye, the rabbit ear chamber, the dorsal skinfold chamber, the cranial window, the hamster cheek pouch window, the rat mesenteric window, the chick embryo chorioallantoic membrane, the air sac in mice and rats, and finally the sponge, fibrin clot, sodium alginate beads and matrigel implants (Auerbach et al., 1991; Jain et al., 1997). However, among these tests, three main models are the most frequently used: assays in the CAM (chorioallantoic membrane), in the cornea, and in subcutaneous implants. This chapter reviews the use of specific protocols for each of these approaches.

9.1. *The corneal assay for angiogenesis*

The corneal assay consists in positioning an angiogenesis inducer (tumor tissue, cell suspension, growth factor) into a pocket made in the cornea in order to evoke vessel outgrowth from the peripherally located limbal vasculature. This assay, with respect to the other in vivo assays, has the advantage of measuring only new blood vessels, since the cornea is initially avascular. Both eyes of each rabbit can be used and the implantation of two or more sponges in the same cornea is a suitable tool for studying synergistic or competitive effects.

9.1.1. Experimental procedure

The corneal assay performed in New Zealand white rabbits was first described by Gimbrone et al., (Gimbrone et al., 1974a). It was chosen for the absence of a vascular pattern and for the ease of manipulation and monitoring of the neovascular growth. This technique has been extensively modified to fulfill different experimental requirements.

Surgery. After being anaesthetized with sodium pentothal (30 mg/kg, i.v.), a micropocket (1.5 × 3 mm) is surgically produced under aseptic conditions using a pliable iris spatula 1.5 mm wide in the lower half of the cornea (Fig. 9.1). A small amount of the aqueous humor can be drained from the anterior chamber when reduced corneal tension is required. The implant is positioned at 2.5–3 mm from the limbus to avoid false positives due to the mechanical procedure, and to allow the diffusion of test substances in the tissue forming a gradient toward the endothelial cells of the limbal vessels. Implants sequestering the test materials and the controls are encoded and implanted by different operators, providing a 'blind' assay.

Sample preparation. The material to be tested can be in the form of slow-release pellets incorporating recombinant growth factors, cell suspensions, or tissue samples (Gallo et al., 1998; Langer et al., 1976).

Slow release preparations: Recombinant growth factors are prepared as slow-release pellets by incorporating the test molecule into an ethylene-vinyl-acetate copolymer (Elvax-40) (DuPont de

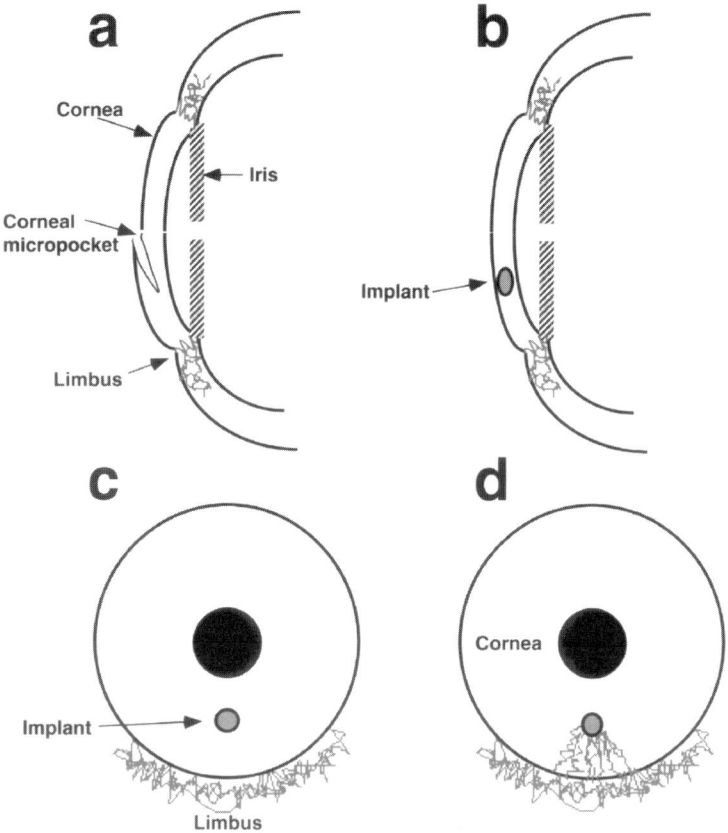

Fig. 9.1. The corneal micropocket assay: a micropocket is surgically produced in the corneal stroma by a pliable spatula (a) and the sample is inserted in the micropocket (b). The newly formed vessels protrude from the limbal vasculature (c, t=0) and progress toward the implanted stimulus (d, t = 7 days).

Nemours, Wilmington, DE). A predetermined volume of Elvax-40 casting solution is mixed with a given amount of test molecule on a flat surface and the polymer is allowed to dry under a laminar flow hood. After drying, the film sequestering the compound is cut into $1 \times 1 \times 0.5$ mm pieces. Empty pellets of Elvax-40 are used as controls.

Cell and tissue implants: Cell suspensions are obtained by trypsinization of confluent cell monolayers. Five microliters containing 2×10^5 cells in medium supplemented with 10% serum are introduced in the corneal micropocket. When the over expression of growth factors by stable transfection of a specific cDNA is studied, one eye is implanted with transfected cells and the other with the wild type cell line. When tissue samples are tested, samples of 2–3 mg are obtained by cutting the original fragments under sterile conditions. The angiogenic activity of tumor samples is compared with macroscopically healthy tissue.

Quantitation. Subsequent daily observation of the implants is made with a slit lamp stereomicroscope without anaesthesia. Angiogenesis, edema and cellular infiltrate are recorded daily with the aid of an ocular grid by an independent operator who did not perform the surgery. An angiogenic response is scored positive when budding of vessels from the limbal plexus occurs after 3–4 days and capillaries progress to reach the implanted pellet in 7–10 days. Implants that fail to produce a neovascular growth within 10 days are considered negative, while implants showing an inflammatory reaction are discarded (Ziche et al., 1989). The number of positive implants over the total implants performed is scored during each observation. The potency of angiogenic activity is evaluated on the basis of the number and growth rate of newly formed capillaries, and an angiogenic score is calculated by the formula *vessel density × distance from limbus* (Fig. 9.2) (Ziche et al., 1997) calculated as follows: A density value of 1 corresponds to 0–25 vessels per cornea, 2 for 25–50, 3 for 50–75, 4 for 75–100 and 5 for >100 vessels. The distance from the limbus is graded with the aid of an ocular grid.

Histological examination. Corneas are removed at the end of the experiment as well as at defined intervals after surgery and/or treatment and fixed in formalin for histological examination. Newly formed vessels and the presence of inflammatory cells are detected by hematoxylin/eosin staining or specific immunohistochemical

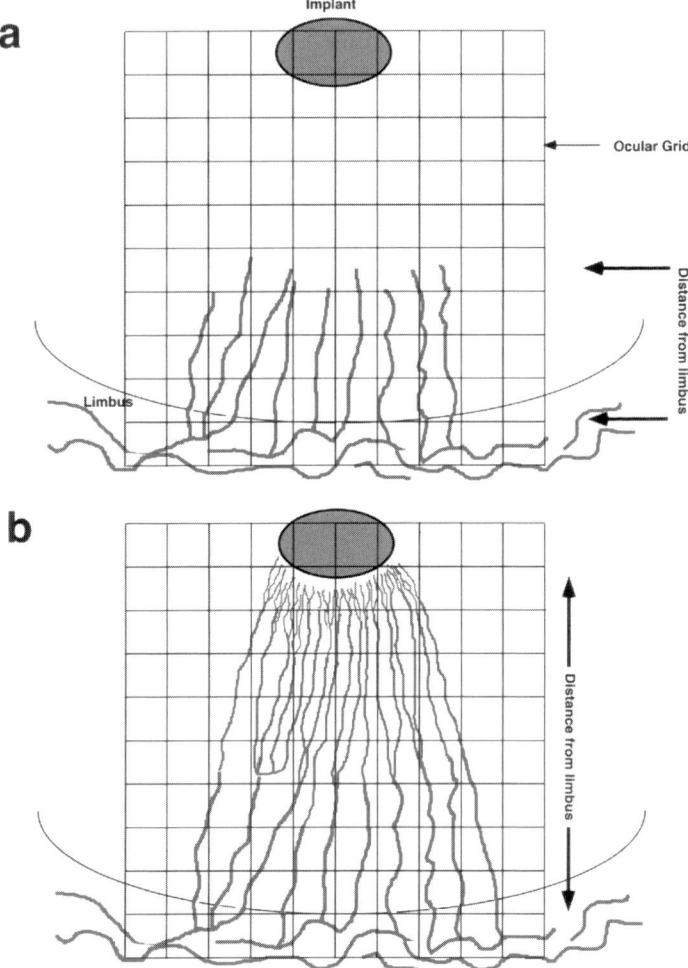

Fig. 9.2. Angiogenic activity is quantified by the formula [number of vessels × distance from the limbus]. The vessel density is calculated counting the number of newly formed capillaries. The distance from the limbus (mm) is calculated using an ocular grid. An angiogenic response is scored positive at day 3–4 when the budding vessels travel approximately 0.5—0.8 mm from the limbal plexus (a); at day 5–7 when capillaries progress into the corneal stroma to 1.0—1.8 mm. and at day 9–12 when the new capillaries reach the pellet containing the angiogenic factor (b).

procedures (i.e. antirabbit macrophage (RAM11) antibodies or anti-CD31 for endothelia). Double staining (i.e. anti-CD31 for vascular endothelium and specific markers for tumor cells) could be useful to label newly formed vessels of the host and proliferating tumor cells implanted in the cornea.

9.1.2. *Features of the rabbit corneal assay for angiogenesis*

Mice and rats can also be used to perform the cornea micropocket assay (Chen et al., 1995; Muthukkaruppan et al., 1979; Voest et al., 1995). However, there is a series of features which lead to prefer rabbits for this technique.

Species: Rabbit cornea has been found avascular in all strains examined so far. In some strains of rats the presence of pre-existing vessels within the cornea and the development of keratitis are serious disadvantages.

Animal size: Rabbits are more docile and amenable to handling and experimentation than mice and rats. The rabbit size (2–3 kg) lets an easy manipulation of both the whole animal and the eye to be easily extruded from its location and to be surgically manipulated. General anaesthesia is required only for surgery while daily examinations occasionally need local anaesthesia. Conversely, the small size of rats and mice allows the use of small amounts of drugs given systematically.

Measurements: In mice and rats it is possible to obtain time-point results. The evolution of the angiogenic response in the same animal is not recommended because each time the cornea is observed the animal has to be anaesthetized. Thus, experiments are made with a large number of animals and vessel growth during time is measured in individual animals. In contrast, multiple observations are possible in rabbits. The use of slit lamp stereomicroscope and of nonanesthetized animals allows the observation of newly formed vessels over time with long term monitoring, even for 1–2 months.

Monitoring of inflammation: Inflammatory reactions are easily detectable in rabbits by stereomicroscopic examination for corneal opacity.

Different experimental procedures: In the rabbit eye, due to its large area, stimuli can be placed in different forms. In particular, the activity of specific growth factors can be studied in the form of slow-release pellets and of tumor or nontumor cell lines stably transfected for the overexpression of angiogenic factors (Gualandris et al., 1996) (Fig. 9.3). Modulation of angiogenic responses by different stimuli can be assessed in the rabbit cornea assay through the implant and/or removal of multiple pellets placed in parallel micropockets produced in the same cornea (Albini et al., 1996; Ziche et al., 1989). The implant of tumor samples from different locations can be performed both in corneal micropockets and in the anterior chamber of the eye to monitor angiogenesis produced by hormone-dependent tissues or tumors (i.e. human breast or ovary carcinoma in female rabbits) and it allows the detection of both iris and corneal neovascular growth (Gallo et al., 1998). Moreover the activity of antiangiogenic molecules released by the tumors can be simultaneously quantified by monitoring in the cornea the angiogenic response to angiogenic stimuli.

9.1.3. Concluding remarks

Continuous monitoring of angiogenesis in vivo is required for the development and evaluation of drugs acting as suppressors or stimulators of angiogenesis. In this respect, the avascular cornea of New Zealand albino rabbits offers a unique model, since progression of neovascularization can be monitored for extended periods of time with a noninvasive approach, and the comparison in the same animal of distinct effectors is possible. Measurements of corneal angiogenesis is useful for quantitating the effects of angiogenic stimuli and for evaluating the efficacy of potential inhibitors of neovascularization. Because accurate methods suitable for recording the entire

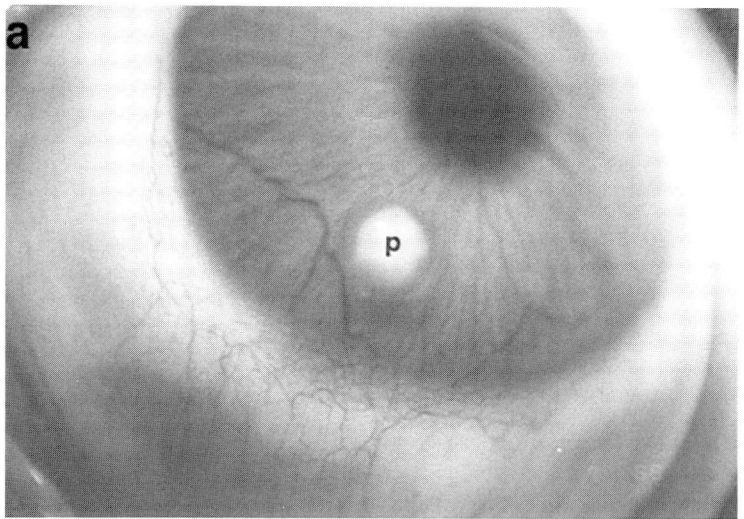

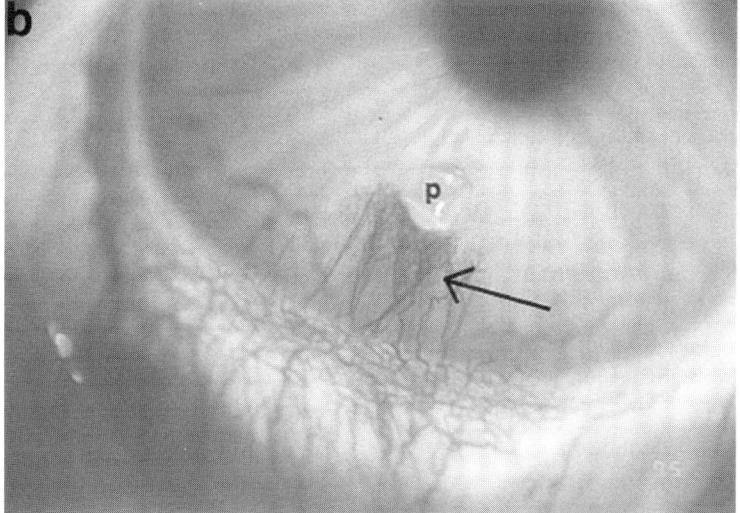

Fig. 9.3. Representative pictures of a negative (a) and a positive (b) angiogenic response. Note in panel b the strong neovascularization caused by the pellet containing VEGF121 (200 ng/pellet) in comparison with the vehicle alone. Pictures are taken 12 days after pellet implant. Arrows indicate the newly formed vessels. Asterisks indicate flash artefacts. p = pellet implant. (18× magnification).

pattern of corneal neovascularization over time and for obtaining a quantitative evaluation of the process in individual living animals do not exist, a noninvasive method to achieve this goal by the use of a computerized image analysis system is in progress.

9.2. The chick embryo chorioallantoic membrane assay

The avian chorioallantoic membrane (CAM) is a useful model to study angiogenesis and its regulation in vivo (Ribatti et al., 1996). Even if this model is based on avian systems, thus phylogenetically distant from mammals, it has been proven to be one of the most frequently successfully used models.

9.2.1. Histogenesis and structure of CAM

The allantois is an extraembryonic membrane derived from the mesoderm in which primitive blood vessels begin to take shape on day 3 of incubation. On day 4, the allantois fuses with the chorion epithelium, derived from the ectoderm, and forms the chorioallantois. The primitive vessels continue to proliferate and to differentiate into an arteriovenous system until the eighth day of incubation producing a network of capillaries that occupy the area next to the chorion epithelium and mediate gas exchange with the outer environment. The CAM vessels grow rapidly up to day 11, after which the endothelial cell mitotic index decreases just as rapidly, and the vascular system attains its final organization on day 18 of incubation, just before hatching (Ausprunk et al., 1974).

9.2.1.1. Utilization of CAM
Fertilized chick eggs can be conserved at room temperature up to 48 h and are incubated at 37°C in humidified atmosphere. On day 3 of incubation, 2–3 ml of albumen are aspirated at the more pointed end of the egg, so that the CAM can be detached from

the shell itself. A window is then cut into the shell with the aid of scissors (Fig. 9.4) which exposes the underlying CAM vessels. These vessels were first used to study tumor angiogenesis by grafting tumors, or fractions of tumors, onto the CAM surface on day 8 of incubation; 48–72 h after the grafting, tumor-induced vasoproliferative response is visible as newly formed vessels converge towards the graft, and can be evaluated in vivo by means of a stereomicroscope (Ausprunk et al., 1975; Knighton et al., 1977a). On day 12 the CAMs are processed for light or electron microscopy. Briefly, the embryos and their membranes are fixed in ovo in 3% phosphate-buffered glutaraldehyde, dehydrated in serial alcohols, post-fixed in 1% phosphate-buffered OsO_4, and embedded in Epon 812. One-micrometer semi- and ultrathin sections are cut on a ultramicrotome. The semithin sections are stained with a 0.5% aqueous solution of toluidine blue and observed under a light microscope. The ultrathin sections are stained with uranyl acetate followed by lead citrate and examined under a transmission electron microscopy.

It should be noted that the CAM was found to be an ideal medium to investigate the tumor-induced vasoproliferative response (Marzullo et al., 1998), because the host's immune system is not yet fully developed, thus avoiding potential problems with rejection (Leene et al., 1973).

The CAM is also used to study different macromolecules displaying angiogenic or antiangiogenic activities. Inert synthetic polymers similar to those used in the rabbit corneal assay are soaked with the macromolecule of interest and laid on the CAM surface. Elvax 40 and hydron are commonly used polymers, originally described and validated by Langer and Folkman (1976): both proved to be biologically inert when implanted onto the CAM and both were found to polymerize in the presence of test substances, allowing sustained release during the assay. When polymers are used in combination with an angiogenic substance, a vasoproliferative response will be recognizable 72–96 h after implantation: the response takes the form of increased vessel density around the implant, with the vessel radially converging towards the center like

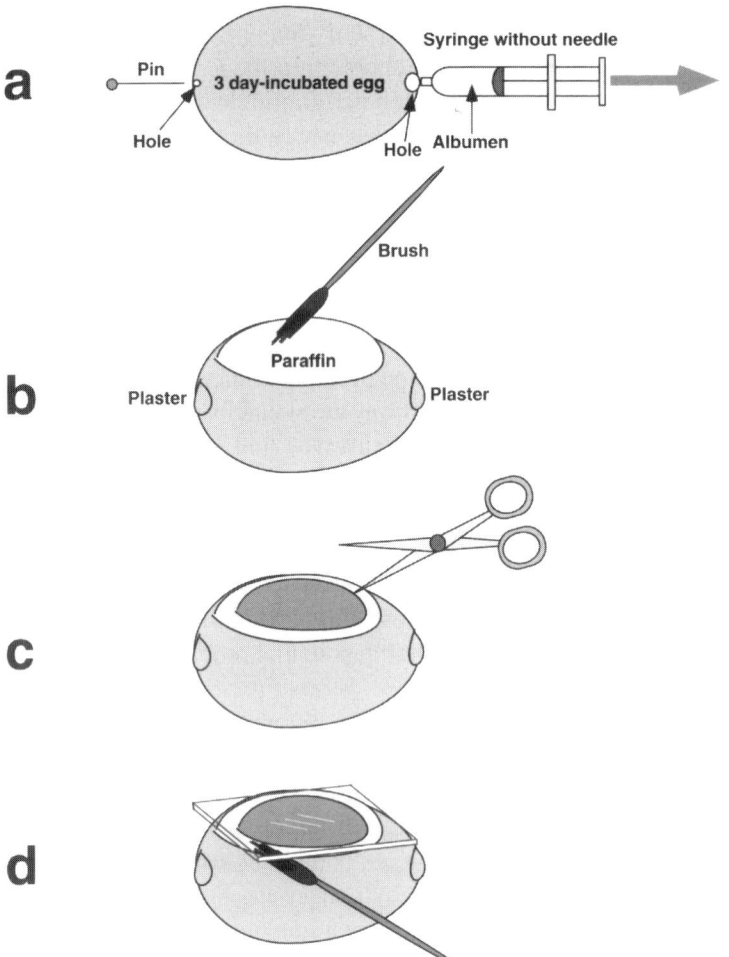

Fig. 9.4. Preparation of the egg for the CAM assay. On day 3 of incubation, 2–3 ml of albumen are aspirated at the acute pole of the egg (a) to detach the developing CAM from the shell; the holes are closed with plaster. The upper surface of the egg is brushed on with paraffin (b) and cut with scissors kept parallel to the surface so as not to damage the embryo (c). The window is covered with a glass slide and sealed with paraffin (d).

spokes in a wheel (Ribatti et al., 1995). Conversely, when polymers combined with an antiangiogenic substance are tested, the vessels become less dense around the implant after 72–96 h, and eventually disappear (Ribatti et al., 1995). Alternatively, fluid substances or cell suspensions can be directly inoculated into the cavity of the allantoic vesicle so that their activity covers the whole vascular area in a uniform manner (Gualandris et al., 1996).

Wilting et al. (1991) used culture coverslips 4-5 mm in diameter on which 5 μl of several angiogenic factors were placed. The coverslips were turned over and placed onto the CAM on day 9 of incubation and the angiogenic response were evaluated 96 h later. Another method has been proposed by Nguyen and colleagues (1994): the test substance is placed into a collagen gel polymerized between two parallel nylon meshes. This implant is useful to align the capillaries on a single plane for counting. The resulting 'sandwich' is then placed on the CAM on day 8 of incubation. These different implants are illustrated in Fig. 9.5.

Yet other methods have been proposed (Fig. 9.6) whereby the CAM vascular networks can be displayed in greater detail, except that the embryo with the extra-embryonic membranes and yolk must be transferred to an in vitro system during the early stages of development (day 3 or 4 of incubation). The system consists of a Petri dish (Auerbach et al., 1974), or an inert plastic container, previously equipped with a 'parafilm' ring (4–5 cm inside depth) to provide a support for the embryo and adnexa (Dugan et al., 1991). The embryo in the petri dish or in the container is then incubated in an humidified CO_2 atmosphere at 37°C. This CAM in petri dish, or 'cracked eggs' method is among the most frequently used CAM assays.

Recently, a new method for the quantitation of angiogenesis and antiangiogenesis in the CAM has been developed based on the implantation of dehydrated gelatin sponges (Gelfoam, Upjohn company) treated with stimulators or inhibitors of blood vessel formation on the top of growing CAM on day 8 of incubation (Ribatti et al., 1997) (Fig. 9.7). Blood vessels growing vertically into the

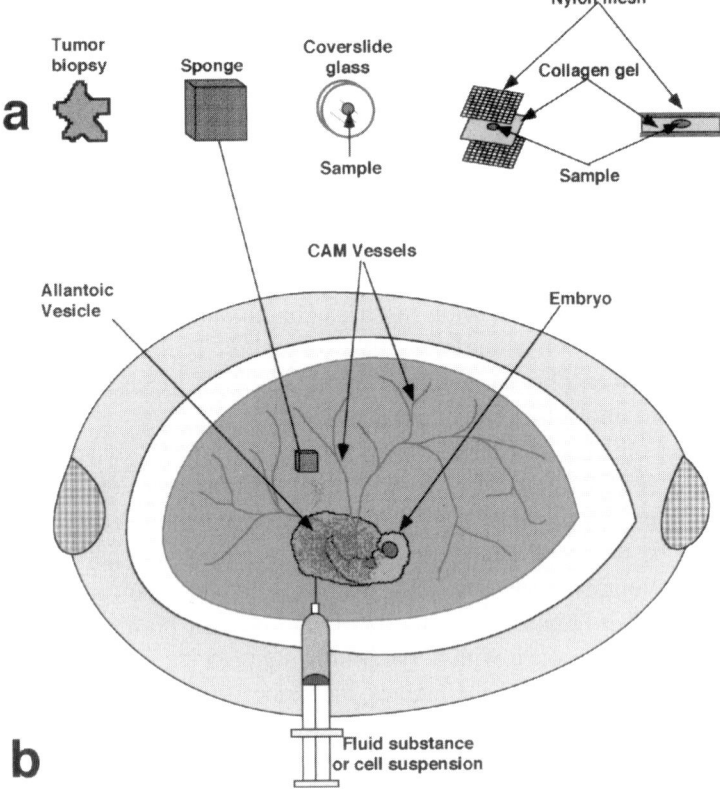

Fig. 9.5. Implants for CAM assay. The sample (a) can be a solid mass (e.g. tumor biopsy), or a solution trapped within a solid substrate (natural or synthetic sponge, glass coverslide, nylon mesh/collagen sandwich). Solid samples are simply laid on the CAM in an avascular area (shown for the sponge implant). As an alternative, fluids can be injected directly into the allantoic vesicle (b) so that their activity is extended to the whole vascular area.

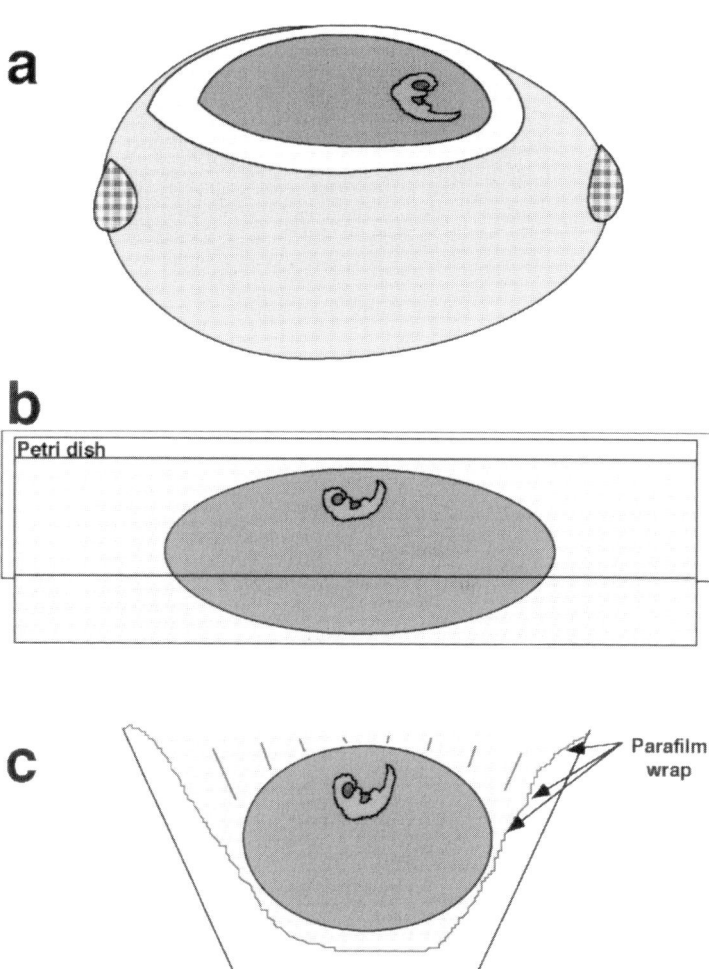

Fig. 9.6. Different methods for the CAM assay: (a) CAM assay; (b) CAM assay in a petri dish; and (c) CAM assay on a parafilm ring.

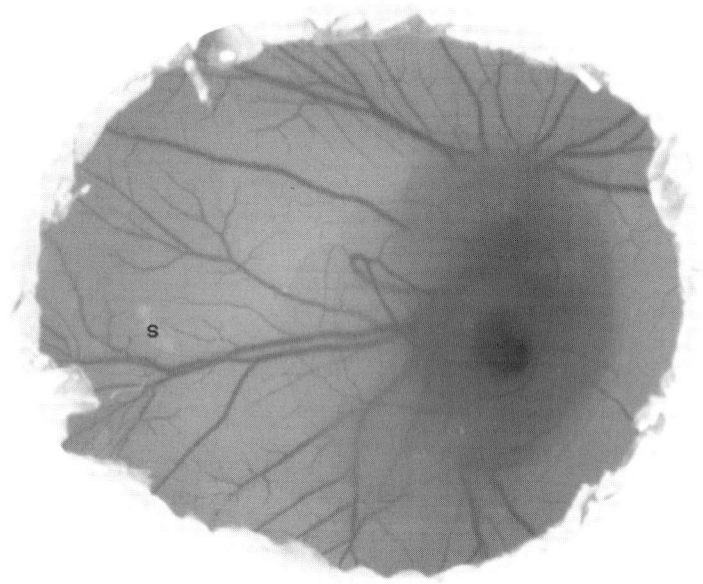

Fig. 19.7. Gelatin implant (S) in an 8-days-old CAM.

sponge and at the boundary between sponge and surrounding CAM mesenchyme are counted by a morphometric method on day 12. The gelatin sponge appears also to be suitable for the delivery of cell suspensions onto the CAM surface and for the evaluation of the associated angiogenic potential (Ribatti et al., 1997, 1998; Vacca et al., 1998). Many techniques can be applied on samples embedded in paraffin and plastic, including histochemistry and immunohistochemistry. Electron microscopy can also be used in combination with light microscopy. Moreover, unfixed sponges can be utilized for chemical studies, such as the determination of DNA, protein and collagen content, as well as for RT-PCR analysis of gene expression by infiltrating cells, including endothelial cells.

9.2.1.2. *Evaluating the vasoproliferative response by semiquantitative methods*

Several semiquantitative methods have been used to evaluate the extent of the vasoproliferative response. One method considers variations in the distribution and density of CAM vessels next to the graft site: these are evaluated in vivo by means of a stereomicroscope at regular intervals following the grafting procedure. The response is rated 0 when no change with respect to controls can be appreciated, in this case the new vessels, developed from the time of grafting, do not show any convergence toward the implant. The score is +1 when few neovessels converging towards the implant are observed; +2 when a considerable change in the number and distribution of the converging neovessels can be appreciated (Knighton et al., 1977b).

In another method (Fig. 9.8, panel a) the degree of vasoproliferative response is defined as a vascular index based on photographic reconstruction. All the vessels converging toward the implant and contained inside a 1 mm diameter ring superimposed on the CAM are counted: the ring is drawn around the implant in such a way that it will form an angle of less than 45 with a straight line drawn starting from the implant's center. Vessels branching dichotomically outside the ring are counted as 2, while those branching inside the ring are counted as 1 (Barnhill et al., 1983; Dusseau et al., 1986).

A third method (Fig. 9.8, panel b) expresses the degree of the vasoproliferative response as evaluated in vivo under the stereomicroscope by means of a 0–5 scale of arbitrary values. Zero describes a condition of the vascular network that is unchanged with respect to the time of grafting; 1 marks a slight increment in vessel density associated to occasional changes in the course of vessels converging towards the grafting site; 2, 3, 4 and 5 correspond to a gradual increase in vessel density associated with increased irregularity in their course: a 5 rating also highlights strong hyperemia. A coefficient describing the degree of angiogenesis can also be derived from the ratio of the calculated value to the highest attainable value; thus,

Semiquantitative methods for the evaluation of the vasoproliferative response in the CAM

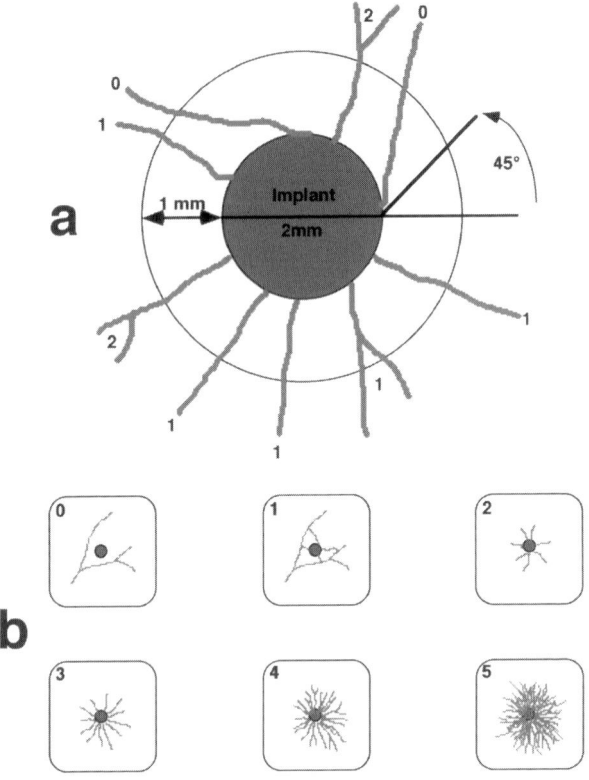

Fig. 9.8. Semiquantitative methods for the evaluation of the vasoproliferative response in the CAM. (a) Vascular index: based on the analysis performed using photos of the CAM; the vessels converging toward the implant and contained within the 1-mm external-diameter ring superimposed on the CAM are counted. Vessels forming an angle bigger than 45 with the diameter of the ring are not included (0), single vessels or those branching dichotomically inside the ring are counted as 1, vessels branching outside the ring are counted as 2. (b) The vasoproliferative response can be evaluated directly under a stereomicroscope using a 0–5 scale of arbitrary values. Zero describes no vascularization of the implant, while 1–5 are associated to the increase of vascular density and the conversion of vessels towards the grafting site; 5 is also representative of strong hyperemia.

the coefficient's lowest value is equal to 0 and the highest value is 1 (Folkman et al., 1976).

9.2.1.3. Evaluating the vasoproliferative response by quantitative methods

Quantitative evaluation of vascular density can be obtained by applying morphometric and planimetric methods to the observation of histologic CAM specimens. The angiogenic response can be evaluated as microvessel area by using the morphometric method of 'point counting' (Ribatti et al., 1997). The sponges and the underlying and immediately adjacent CAM portions are fixed, removed and processed for paraffin embedding. Eight-micrometer serial sections parallel to the surface of the CAM are cut, stained with a 0.5% aqueous solution of toluidine blue and observed with a double-headed photomicroscope: two investigators simultaneously identify the transversely cut microvessels (diameter ranging from 3 to 10 m), and each identification is agreed on in turn. Microvessels are studied at a magnification of $250\times$ with a square mesh inserted in the eyepiece. The mesh has 12 lines per side giving 144 intersection points. Six randomly chosen microscopic fields of each section (every third section within 30 serial slides from an individual specimen are analyzed) are evaluated for the total number of intersection points which are occupied by microvessels. For vessel counts, mean values +1 standard deviation are determined for each analysis. The microvessel area is indicated by the final mean number of the occupied intersection points, expressed again as percentage of the total number of intersection points. Statistically significant differences between the mean values of the intersection points in the experimental CAMs and controls are determined by Student's t test for unpaired data. Just like for the cornea, automatic image analyses have also been suggested for CAM (Jakob et al., 1984).

9.2.1.4. The limitations of CAM assay
The main limitation of CAM is represented by nonspecific inflammatory reactions which may develop as a result of grafting and in turn induce a secondary vasoproliferative response, sometime making it difficult to quantify the primary response being investigated (Jakob et al., 1978; Spanel-Burowski et al., 1988). The study of histological CAM sections would help to detect the possible presence of a perivascular inflammatory infiltrate together with a hyperplastic reaction, if any, of the chorion epithelium. However, the possibilities of causing a nonspecific inflammatory response are much lower when the test material is grafted as soon as CAM begins to develop, since then the host's immune system is relatively immature (Leene et al., 1973).

Another limit of CAM could be the presence of receptors with a different affinity for some molecules derived from mammals. Taking into account this possibility would require to perform a dose-dependence test in order to find the optimal concentration of the molecule that gives a positive response.

There are two more drawbacks to the CAM assay: first, the test material is placed on pre-existing vessels and newly formed blood vessels grow within the CAM mesenchyme. Thus, the actual neovascularization can hardly be distinguished from a falsely increased vascular density due to the rearrangement of pre-existing vessels that follows contraction of the membrane (Knighton et al., 1991). Second, the timing of the CAM angiogenic response is essential. Many studies determine angiogenesis after 24 h of incubation, a time at which there is no angiogenesis, but only vasodilation. Measurements of vessel density are really measurements of visible vessel density, and the distinction between vasodilation and neovascularization is not easy to make. To circumvent this drawback it is useful to utilize sequential photography to document new vessel formation.

9.3. Subcutaneous implant assay

The general goal of most subcutaneous implant models is to trap a putative angiogenic substance into a suitable carrier, mostly an avascualar sponge-like structure, which slowly releases the factor at the site of implant causing the recruitment of new vessels into the implant. Several assays have been established to date: the dorsal skin flaps (Arranz et al., 1995), the sponge implant assay (Andrade et al., 1987, 1997), the sodium alginate beads (Hoffmann et al., 1997), and the air sac model (Lichtenberg et al., 1997); a recent review has compared these different approaches showing progress and problems for each method (Jain et al., 1997).

Passaniti et al described a simple adaptation of the cutaneous implant assays using reconstituted BMs (Passaniti et al., 1992). The group of Albini in Genova (I) has extensively used reconstituted BMs to study specific steps of the angiogenic process in vitro and assess angiogenesis in vivo (Albini et al., 1994, 1995; Benelli et al., 1994; Iurlaro et al., 1998). The use of reconstituted BMs has several advantages both in vitro and in vivo.

9.3.1. The reconstituted basement membrane: one tool for several tests

Three-dimensional matrix substrates have been extensively used in vitro to study differentiation of endothelial cells into vessel-like structures: fibrin and collagen gels have proven good substrates (Montesano et al., 1983; Montesano et al., 1992); however, the extensive time required for these assays is a major limiting factor.

These problems can be circumvented by using a reconstituted BM. The most widely used material is known as Matrigel^{TM}, an extract of a murine tumor which produces massive amounts of BM (Kleinman et al., 1986). In fact, Matrigel^{TM} is a mixture of BM components (mostly laminin) prepared from the murine EHS tumor of C57/black mice grown in vivo. Once extracted from the tumors with urea-containing buffer, the matrix mixture is dialyzed in DMEM at 4°C and stored frozen (−20°C) at concentrations ranging from 12–18 mg/ml. Commercially prepared Matrigel^{TM} is generally supplied in aliquots at 10–12 mg/ml (Becton Dickinson, Bedford, MA). A commercial growth factor-reduced Matrigel^{TM} (Becton Dickinson, Bedford, MA) is also available, but this preparation is obtained reducing the content of heparan sulfate proteoglycan, thus altering Matrigel^{TM} composition. Matrigel^{TM} can be frozen and thawed several times without any problem, it is convenient to prepare small aliquots to shorten the thawing procedure in ice/water bath. As Matrigel^{TM} quickly polymerizes at room temperature, all the material used for its manipulation (pipettes, tips, tubes, multiwells) must be prechilled.

Matrigel^{TM} is often used in three different tests related to angiogenesis: the chemoinvasion assay, the morphology assay and the in vivo sponge model. The in vitro models are briefly described from an 'angiogenic' point of view and a detailed protocol for the in vivo test given.

9.3.1.1. The chemoinvasion assay
This is a modification of the classic Boyden chemotaxis assay in Boyden chambers (Albini et al., 1987). PVP-free polycarbonate filters (12 μm pore size for endothelial cells, Nuclepore, Concorezzo, Milan, Italy) are coated with Matrigel^{TM} diluted with cold water, on ice (15–40 μg/50 ml/13 mm diameter filter for endothelial cells). Filters are placed on tissue culture plates and the liquid Matrigel^{TM} is pipetted onto the filter. During this step care must be taken to insure that the Matrigel^{TM} solution is homogeneously applied on the filter surface, leaving a small border of uncoated membrane to prevent leakage off the filter. The Matrigel^{TM} coated filters are dried under a laminar flow and the plate tightly closed with parafilm (they can be maintained in this form up to 48 h at 4°C). The filters are rehydrated just before performing the assay using serum free DMEM at 4°C. Numerous substances can be placed as chemoattractants in the lower chamber. Kaposi's Sarcoma cell conditioned

medium is a good chemoattractant for endothelial cells, mimicking a highly angiogenic environment (Albini et al., 1992; Benelli et al., 1996); purified angiogenic growth factors are also frequently used, i.e. vascular endothelial growth factor (VEGF) or bFGF (Dvorak et al., 1995; Potgens et al., 1995). Both chemoattractants and cells are suspended in serum-free medium containing 0.1% BSA (SFM). Medium with BSA alone in the lower chamber serves as a negative control. Chemotaxis and chemoinvasion assays are often done in parallel to determine if treatments affect chemotaxis itself or if they are specific for invasion. A control of chemotaxis in a chemoinvasion assay is also useful to verify that Matrigel^{TM} concentration is sufficient to act as a real barrier to cell migration. An invasion index (invasion/chemotaxis) can be calculated and indicates the specific contribution of matrix degradation using the following formula:

$$\text{Invasion index} = (\text{invasive cells/migrating cells}) \times 100$$

The upper chamber is filled with an endothelial cell suspension (usually a 1.2×10^5 cells/800 μl/chamber). The chambers are then incubated at 37°C in 5% CO_2 for 6 h. While tumor cells usually do not migrate in response to SFM alone, endothelial cells show a variable degree of random unstimulated migration due to the loss of cell-cell contact; once determined, this value is usually subtracted to the final data as 'background migration'. Invasion inhibitors acting on the chemoattractant are added to the lower compartment of the Boyden chamber, while inhibitors acting on the cells or cell products are generally added along with the cells. At the end of the incubation time, cells remaining on the upper surface of the filter are mechanically removed by wiping them with a cotton swab or stripping on a glass slide. The cells migrated to the under surface are quantitated after staining (with toluidine blue, hematoxylin/eosin or others). Assays are generally performed in triplicate and repeated two or three times. In the original assay quantitation was performed at the microscope, counting five to ten random fields for each filter.

Various alternatives have been proposed such as colorimetric detection of the staining, image analysis, and metabolic labeling with MTT or similar compounds (Albini, 1998).

9.3.1.2. Morphogenesis on Matrigel^{TM}

Matrigel^{TM} can also be used for morphological studies on vascular cell organization (Baatout et al., 1996). Matrigel^{TM} is thawed at 4°C in an ice/water bath, and 0.3–0.4 ml of a concentrated solution (10 mg/ml) carefully pipetted into 13 mm/diameter tissue culture wells (48-wells chambers), avoiding even the smallest bubbles. The matrigel is then polymerized for 1 h at 37°C. Once polymerization has occurred, 7×10^4 endothelial cells in 1 ml of medium are carefully pipetted on top of the gel. The plates are then incubated at 37°C in a 5% CO_2, humidified atmosphere. The assays are photographed and monitored with an inverted microscope. A modification using MTT as a dye has been recently described (Sasaki et al., 1998). HUVE cells suspended in 10% FCS-containing medium with heparin and ECGS organize into capillary-like networks within 6 h, this structure is maintained for about 48 h and then lost. The kinetics of network organization and cell viability is modulated by angiogenic factors and antiangiogenic drugs.

9.3.1.3. The in vivo sponge model

Although some synthetic polymers have proven suitable systems to trap angiogenic factors in vivo, they are an extraneous material implanted in a living organism. In addition, these polymers frequently need protein enrichment to obtain a slow release of the trapped material (Langer and Folkman, 1976). The subcutaneous Matrigel^{TM} implant model has several advantages over many other in vivo implant models. Matrigel^{TM} is fluid at 4°C but quickly polymerizes at 37°C, thus the material is readily implanted by a simple injection, avoiding slow and complex surgical procedures. For this reason it is an ideal tool to trap cytokines, cells or cell supernatants subcutaneously in mice by simple injection. In addition the EHS tumor which produces Matrigel^{TM} is syngeneic for the C57/black mice generally used for the experiments. This fact allows the use of a biological matrix in immunologically competent mice without any immunologic reaction. Another important observation regarding the use of Matrigel^{TM} is its content of Perlecan, the heparan-sulphate proteoglycan of the BM, which is involved in the storage and correct presentation to the receptor of angiogenic growth factors (Aviezer et al., 1997; Sharma et al., 1998).

Procedure. The assay was originally described by Passaniti (Passaniti et al., 1992) who studied the angiogenesis induced by bFGF and heparin. This method has been extensively used to study the angiogenic potential of Kaposi's Sarcoma cell (KS) supernatants and HIV Tat protein (Albini et al., 1994) and to test numerous antiangiogenic compounds (TIMP-2, IFNs, Somatostatin ...) (Albini et al., 1994, 1999; Benelli et al., 1997; Iurlaro et al., 1998).

The test is performed using at least 5–6 C57/black mice per point (two implants/mouse). Matrigel^{TM} should have a concentration ranging from 12 to 14 mg/ml to allow a further dilution by the addition of cytokines or cell supernatants (in PBS or DMEM) to reach a final concentration of 11 mg/ml. More dilute Matrigel^{TM} does not polymerize well in vivo. The ideal final volume to be injected is 500–600 μl. Cell supernatants need to be concentrated 10× so that once diluted 1 : 10 with Matrigel^{TM} the original concentration is restored. Serum-free KS supernatants are concentrated using the Centriprep centrifuge concentrators (Amicon) with a size cut-off of 3 kDa, and used as a 10x stock solution stored at $-80°$C. Most of the angiogenic growth factors (AGF) are heparin binding (HB), and often use heparin/heparan-sulphates for a correct presentation to the receptor. Matrigel^{TM} contains Perlecan (Noonan et al., 1991, 1993) which is able to bind these factors. This interaction constitutes a perfect angiogenic environment when the supernatants tested have high titers of such cytokines, as mouse endothelial cells are easily exposed to this stimulus. The AGF or the cell supernatant should be mixed with heparin (use concentrated stock to reach a final concentration of 20–30 U/ml) to favor the diffusion of the AGF and its presentation to the low and high affinity receptors of endothelium. The concentration of heparin is crucial and should be assessed for each new commercial stock and Matrigel^{TM} batch used, since low heparin concentration does not allow for HB activity, while high heparin can be angiogenic per se.

It is convenient to prepare the samples (i.e. Matrigel^{TM}+ AGF+heparin, total volume 600 μl) in single 1.5 ml sterile tubes (eppendorf) and use one insulin syringe for each injection (keeping everything in ice): the sample needs to be mixed with the syringe 3–4 times, just before the injection, avoiding bubbles. The mix is then slowly injected subcutaneously in the dorsal area of C57Bl/6 mice. Matrigel^{TM} takes about 1 min to polymerize, after this time the animal can be returned to the cage. After 4 days the animals are sacrificed, the implants collected, and immediately photographed.

Quantitation of the angiogenic response. Matrigel^{TM} sponges can be processed by two different ways: (1) by histologic examination; and (2) for hemoglobin content. The first approach provides qualitative data regarding the presence and organization of vessels and the immunohistochemical characterization of recruited cells. The second gives a numeric value proportional to the extent of vascularization. Each method excludes the other, because for hemoglobin quantitation the Matrigel^{TM} pellet needs to be minced.

Histologic evaluation: the gel is routinely processed for paraffin embedding by para-formaldehyde fixation (4% in PBS) progressive dehydration with ethanol/water solutions (30, 50, 70, 80, 90 and 100% twice, 30′ each step), xylol treatment (20', twice), xylol-paraffin (1:1) incubation (1 h, 52°C), double paraffin incubation (40′, 52°C) and finally embedded. Note that for the high content of water the Matrigel^{TM} pellet usually shrinks so that the number of potential sections is limited.

Hemoglobin Content: Hemoglobin content is determined with the Sigma Drabkin solution Kit 525. The principle of the assay is based on the oxidation of hemoglobin to cyan-meta-hemoglobin which is detectable spectrophotometrically at 540 nm.

The Matrigel™ pellets are weighed and minced in 1 ml of distilled water; after centrifugation (13,000 g/5′) 200 μl of each sample is diluted in 800 ml of Drabkin solution. The kit contains a standard hemoglobin solution to prepare a titration curve. The spectrophotometric measure at 540 nm indicates the relative hemoglobin concentration (RHC) of the sample in g/dl. The absolute hemoglobin concentration (AHC) can be calculated using the following formula:

AHC (g/dl) = [RHC (g/dl)/sample weight (g)] $\times$ 100

If it is necessary to inject allogenic cells along with matrigel the test can be modified using nude mice. Nude mice usually exhibit a lower angiogenic response as compared to C57/black mice; it is necessary to titrate the cytokines and heparin concentration.

References

Aaronson, S. A. (1991). Growth factors and cancer. Science *254*, 1146–1153.

Abercrombie, M. and Heaysman, J. E. M. (1953). Observations on the social behavior of cells in tissue culture. Exp. Cell Res. *5*, 111–131.

Abercrombie, M., Heaysman, J. E. M. and Karthauser, H. M. (1957). Social behavior of cells in tissue culture. Exp. Cell Res. *13*, 276–291.

Ager, A. (1987). Isolation and culture of high endothelial cells from rat lymph nodes. J. Cell Sci. *87*, 133–144.

Aharoni, D., Meiri, I., Atzmon, R., Vlodavsky, I. and Amsterdam, A. (1997). Differential effects of components of the extracellular matrix on differentiation and apoptosis. Curr. Biol. *7*, 43–51.

Ahmad, A., Marshall, J. F., Basset, P., Anglard, P. and Hart, I. R. (1997). Modulation of human stromelysin 3 promoter activity and gene expression by human breast cancer cells. Int. J. Cancer *73*, 290–296.

Akhurst, R. J. and Balmain, A. (1999). Genetic events and the role of TGFβ in epithelial tumour progression. J. Pathol. *187*, 82–90.

Akiyama, S. K., Yamada, S. S., Chen, W. T. and Yamada, K. M. (1989). Analysis of fibronectin receptor function with monoclonal antibodies: roles in cell adhesion, migration, matrix assembly, and cytoskeletal organization. J. Cell Biol. *109*, 863–875.

Akiyama, S. K., Olden, K. and Yamada, K. M. (1995). Fibronectin and integrins in invasion and metastasis. Cancer Metastasis Rev. *14*, 173–189.

Al-Mondhiry, H. (1984). Tumor interaction with hemostasis: the rationale for the use of platelet inhibitors and anticoagulants in the treatment of cancer. Am. J. Hematol. *16*, 193–202.

Albelda, S. M. (1993). Role of integrins and other cell adhesion molecules in tumor progression and metastasis. Lab. Invest. *68*, 4–17.

Albert, D. M., Wagoner, M. D., Moazed, K., Kimball, G. P. and Gonder, J. R. (1980). Heterotransplantation of human choroidal melanoma into the athymic nude mouse. Invest. Ophthal. Vis. Sci. *19*, 555–559.

Albini, A. (1998). Tumor and endothelial cell invasion of basement membranes. Pathol. Oncol. Res. *4*, 1–12.

Albini, A., Iwamoto, Y., Kleinman, H. K., Martin, G. R., Aaronson, S. A., J. M., K. and McEwan, R. N. (1987). A rapid in vitro assay for quantitating the invasive potential of tumor cells. Cancer Res. *47*, 3239–3245.

Albini, A., Repetto, L., Carlone, S., Benelli, R., Gendelman, R., Filippi, P. D., Bussolino, F., Monaco, L., Soria, M. and Parravicini, C. (1992). Characterization of Kaposi's sarcoma-derived cell cultures from an epidemic and a classic case. Int. J. Oncol. *1*, 723–730.

Albini, A., Fontanini, G., Masiello, L., Tacchetti, C., Bigini, D., Luzzi, P., Noonan, D. M. and Stetler-Stevenson, W. G. (1994). Angiogenic potential in vivo by Kaposi sarcoma cell-free supernatants and HIV1-tat product: inhibition of KS-like lesions by TIMP–2. AIDS *8*, 1237–1244.

Albini, A., Barillari, G., Benelli, R., Gallo, R. C. and Ensoli, B. (1995). Angiogenic properties of human immunodeficiency virus type 1 Tat protein. Proc. Natl. Acad. Sci. USA *92*, 4838–4842.

Albini, A., Benelli, R., Presta, M., Rusnati, M., Ziche, M., Rubartelli, A., Paglialunga, G., Bussolino, F. and Noonan, D. (1996). HIV-tat protein is a heparin-binding angiogenic growth factor. Oncogene *12*, 289–297.

Albini, A., Florio, T., Giunciuglio, D., Masiello, L., Carlone, S., Corsaro, A., Thellung, S., Cai, T., Noonan, D. M. and Schettini, G. (1999). Somatostatin controls Kaposi's sarcoma tumor growth through inhibition of angiogenesis. FASEB J. *13*, 647–655.

Albrecht-Buehler, G. (1977a). The phagokinetic tracks of 3T3 cells. Cell *11*, 395–404.

Albrecht-Buehler, G. (1977b). Phagokinetic tracks of 3T3 cells: parallels between the orientation of track segments and of cellular structures which contain actin and tubulin. Cell *12*, 333–339.

Alby, L. and Auerbach, R. (1984). Differential adhesion of tumor cells to capillary endothelial cells in vitro. Proc. Natl. Acad. Sci. USA *81*, 5739–5743.

Almasio, P., Piovella, F., Ricetti, M. M., Samaden, A. and Semino, G. (1984). ECMs produced in the presence of dexamethasone: a natural growth agent for human cultured endothelium. Boll. Soc. Ital. Biol. Sperim. *60*, 225–230.

Alterman, A. L., Fornabaio, D. M., Kim, Y. S. and Stackpole, C. W. (1989). The role of intratumor environment in determining spontaneous metastatic activity of a B16 melanoma clone. Invasion Metastasis *9*, 242–253.

Altevogt P., Heckl-Ostreicher B., Schmitt H. P. and Schirrmacher V. (1985). Biochemical identification of membrane proteins and molecular cloning of encoding genes that determine the metastatic phenotype of ESb lymphoma cells. In: Treatment of Metastasis: Problems and Prospects (Hellmann, K. and Eccles, S. A., eds.). Taylor and Francis, London, pp. 381–384.

Ambartsumian, N. S., Grigorian, M. S., Larsen, I. F., Karlstrom, O., Sidenius, N., Rygaard, J., Georgiev, G. and Lukanidin, E. (1996). Metastasis of mammary carcinomas in GRS/A hybrid mice transgenic for the mts1 gene. Oncogene *13*, 1621–1630.

An, Z. L., Wang, X. E., Kubota, T., Moossa, A. R. and Hoffman, R. M. (1996). A clinical nude mouse metastatic model for highly malignant human pancreatic cancer. Anticancer Res. *16*, 627–631.

An, Z. L., Wang, X. E., Willmott, N., Chander, S. K., Tickle, S., Docherty, A. J. P., Mountain, A., Millican, A. T., Morphy, R., Porter, J. R., Epemolu, R. O., Kubota, T., Moossa, A. R. and Hoffman, R. M. (1997). Conversion of highly malignant colon cancer from an aggressive to a controlled disease by oral administration of a metalloproteinase inhibitor. Clin. Exptl. Metastasis *15*, 184–195.

Andrade, S. P., Fan, T. P. and Lewis, G. P. (1987). Quantitative in-vivo studies on angiogenesis in a rat sponge model. Br. J. Exp. Pathol. *68*, 755–766.

Andrade, S. P., Machado, R. D., Teixeira, A. S., Belo, A. V., Tarso, A. M. and Beraldo, W. T. (1997). Sponge-induced angiogenesis in mice and the pharmacological reactivity of the neovasculature quantitated by a fluorimetric method. Microvasc. Res. *54*, 253–261.

Anklesaria, P., Teixidó, J., Marikki, L., Pierce, J. H. and Greenberger, J. S. (1990). Cell-cell adhesion mediated by binding of membrane- anchored transforming groath factor alpha to epidermal growth factor receptors promotes cell proliferation. Proc. Natl. Acad. Sci. USA *87*, 3289–3293.

Antonia, S. J., Uchida, J., Cohen, S. and Cohen, M. C. (1989). Attachment of tumor cells to endothelial monolayers: detection of surface molecules involved in cell-cell binding. Clin. Immunol. Immunopathol. *53*, 281–296.

Araki, M., Araki, K., Biancone, L., Stamenkovic, I., Izui, S., Yamamura, K. and Vassalli, P. (1997). The role of E-selectin for neutrophil activation and tumor metastasis in vivo. Leukemia Suppl. *3*, 209–212.

Arisawa, Y., Kubota, T., Suto, A., Kodaira, S., Ishibiki, K. and Abe, O. (1990). Nude mouse resists hepatic metastasis of the allogeneic tumor, colon 26. Jpn. J. Surg. *20*, 487–490.

Arranz, L. J., Suarez, N. C., Barthe, G. P. and Rojo, O. J. (1995). Evaluation of angiogenesis in delayed skin flaps using a monoclonal antibody for the vascular endothelium. Br. J. Plast. Surg. *48*, 479–486.

Aruffo, A., Dietsch, M. T., Wan, H., Hellstrom, K. E. and Hellstrom, I. (1992). Granule membrane protein 140 (GMP 140). binds to carcinomas and carcinoma-derived cell lines. Proc. Natl. Acad. Sci. USA *89*, 2292–2296.

Asai, M., Kato, M., Asai, N., Iwashita, T., Murakami, H., Kawai, K., Nakashima, I. and Takahashi, M. (1999). Differential regulation of MMP-9 and TIMP-2 expression in malignant melanoma developed in Metallothionein/RET transgenic mice. Jap. J. Cancer Res. *90*, 86–92.

Atherton, A. and Born, G. V. R. (1973). Relationship between the velocity of rolling granulocytes and that of blood flow in venules. J. Physiol. *233*, 157–165.

Auerbach, R., Kubai, L., Knighton, D. R. and Folkman, J. (1974). A simple procedure for the long-term cultivation of chicken embryos. Dev. Biol. *41*, 391–394.

Auerbach, R., Alby, J., Grieves, J., Joseph, J., Morrissey, L., Sidky, Y. and Watt, L. (1982). A monoclonal antibody against angiotensin converting enzyme: its use as a marker for murine, bovine and human endothelial cells. Proc. Natl. Acad. Sci. USA *79*, 7891–7895.

Auerbach, R., Lu, W., Pardon, E., Gumkowski, F., Kaminski, G. and Kaminski, M. (1987). Specificity of adhesion between murine tumor cells and capillary endothelium: an in vitro correlate of preferential metastasis in vivo. Cancer Res. *47*, 1492–1496.

Auerbach, R., Auerbach, W. and Polakowsky, I. (1991). Assays for angiogenesis: a review. Pharmacol. Ther. *51*, 1–11.

Ausprunk, D. H., Knighton, D. R. and Folkman, J. (1974). Differentiation of vascular endothelium in the chick chorioallantois: a structural and autoradiographic study. Dev. Biol *38*, 237–249.

Ausprunk, D. H., Knighton, D. R. and Folkman, J. (1975). Vascularization of normal and neoplastic tissues grafted to the cick chorioallantois. Am. J. Pathol. *79*, 597–610.

Aviezer, D., Iozzo, R. V., Noonan, D. M. and Yayon, A. (1997). Suppression of autocrine and paracrine functions of basic fibroblast growth factor by stable expression of perlecan antisense cDNA. Mol. Cell Biol. *17*, 1938–1946.

Avis, I. L., Kovacs, T. O., Kasprzyk, P. G., Treston, A. M., Bartholomew, R., Walsh, J. H., Cuttitta F. and Mulshine, J. L. (1991). Preclinical evaluation of an anti-autocrine growth factor monoclonal antibody for treatment of patients with small-cell lung cancer. J. Natl. Cancer Inst. *83*, 1470–1476.

Aznavoorian, S., Stracke, M. L., Krutzsch, H. C., Schiffman, E. and Liotta, L. A. (1990). Signal transduction for chemotaxis and haptotaxis by matrix macromolecules in tumor cells. J. Cell Biol. *110*, 1427–1438.

Baatout, S. and Cheta, N. (1996). Matrigel: a useful tool to study endothelial differentiation. Rom J. Intern. Med. *34*, 263–269.

Bade, E. G. and Feindler, S. (1988). Liver epithelial cell migration induced by EGF or TGF alpha is associated with changes in the gene expression of secreted proteins. In Vitro Cell. Dev. Biol. *24*, 149–154.

Bade, E. G. and Nitzgen, B. (1985). ECM modulates the EGF-induced migration of liver epithelial cells in serum-free, hormone-supplemented medium. In Vitro Cell. Dev. Biol. *21*, 245–248.

Baik, M. G., Lee, M. J. and Choi, Y. J. (1998). Gene expression during involution of mammary gland (review). Int. J. Mol. Med. *2*, 39–44.

Bailly, M. and Doré, J.-F. (1991). Human tumor spontaneous metastasis in immunosuppressed newborn rats. II. Multiple selections of human melanoma metastatic clones and variants. Intl. J. Cancer *49*, 750–757.

Balconi, G. and Dejana, E. (1986). Cultivation of endothelial cells: limitations and persectives. Med. Biol. *64*, 231–245.

Bammer, K. (1981). Stress, spread and cancer. In: Stress and Cancer (Bammer, K. and Newberry B. H. eds,). C. J. Hogrefe, Toronto, pp. 137–163.

Bani, M. R., Garofalo, A., Scanziani, E. and Giavazzi, R. (1991). Effect of IL-1b on metastasis formation in different tumor systems. J. Natl. Cancer Inst. *83*, 119–123.

Bani, M. R., Rak, J., Adachi, D., Wiltshire, R., Trent, J. M., Kerbel, R. S. and Ben-David, Y. (1996). Multiple features of advanced melanoma recapitulated in tumorigenic variants of early stage (radial growth phase) human melanoma cell lines: evidence for a dominant phenotype. Cancer Res. *56*, 3075–3086.

Bao, L., Matsumura, Y., Baban, D., Sun, Y. and Tarin, D. (1994). Effects of inoculation site and Matrigel on growth and metastasis of human breast cancer cells. Br. J. Cancer *70*, 228–232.

Barnhill, R. L. and Ryan, T. J. (1983). Biochemical modulation of angiogenesis in the chorioallantoic membrane of the chick embryo. J. Invest. Dermatol. *81*, 485–488.

Barrack, E. R. (1997). TGF beta in prostate cancer: a growth inhibitor that can enhance tumorigenicity. Prostate *31*, 61–70.

Baselga, J., Tripathy, D., Mendelsohn, J., Baughman, S., Benz, C. C., Dantis, L., Sklarin, N. T., Seidman, A. D., Hudis, C. A., Moore, J., Rosen, P. P., Twaddell, T., Henderson, I. C. and Norton, L. (1996). Phase II study of weekly intravenous recombinant humanized anti-p185HER2 monoclonal antibody in patients with HER2/neu-overexpressing metastatic breast cancer. J. Clin. Oncol. *14*, 737–744.

Bashkin, P., Doctrow, S., Klagsbrun, M., Svahn, C. M., Folkman, J. and Vlodavski, I. (1989). Basic FGF binds to subendothelial ECM and is released by heparitinase and heparin-like molecules. Biochemistry *28*, 1737–1743.

Basset, P., Bellocq, J. P., Wolf, C., Stoll, I., Hutin, P., Limacher, J. M., Podhajcer, O. L., Chenard, M. P., Rio, M. C. and Chambon, P. (1990). A novel metalloproteinase gene specifically expressed in stromal cells of breast carcinomas. Nature *348*, 699–704.

Bastida, E., Ordinas, A., Giardina, S. and Jamieson, G. A. (1982). Differentiation of platelet aggregating effects on human tumor cell lines based on inhibition studies with apyase, hirudin and phospholypase. Cancer Res. *42*, 4348–4352.

Bastida, E., Almirall, L., Bertomeu, M. C. and Ordinas, A. (1989). Influence of shear stress on tumor cell adhesion to endothelial cell extracellular matrix and its modulation by fibronectin. Int. J. Cancer *43*, 1174–1178.

Becker, K. F., Atkinson, M. J., Reich, U., Becker, I., Nekarda, H., Siewert, J. R. and Hoefler, H. (1995). E-cadherin gene mutations provide clues to diffuse type gastric carcinoma. Cancer Res. *54*, 3845–3852.

Behrens, J. (1993). The role of cell adhesion molecules in cancer invasion and metastasis. Breast Cancer Res. Treat. *24*, 175–184.

Behrens, J. (1994–95). Cell contacts, differentiation, and invasiveness of epithelial cells. Invasion Metastasis *14*, 61–70.

Beilke, M. A. (1989). Vascular endothelium in immunology and infectious disease. Rev. Infect. Dis. *11*, 273–283.

Bell, C., Frost, P., Kerbel, R. S. (1991). Cytogenetic heterogeneity of genetically marked and metastatically competent 'dominant' tumor cell clones. Cancer Genet. Cytogenet. *1154*, 153–161.

Bellinger, D. L, Madden, K. S., Felten, S. Y. and Felten, D. L. (1994). Neural and endocrine links between the brain and the immune system. In: The Psychoimmunology of Cancer (Lewis, C. E., O'Sullivan C. and Barraclough J., eds.). Oxford University Press, Oxford, pp. 55–106.

Belloc, F., Dumain, P., Boisseau, M. R., Jalloustre, C., Reiffers, J., Bernard, P. and Lacombe, F. (1994). A flow cytometric method using Hoechst 33342 and propidium iodide for simultaneous cell cycle analysis and apoptosis determination in unfixed cells. Cytometry *17*, 59–65.

Belloc, C., Lu, H., Soria, C., Fridman, R., Legrand, Y. and Menashi, S. (1995). The effect of platelets on invasiveness and protease production of human mammary tumor cells. Int. J. Cancer *60*, 413–417.

Belloni, P. N. and Nicolson, G. L. (1988). Differential expression of cell surface glycoproteins on various organ-derived microvascular endothelia and endothelial cell cultures. J. Cell Physiol. *136*, 398–410.

Belloni, P. N. and Tressler, R. J. (1989/90). Microvascular endothelial cell heterogeneity: interactions with leukocytes and tumor cells. Cancer Metastasis Rev. *8*, 353–389.

Belloni, P. N., Carney, D. H. and Nicolson, G. L. (1992). Organ-derived microvessel endothelial cells exhibit differential responsiveness to thrombin and other growth factors. Microvasc. Res. *43*, 20–45.

Bellusci, S., Moens, G., Thiery, J. P. and Jouanneau, J. (1994). A scatter factor-like factor is produced by a metastatic variant of a rat bladder carcinoma line. J. Cell Sci. *107*, 1277–1287.

Benelli, R., Adatia, R., Ensoli, B., Stetler-Stevenson, W. G., Santi, L. and Albini, A. (1994). Inhibition of AIDS-Kaposi's sarcoma cell induced endothelial cell invasion by TIMP-2 and a synthetic peptide from the metalloproteinase propeptide: implications for an anti-angiogenic therapy. Oncol. Res. *6*, 251–257.

Benelli, R., Albini, A., Parravicini, C., Carlone, S., Repetto, L., Tambussi, G. and Lazzarin, A. (1996). Isolation of spindle-shaped cell populations from primary cultures of Kaposi's Sarcoma of different stage. Cancer Lett. *100*, 125–132.

Benelli, R., Masiello, L., Paglialunga, G., Noonan, D. M., Repetto, L., Carlone, S., Orengo, G., Fontanini, G. and Albini, A. (1997). IFN-alpha Inhibits Tat-induced angiogenesis. HIV and cytokines. Paris, Les Editions INSERM, pp. 447–458.

Bennett, M., Cudkowicz, G., Foster, R. S. jr. and Metcalf, D. (1968). Hemopoietic progenitor cells of W anemic mice studied in vivo and in vitro. J. Cell. Physiol. *71*, 211–226.

Bereta, M., Bereta, J., Cohen, S., Zaifert, K. and Cohen, M. C. (1991). Effect of inflammatory cytokines on the adherence of tumor cells to endothelium in amurine model. Cell. Immunol. *136*, 263–277.

Berlin, O., Samid, D., Donthineni-Rao, R., Akeson, W., Amiel, D. and Woods, V. L. (1993). Development of a novel spontaneous metastasis model of human osteosarcoma transplanted orthotopically into bone of athymic nude mice. Cancer Res. *53*, 4890–4895.

Berliner, J. A. (1981). Regulation of endothelial cell DNA synthesis and adherence. In Vitro *17*, 985–992.

Bertomeu, M. C., Gallo, S., Lauri, D., Haas, T. A., Orr, F. W., Bastida, E. and Buchanan, M. R. (1993). IL-1-induced cancer cell/endothelial cell adhesion in vitro and its relationship to metastasis in vivo: role of vessel wall 13-HODE synthesis and integrin expression. Clin Exp. Metastasis *11*, 243–250.

Bevilacqua, M. P. and Nelson, R. M. (1993). Selectins. J. Clin. Invest. *91*, 379–387.

Biancone, L., Araki, M., Araki, K., Vassalli, P. and Stamenkovic, I. (1996). Redirection of tumor metastasis by expression of E-selectin in vivo. J. Exp. Med. *183*, 581–587.

Billadeau, D., Liu, P., Jelinek, D., Shah, N., LeBien, T. W. and van Ness, B. (1997). Activating mutations in the N- and K-ras oncogenes differentially affect the growth properties of the IL-6-dependent myeloma cell line ANBL6. Cancer Res. *57*, 2268–2275.

Birchmeier, W. and Behrens, J. (1994). Cadherin expression in carcinomas: role in the formation of cell junctions and the prevention of invasiveness. Biochem. Biophys. Acta Reviews on Cancer *1198*, 11–26.

Bishop, J. M. (1991). Molecular themes in oncogenesis. Cell *64*, 235–248.

Bissell, M. J., Weaver, V. M., Lelievre, S. A., Wang, F., Petersen, O. W. and Schmeichel, K. L. (1999). Tissue structure, nuclear organization, and gene expression in normal and malignant breast. Cancer Res. *59*, 1757s–1763s.

Blood, C. H. and Zetter, B. R. (1990). Tumor interactions with the vasculature: angiogenesis and tumor metastasis. Biochem. Biophys. Acta *591032*, 89–118.

Boeryd, B., Ganelius, T., Lundin, P. and Mellgren, J. (1966). Counting and sizing of tumor metastases in experimental oncology. Intl. J. Cancer *1*, 497–502.

Bonmassar, E., Houchens, D. P., Fioretti, M. C. and Goldin, A. (1975). Uptake of 5-iododeoxyuridine as a measure of tumor growth and tumor inhibition. Chemotherapy *21*, 321–329.

Booyse, F. M., Sedlak, B. J. and Rafelson Jr., M. E. (1975). Culture of artherial endothelial cells: characterization and growth of bovine aortic cells. Thromb. Diath. Haemorrh. *5934*, 825–839.

Borchers, A. H., Sanders, L. A., Powell, M. B. and Bowden, G. T. (1997). Melanocyte mediated paracrine induction of extracellular matrix degrading proteases in squamous cell carcinoma cells. Exp. Cell Res. *231*, 61–65.

Bosenberg, M. W. and Massagué, J. (1993). Juxtacrine cell signaling molecules. Curr. Opin. Cell Biol. *5*, 832–838.

Bottaro, D. P., Rubin, J. S., Faletto, D. L., Chan, A. M., Kmiecik, T. E., Vande Woude, G. F. and Aaronson, S. A. (1991). Identification of the hepatocyte growth factor receptor as the c-met proto-oncogene product. Science *59251*, 802–804.

Boudreau, N. and Bissell, M. J. (1998). Extracellular matrix signaling: integration of form and function in normal and malignant cells. Curr. Opin. Cell Biol. *10*, 640–646.

Bourguignon, L. Y. W., Gunja-Smith, Z., Iida, N., Zhu, H. B., Young, L. J. T., Muller, W. J. and Cardiff, R. D. (1998). CD44v3,8–10 is involved in cytoskeleton-mediated tumor cell migration and matrix metalloproteinase (MMP-9). association in metastatic breast cancer cells. J. Cell Physiol. *59176*, 206–215.

Bowman, P. D., Betz, A. L., Ar, D., Wolinsky, J. S., Penny, J. B., Shivers, R. R. and Goldstein, G. W. (1981). Primary culture of capillary endothelium from rat brain. In Vitro *5917*, 353–362.

Boxberger, H. J., Paweletz, N., Spiess, E. and Kriehuber, R. (1989). An in vitro model study of BSp73 rat tumor cell invasion into endothelial monolayer. Anticancer Res. *9*, 1777–1786.

Boyden, S. (1962). The chemotactic effect of mixtures of antibody and antigen on polymorphonuclear leukocytes. J. Exp. Med. *115*, 453–466.

Bracke, M. E., van Roy, F. M. and Mareel, M. M. (1996). The E-cadherin/catenin complex in invasion and metastasis. Curr. Top. Microbiol. Immunol. *213(1)*, 123–161.

Brandley, B. K. and Schnaar, R. L. (1989). Tumor cell haptotaxis on covalently immobilized linear and exponential gradients of a cell adhesion peptide. Dev. Biol. *135*, 74–86.

Brandley, B. K., Shaper, J. H. and Schnaar, R. L. (1990). Tumor cell haptotaxis on immobilized N-acetylglucosamine gradients. Dev. Biol. *140*, 161–171.

Brennan, M. J., Oldberg, A., Hayman, E. G. and Ruoslahti, E. (1983). Effect of a proteoglycan produced by rat tumor cells on their adhesion to fibronectin-collagen substrata. Cancer Res. *43*, 4302–4307.

Bresalier, R. S., Hujanen, E. S., Raper, S. E., Roll, F. J., Itzkowitz, S. H., Martin, G. R. and Kim, Y. S. (1987). An animal model for colon cancer metastasis: establishment and characterization of murine cell lines with enhanced liver-metastasizing ability. Cancer Res. *47*, 1398–1406.

Brinkmann, V., Foroutan, H., Sachs, M., Weidner, M. K. and Birchmeier, W. (1995). HGF/SF induces a variety of tissue-specific morphogenic programs in epithelial cells. J. Cell Biol. *131*, 1573–1586.

Brodt, P. (1986). Characterization of two highly metastatic variants of Lewis lung carcinoma with different organ specificities. Cancer Res. *46*, 2442–2448.

Brodt, P. (1989). Selection of a highly metastatic liver-colonizing subpopulation of Lewis lung carcinoma variant H-59 using murine hepatocyte monolayers. Clin. Exp. Metastasis *7*, 525–539.

Brooks, D. E. (1984). The biorheology of tumor cells. Biorheology, *21*, 85–91.

Browder, T. M., Dunbar, C. E. and Nienhuis, A. W. (1990). Private and public autocrine loops in neoplastic cells. Cancer Cells *1*, 9–17.

Brundage, R. A., Fogarty, K. E., Tuft, R. A. and Fay, F. S. (1991). Calcium gradients underlying polarization and chemotaxis of eosinophils. Science *254*, 703–706.

Brunner, N., Thompson, E. W., Spang-Thomsen, M., Rygaard, J., Dano, K. and Zwiebel, J. A. (1992). lacZ transduced human breast cancer xenografts as an in vivo model for the study of invasion and metastasis. Eur. J. Cancer *28A*, 1989–1995.

Brunner, N., Boysen, B., Romer, J. and Srang-Thomsen, M. (1993). The nude mouse as an in vivo model for human breast cancer invasion and metastasis. Breast Cancer Res. Treat. *24*, 257–264.

Brunschwig, A., Southam, C. M. and Levin, A. G. (1965). Clinical experiments by homotransplants, autotransplants and admixture of autologous leukocytes. Ann. Surg. *162*, 416–425.

Brunson, K. W. and Nicolson, G. L. (1978). Selection and biologic properties of malignant variants of a murine lymphosarcoma. J. Natl. Cancer Inst. *61*, 1499–1503.

Brunson, K. W. and Nicolson, G. L. (1979). Selection of malignant melanoma variant cell lines for ovary colonization. J. Supramolec. Struct. Cell. Biochem. *11*, 517–528.

Brunson, K. W., Beattie, G. and Nicolson, G. L. (1978). Selection and altered properties of brain-colonising metastatic melanoma. Nature *272*, 543–545.

Bryant, P. J., Watson, K. L., Justice, R. W. and Woods, D. F. (1993). Tumor suppressor genes encoding proteins required for cell interactions and signal transduction in Drosophila. Dev. *Suppl.*, 239–249.

Burger, M. M. and Madnick, H. M. (1983). General and liver-specific metastasis: possible mechanisms. In: Liver in Metabolic Diseases: Proceedings of the 35th Falk Symposium, Basel, October 15–17, 1982 (Bianchi, L., Landmann, L., Gerok, W., Sickinger, K. and Stalder, G. A., eds.). MTP Press, Boston, pp. 351–365.

Bürk, R. R. (1973). A factor from a transformed cell line that affects cell migration. Proc. Natl. Acad. Sci. USA *70*, 369–372.

Burke, F., Rozengurt, E. and Balkwill, F. R. (1995). Measurement of proliferative, cytolytic, and cytostatic activity of cytokines. In: Cytokines: A Practical Approach (Balkwill, F. R., ed.). IRL Press, Oxford.

Bussolino, F., DiRenzo, M. F., Ziche, M., Bocchietto, E., Olivero, M., Naldini, L., Gaudino, G., Tamagnone, L., Coffer, A. and Comoglio, P. M. (1992). HGF is a potent angiogenic factor which stimulates endothelial cell motility and growth. J. Cell Biol. *119*, 629–641.

Butcher, E. C., Scollay, R. G. and Weissman, I. L. (1979). Lymphocyte adherence to high endothelial venules: characterization of a modified in vitro assay, and examination of the binding of syngeneic and allogeneic lymphocyte populations. J. Immunol. *123*, 1996–2003.

Butler, T. P. and Gullino, P. M. (1975). Quantitation of cell shedding into efferent blood of mammary adenocarcinoma. Cancer Res. *35*, 512–516.

Caenazzo, C., Onisto, M., Sartor, L., Scalerta, R., Giraldo A., Nitti, D. and Garbisa, S. (1998). Augmented membrane type 1 matrix metalloproteinase (MT1-MMP):MMP-2 messenger RNA ration in gastric carcinomas with poor prognosis. Clin. Cancer Res. *4*, 2179–2186.

Caldwell, P. R., Segal, B. C. and Hsu, K. C. (1976). Angiotensin-converting enzyme: vascular endothelial location. Science *191*, 1050–1051.

Calorini, L., Marozzi, A., Byers, H. R., Waneck, G. L., Lee, K. W., Isselbacher, K. J. and Gattoni-Celli, S. (1992). Expression of a transfected H–2Kb gene in B16 cells correlates with suppression of liver metastases in triple immunodeficient mice. Cancer Res. *52*, 4036–4041.

Cannistra, S. A. (1993). Cancer of the ovary. N. Engl. J. Med. *329*, 1550–1559.

Carmichael, D. F., Sommer, A., Thompson, R. C. et al. (1986). Primary structure and cDNA cloning of human fibroblast collagenase inhibitor. Proc. Natl. Acad. Sci. USA *83*, 2407–2411.

Carpenter, R. R. (1963). In vitro studies of cellular hypersensitivity. I. Specific inhibition of migration of cells from adjuvant immunized animals by purified protein derivative and other protein antigens. J. Immunol. *91*, 803–818.

Castronovo, V., Taraboletti, G. and Sobel, M. E. (1991). Laminin receptor cDNA-deduced synthetic peptide inhibits cancer cell attachment to endothelium. Cancer Res. *51*, 5672–5678.

Cavanaugh, P. G. and Nicolson, G. L. (1991). Lung-derived growth factor that stimulates the growth of lung-metastasizing tumor cells: identification as transferrin. J. Cell. Biochem. *47*, 261–271.

Cavanaugh, P. G. and Nicolson, G. L. (1998). Selection of highly metastatic rat MTLn2 mammary adenocarcinoma cell variants using in vitro growth response to transferrin. J. Cell Physiol. *174*, 48–57.

Chackal-Roy, M., Niemeyer, C., Moore, M. and Zetter, B. R. (1989). Stimulation of human prostatic carcinoma cell growth by factors present in human bone marrow. J. Clin. Invest. *84*, 43–50.

Chalfie, M., Tu, Y., Euskirchen, G., Ward, W. W. and Prasher, D. C. (1994). Green fluorescent protein as a marker for gene expression. Science *263*, 802–805.

Chambers, A. F. and Tuck, A. B. (1988). Oncogene transformation and the metastatic phenotype. Anticancer Res. *8*, 861–872.

Chambers, A. F. and Wilson, S. (1988). Use of neor B16-F1 murine melanoma cells to assess clonality of experimental metastases in the immune-deficient chick embryo. Clin. Exptl. Metastasis *6*, 171–182.

Chambers, A. F., Shafir, R. and Ling, V. (1982). A model system for studying metastasis using the embryonic chick. Cancer Res. *42*, 4018–4025.

Chambers, A. F., MacDonald, I. C., Schmidt, E. E., Koop, S., Morris, V. L., Khokha, R. and Groom, A. C. (1995). Steps in tumor metastasis: new concepts from intravital videomicroscopy. Cancer Metastasis Rev. *14*, 279–301.

Chan, W. -S., Page, C. M., Maclellan, J. R. and Turner, G. A. (1988). The growth and metastasis of four commonly used tumour lines implanted into eight different sites: evidence for site and tumour effects. Clin. Exptl. Metastasis *6*, 233–244.

Chen, P., Gupta, K. and Wells, A. (1994). Cell movement elicited by EGFr requires kinase and autophosphorylation but is separable from mitogenesis. J. Cell Biol. *124*, 547–555.

Chen, C., Parangi, S., Tolentino, M. and Folkman, J. (1995). A strategy to discover circulating angiogenesis inhibitors generated by human tumors. Cancer Res. *55*, 4230–4233.

Cheng, K. C. and Loeb, L. A. (1993). Genomic instability and tumor progression: Mechanistic considerations. Adv. Cancer Res. *60*, 121–156.

Cheng, L., Fu, J., Tsukamoto, A. and Hawley, R. G. (1996). Use of green fluorescent protein variants to monitor gene transfer and expression in mammalian cells. Nat. Biotechnol. *14*, 606–609.

Chew, E. C., Josephson, R. L. and Wallace, J. (1976). Morphological aspects of the arrest of circulating cancer cells. In: Fundamental Aspects of Metastasis (Weiss, L., ed.). Elsevier-North Holland, Amsterdam, p. 121.

Chiquet-Ehrismann, R., Kalla, P., Pearson, C., Beck, K. and Chiquet, M. (1988). Tenascin interferes with fibronectin action. Cell *53A*, 383–390.

Chishima, T., Miyagi, Y., Wang, X., Yamaoka, H., Shimada, H., Moossa, A. R. and Hoffman, R. M. (1997a) Cancer invasion and micrometastasis visualized in live tissue by green fluorescent protein expression. Cancer Res. *57*, 2042–2047.

Chishima, T., Miyagi, Y., Wang, X., Tan, Y., Shimada, H., Moossa, A. and Hoffman, R. M. (1997b) Visualization of the metastatic process by green fluorescent protein expression. Anticancer Res. *17*, 2377–2384.

Chishima, T., Miyagi, Y., Wang, X., Baranov, E., Tan, Y., Shimada, H., Moossa, A. R. and Hoffman, R. M. (1997c) Metastatic patterns of lung cancer visualized live and in process by green fluorescence protein expression. Clin. Exp. Metastasis *15*, 547–552.

Chishima, T., Yang, M., Miyagi, Y., Li, L., Tan, Y., Baranov, E., Shimada, H., Moossa, A. R., Penman, S. and Hoffman RM (1997d) Governing step of metastasis visualized in vitro. Proc. Natl. Acad. Sci. USA *94*, 11573–11576.

Chishima, T., Miyagi, Y., Li, L., Tan, Y., Baranov, E., Yang, M., Shimada, H., Moossa, A. R. and Hoffman, R. M. (1997e) Use of histoculture and green fluorescent protein to visualize tumor cell host interaction. In Vitro Cell Dev. Biol. Anim. *33*, 745–747.

Christofori, G. and Semb, H. (1999). The role of the cell-adhesion molecule E-cadherin as a tumour-suppressor gene. Trends Biochem. Sci. *24*, 73–76.

Christov, K., Chew, K. L., Ljung, B. M., Waldman, F. M., Duarte L. A., Goodson, W. H. 3rd, Smith, H. S. and Mayall, B. H. (1991). Proliferation of normal breast epithelial cells as shown by in vivo labeling with bromodeoxyuridine. Am. J. Pathol. *138*, 1371–1377.

Christov, K., Chew, K. L., Ljung, B. M., Waldman, F. M., Goodson, W. H. 3rd, Smith, H. S. and Mayall, B. H. (1994). Cell proliferation in hyperplastic and in situ carcinoma lesions of the breast estimated by in vivo labeling with bromodeoxyuridine. J. Cell. Biochem. *Suppl. 19*, 165–172.

Chung, A. Y. K., Ryan, J. W., Ryan, J. P. A. and Ryan, U. S. (1986). Radiolabelled substrates for angiotensin converting enzyme. Adv. Exp. Med. Biol. *198A*, 427–434.

Clark, W. H., Elder, D. E., Guerry, D., Braitman, L. E., Trock, B. J., Schultz, D., Synnestvedt, M. and Halpern, A. C. (1989). Model predicting survival in Stage I melanoma based on tumor progression. J. Natl. Cancer Inst. *81*, 1893–1904.

Clarke, R. (1996). Human breast cancer cell line xenografts as models of breast cancer—The immunobiologies of recipient mice and the characteristics of several tumorigenic cell lines. Breast Cancer Res. Treat. *39*, 69–86.

Clayman, R. V., Figenshau, R. S. and Bear, A. (1985). Transplantation of human renal carcinomas into nude mice. Cancer Res. *45*, 2650–2653.

Codington, J. F., Cooper, A. G., Miller, D. K., Slayter, H. S., Brown, M. C., Silber, C. and Jeanloz, R. W. (1979). Isolation and partial characterization of an epiglycanin-like glycoprotein from a new non-strain-specific subline of TA_3 murine mammary adenocarcinoma. J. Natl. Cancer Inst. *63*, 153–161.

Cohnheim, J. (1889). Lectures on general pathology: a handbook for practitioners and students. The New Sydenham Society, London.

Collard, J. G., Habets, G. G., Michiels, F., Stam, J., van der Kammen, R. A. and van Leeuwen F (1996). Role of Tiam 1 in Rac-mediated signal transduction pathways. Curr. Top. Microbiol. Immunol. *213*, 253–265.

Collier, I. E. Wilhelm, S. M., Eisen, A. Z., Marmer, D. L., Grant, G. A., Seltzer, J. L., Cronberger, A., He, C., Bauer, G. A. and Goldberg, G. I. (1988). H-ras-oncogene-transformed human bronchiolar epithelial cells (TBE-1). secrete a single metalloproteinase capable of degrading basement membrane collagen. J. Biol. Chem. *263*, 6579–6587.

Coman, D. R. (1953). Mechanisms responsible for the origin and distribution of blood-borne tumor metastasis: a review. Cancer Res. *13*, 397–404.

Condeelis, J. (1993). Life at the leading edge: the formation of cell protrusions. Ann. Rev. Cell Biol. *9*, 411–444.

Constantini, V., Zacharski, L. R., Moritz, T. E. and Edwards, R. L. (1990). The platelet count in carcinoma of the lung and the colon. Thromb. Haemost. *64*, 501–505.

Cooper, D. L. and Dougherty, G. J. (1995). To metastasize or not? Selection of CD44 splice sites. Nat. Med. *1*, 635–637.

Cordon-Cardo, C., Koff, A., Drobnjak, M., Capodieci, P., Osman, I., Millard, S. S., Gaudin, P. B., Fazzari M., Zhang Z. F., Massague J. and Scher H. I. (1998). Distinct altered patterns of p27KIP1 gene expression in benign prostatic hyperplasia and prostatic carcinoma. J. Natl. Cancer Inst. *90*, 1284–1291.

Cornil, I., Theodorescu, D., Man, S., Herlyn, M., Jambrosic, J. and Kerbel, R. S. (1991). Fibroblast cell interactions with human melanoma cells affect tumor cell growth as a function of tumor progression. Proc. Natl. Acad. Sci. USA *88*, 6028–6032.

Cortner, J., Vande Woude, G. F. and Rong, S. (1995). The Met-HGF/SF autocrine signaling is involved in sarcomagenesis. EXS *74*, 89–121.

Cory, A. H, Owen, T. C., Barltrop, J. A. and Cory, J. G. (1991). Use of an aqueous soluble tetrazolium/formazan assay for cell growthassays in culture. Cancer Commun. *3*, 207–212.

Cotte, C., Raghavan, D., McIlhinney, R. A. and Monaghan, P. (1982). Characterization of a new human cell line derived from a xenografted embryonal carcinoma. In Vitro *18*, 739–749.

Coutts, A. S. and Murphy, L. C. (1998). Elevated mitogen-activated protein kinase activity in estrogen-nonresponsive human breast cancer cells. Cancer Res. *58*, 4071–4074.

Crissman, J. D., Hatfield, J., Schaldenbrand, M., Sloane, B. F. and Honn, K. V. (1985). Arrest and extravasation of B16 amelanotic melanoma in murine lungs. A light and electron microscopy study. Lab. Invest. *53*, 470–478.

Crissman, J. D., Hatfield, J., Menter, D. J., Sloane, B. F. and Honn, K. V. (1988). Morphological study of the interaction of intravascular tumor cells with endothelial cells and subendothelial matrix. Cancer Res. *48*, 4065–4072.

Crnalic, S., Håkansson, I., Boquist, L., Löfvenberg, R. and Broström, L. Å. (1997). A novel spontaneous metastasis model of human osteosarcoma developed using orthotopic transplantation of intact tumor tissue into tibia of nude mice. Clin. Exptl. Metastasis *15*, 164–172.

Cubitt, A. B., Heim, R., Adams, S. R., Boyd, A. E., Gross, L. A. and Tsien, R. Y. (1995). Understanding, improving and using green fluorescent proteins. Trends in Biochem. Sci. *20*, 448–455.

Culp, L. A., Lin, W., Kleinman, N. R., O'Connor, K. L. and Lechner, R. (1998). Earliest steps in primary tumor formation and micrometastasis resolved with histochemical markers of gene-tagged tumor cells. J. Histochem. Cytochem. *46*, 557–568.

Dankort, D. L. and Muller, W. J. (1996). Transgenic models of breast cancer metastasis. Cancer Treat. Res. *83*, 71–88.

Daphna-Iken D., Shankar D. B., Lawshe A., Ornitz D. M., Shackleford G. M. and MacArthur C. A. (1998). MMTV-Fgf8 transgenic mice develop mammary and salivary gland neoplasia and ovarian stromal hyperplasia. Oncogene *17*, 2711–2717.

Davies, M. P. A., Rudland, P. S., Robertson, L., Parry, E. W., Jolicoeur, P. and Barraclough, R. (1996). Expression of the calcium-binding protein S100A4 (p9Ka) in MMTV- neu transgenic mice induces metastasis of mammary tumours. Oncogene *13*, 1631–1637.

Davison, P. M., Bensch, K. and Karasek, M. A. (1980). Growth and morphology of rabbit marginal vessel endothelium in cell culture. J. Cell Biol. *85*, 187–195.

De la Rosa, A., Williams, R. L. and Steeg, P. S. (1995). Nm23/nucleoside diphosphate kinase: toward a structural and biochemical understanding of its biological functions. Bioessays *17*, 53–62.

De Giovanni, C., Palmieri, G., Nicoletti, G., Landuzzi, L., Scotlandi, K., Bontadini, A., Tazzari, P. L., Sensi, M., Santoni, A., Nanni, P., et al (1991).

Immunological and non-immunological influence of H–2Kb gene transfection on the metastatic ability of B16 melanoma cells. Int. J. Cancer 48, 270–276.

De Vries, T. J., Kitson, J. L., Silvers, W. K. and Mintz, B. (1995). Expression of plasminogen activators and plasminogen activator inhibitors in cutaneous melanomas of transgenic melanoma-susceptible mice. Cancer Res. 55, 4681–4687.

Dear, T. N., Ramshaw, I. A. and Kefford, R. F. (1988). Differential expression of a novel gene, WDNM1, in nonmetastatic rat mammary adenocarcinoma cells. Cancer Res. 48, 5203–5209.

Dear, T. N., McDonald, D. A. and Kefford, R. F (1989). Transcriptional down-regulation of a rat gene, WDNM2, in metastatic DMBA–8 cells. Cancer Res. 49, 5323–5328.

DeBault, L., Henriquez, E., Hart, M. and Cancilla, P. (1981). Cerebral microvessels and derived cells in tissue culture. In Vitro 17, 480–494.

Denizot, F. and Lang, R. (1986). Rapid colorimetric assay for cell growth and survival. Modifications to the tetrazolium dye procedure giving improved sensitivity and reliability. J. Immunol. Meth. 89, 271–277.

Dennis, J. W. (1985). Partial reversion of the metastatic phenotype in a wheat germ agglutinin-resistant mutant of the murine tumor cell line MDAY-D2 selected with Bandeiraea simplicifolia seed lectin. J. Natl. Cancer Inst. 74, 1111–1120.

Dennis, J. W. and Laferté, S. (1985). Recognition of asparagine-linked oligosaccharides on murine tumor cells by natural killer cells. Cancer Res. 45, 6034–6040.

Dennis, J. W. and Laferté, S. (1987). Tumor cell surface carbohydrate and the metastatic phenotype. Cancer Metastasis Rev. 5, 185–204.

Dennis, J. W., Donaghue, T. P. and Kerbel, R. S. (1981). Membrane-associated alterations detected in poorly tumorigenic lectin-resistant variant sublines of a highly malignant and metastatic murine tumor. J. Natl. Cancer Inst. 66, 129–137.

Dennis, J. W., Carver, J. P. and Schachter, H. (1984). Asparagine-linked oligosaccharides in murine tumor cells: comparison of a WGA-resistant (WGAr) nonmetastatic mutant and a related WGA-sensitive (WGAs) metastatic line. J. Cell Biol. 99, 1034–1044.

Derynck, R., Jarrett, J. A., Chen, E. Y., Eaton, D. H., Bell, J. R., Assoian, R. K., Roberts, A. B., Sporn, M. B. and Goeddel, D. V. (1985). Human transforming growth factor-beta complementary DNA sequence and expression in normal and transformed cells. Nature 316, 701–705.

Dhingra, K., Sahin, A., Emami, K., Hortobagyi, G. N. and Estrov, Z. (1998). Expression of leukemia inhibitory factor and its receptor in breast cancer: a potential autocrine and paracrine growth regulatory mechanism. Breast Cancer Res. Treat. 48, 165–174.

Di Renzo, M. F., Olivero, M., Giacomini, A., Porte, H., Chastre, E., Mirossay, L., Nordlinger, B., Bretti, S., Bottardi, S., Giordano, S., et al. (1995). Overexpression and amplification of the met/HGF receptor gene during the progression of colorectal cancer. Clin Cancer Res. *1(2)*, 147–154

Diglio, C., Grammas, P., Giaccamelli, F. and Wiener, J. (1982). Primary culture of rat cerebral microvascular endothelial cells. Lab. Invest. *46*, 554–563.

Dodge, A. B., Patton, W. F., Yoon, M. U., Hechtman, H. B. and Shepro, D. (1991). Organ and species-specific differences in cytoskeletal proteins profiles of cultured microvascular endothelial cells. Comp. Biochem. Physiol. *98B*, 461–470.

Doerr, R., Zvibel, I., Chiuten, D., D'Olimpio, J. and Reid, L. M. (1989). Clonal growth of tumors on tissue-specific biomatrices and correlation with organ-site specificity of metastases. Cancer Res. *49*, 384–392.

Dong, Z., Radinsky, R., Fan, D., Tsan, R., Bucana, C. D., Wilmanns, C. D. and Fidler, I. J. (1994). Organ-specific modulation of steady state mdr gene expression and drug resistance in murine colon cancer cells. J. Natl. Cancer Inst. *86*, 913–920.

Dong, Z. Y., Greene, G., Pettaway, C., Dinney, C. P. N., Eue, I., Lu, W. X., Bucana, C. D., Balbay, M. D., Bielenberg D. and Fidler I. J. (1999). Suppression of angiogenesis, tumorigenicity, and metastasis by human prostate cancer cells engineered to produce interferon-β. Cancer Res. *59*, 872–879.

Dorkin, T. J. and Neal, D. E. (1997). Basic science aspects of prostate cancer. Semin. Cancer Biol. *8*, 21–27.

Doyle, G. M., Sharief, Y. and Mohler, J. L. (1992). Prediction of metastatic potential by cancer cell motility in the Dunning R–3327 prostatic adenocarcinoma in vivo model. J. Urol. *147*, 514–518.

Dracopoli, N. C., Houghton, A. N. and Old, L. J. (1985). Loss of polymorphic restriction fragments in malignant melanoma: implications for tumor heterogeneity. Proc. Natl. Acad. Sci. USA *82*, 1470–1474.

Dracopoli, N. C., Alhadeff, B., Houghton, A. N. and Old, L. J. (1987). Loss of heterozygosity at autosomal and X-linked loci during tumor progression in a patient with melanoma. Cancer Res. *47*, 3995–4000.

Driessens, M. H. E., Stroeken, P. J. M., Erena, N. F. R., van der Valk, M. A., van Rijthoven, E. A. M. and Roos, E. (1995). Targeted disruption of CD44 in MDAY-D2 lymphosarcoma cells has no effect on subcutaneous growth or metastatic capacity. J. Cell Biol. *131*, 1849–1855.

Duerst, R. E., Rose, D. and Frantz, C. N. (1991). Complement depletion in vitro limits monoclonal antibody 6–19-dependent complement-mediated killing of tumor cells in bone marrow. Exp. Hematol. *19*, 863–867.

Duffy, M. J. (1996). Proteases as prognostic markers in cancer. Clin. Cancer Res. *2*, 613–618.

Duffy, M. J., Duggan, C., Maguire, T., Mulcahy, K., Elvin, P., Mc Dermott, E., Fennelly, J. J. and O'Higgins, N. (1996). Urokinase plasminogen activator as a predictor of aggressive disease in breast cancer. Enzyme Prot. *49*, 85–93.

Dugan, J. D. J., Lawton, M. T., Glaser, B. and Brem, H. (1991). A new technique for explantation and in vitro cultivation of chicken embryos. Anat Rec, *229*, 125–128.

Dumont, J. A., Jones Jr., W. D. and Bitonti, A. J. (1992). Inhibition of experimental metastasis and cell adhesion of B16F1 melanoma cells by inhibitors of protein kinase C. Cancer Res. *52*, 1195–1200.

Dusseau, J. W., Hutchins, P. M. and Malbasa, D. S. (1986). Stimulation of angiogenesis by adenosine on the chick chorioallantoic membrane. Circ. Res. *59*, 163–170.

Dutt, M. K. (1980). Staining of depolymerized DNA in mammalian tissues with methyl violet 6B and crystal violet. Folia Histochem. Cytochem. (Krakow) *18*, 79–83.

Dvorak, H. F., Detmar, M., Claffey, K. P., Nagy, J. A., van de Water, L. and Senger, D. R. (1995). Vascular permeability factor/vascular endothelial growth factor: an important mediator of angiogenesis in malignancy and inflammation. Int Arch Allergy Immunol. *107*, 233–235.

Eaves, G. (1973). The invasiveness growth of malignant tumors as a purely mechanical process. J. Pathol. *109*: 233–237.

Ebralidze, A., Tulchinsky, E., Grigorian, M., Afanasyeva, A., Senin, V., Revazova, E. and Lukanidin, E. (1989). Isolation and characterization of a gene specifically expressed in different metastatic cells and whose deduced gene product has a high degree of homology to a Ca2+-binding protein family. Genes Dev. *3*, 1086–1093.

Egan, S. E., Jarolim, L., Rogelj, S., Spearman, M., Wright, J. A. and Greenberg, A. H. (1990). Growth factor modulation of metastatic lung colonization. Anticancer Res. *10*, 1341–1346.

Eitzman, D. T., Krauss, J. C., Shen, T. L., Cui, J. S. and Ginsburg, D. (1996). Lack of plasminogen activator inhibitor–1 effect in a transgenic mouse model of metastatic melanoma. Blood *87*, 4718–4722.

El-Fouly, M. H., Trosko, J. E. and Chang, C. (1987). Scrape-loading and dye transfer. Exp. Cell Res. *168*, 422–430.

Elble, R. C., Widom, J., Gruber, A. D., Abdel-Ghany, M., Levine, R., Goodwin, A., Cheng, H. C. and Pauli, B. U. (1997). Cloning and characterization of lung-endothelial cell adhesion molecule–1suggest it is an endothelial chloride channel. J. Biol. Chem. *272*, 27853–27861.

Elgavish, S. and Shaanan, B. (1997). Lectin-carbohydrate interactions: different folds, common recognition principles. Trends Biochem. Sci. *22*, 462–467.

Ellis, L. M. and Fidler, I. J. (1995). Angiogenesis and breast cancer metastasis. Lancet *346*, 388–390.

Ercolani, L., Florence, B., Denaro, M., Alexander, M. (1988). Isolation and complete sequence of a functional human glyceraldehyde–3-phosphate dehydrogenase gene. J. Biol. Chem. *263*, 15335–15341.

Ernst, L. K., Rajan, V. P., Larsen, R. D., Ruff, M. M. and Lowe, J. B. (1989). Stable expression of blood group H determinants and GDP-L-fucose:β-D-galactoside 2-α-L-fucosyltransferase in mouse cells after transfection with human DNA. J. Biol. Chem. *264*, 3436–3447.

Essodaigui, M., Broxterman, H. J. and Garnier-Suillerot, A. (1998). Kinetic analysis of calcein and calcein-acetoxymethylester efflux mediated by the multidrug resistance protein and P-glycoprotein. Biochemistry *37*, 2243–2250.

Evans, C. P., Walsh, D. S. and Kohn, E. C. (1991). An autocrine motility factor secreted by the Dunning R–3327 rat prostatic adenocarcinoma cell subtype AT2. 1. Int. J. Cancer *49*, 109–113.

Ewing, J. (1928). A Treatise on Tumors, 3rd edn. W. B. Saunders, Philadelphia.

Farina, K. L., Wyckoff, J. B., Rivera, J., Lee, H., Segall, J. E., Condeelis, J. S. and Jones, J. G. (1998). Cell motility of tumor cells visualized in living intact primary tumors using green fluorescent protein. Cancer Res. *58*, 2528–2532.

Fearon, E. R., Cho, K. R., Nigro, J. M., Kern, S. E., Simons, J. W., Ruppert, J. M., Hamilton, S. R., Preisinger, A. C., Thomas, G., Kinzler, K. W. and Vogelstein, B. (1990). Identification of a chromosome 18q gene that is altered in colorectal cancer. Science *247*, 49–56.

Fialkow, P. J. (1972). Use of genetic markers to study cellular origin and development of tumors in human females. Adv. Cancer Res. *15*, 191–226.

Ferracini, R., Di Renzo, M. F., Scotlandi, K., Baldini, N., Olivero, M., Lollini, P., Cremona, O., Campanacci, M. and Comoglio, P. M. (1995). The Met/HGF receptor is over-expressed in human osteosarcomas and is activated by either a paracrine or an autocrine circuit. Oncogene *10*, 739–749.

Fialkow, P. J. (1976). Clonal origin of human tumors. Biochim. Biophys. Acta *458*, 283–321.

Fidler I. J. (1970). Metastasis: quantitative analysis of distribution and fate of tumor emboli labeled with ^{125}I-5-iodo-2′-deoxyuridine. J. Natl. Cancer Inst. *45*, 773–782.

Fidler, I. J. (1973a) Selection of successive tumor lines for metastasis. Nat. New Biol. *242*, 148–149.

Fidler, I. J. (1973b) The relationship of embolic heterogeneity, number size and viability to the incidence of experimental metastasis. Eur. J. Cancer *9*, 223–227.

Fidler, I. J. (1986). Rationale and methods for the use of nude mice to study the biology and therapy of human cancer metastasis. Cancer Metastasis Rev. *5*, 29–49.

Fidler, I. J. (1990). Critical factors in the biology of human cancer metastasis. Twenty-eighth G. H. A. Clowes memorial award lecture. Cancer Res. *50*, 6130–6138.

Fidler, I. J. (1991). Orthotopic implantation of human colon carcinomas into nude mice provides a valuable model for the biology and therapy of metastasis. Cancer Metastasis Rev. *10*, 229–243.

Fidler, I. J. and Kripke, M. L. (1977). Metastasis results from preexisting variant cells within a malignant tumor. Science *197*, 893–895.

Fidler, I. J. and Lieber, S. (1972). Quantitative analysis of the mechanism of glucocorticoid enhancement of experimental metastasis. Res. Comm. Chem. Path. Pharmacol. *4*, 607–613.

Fidler, I. J. and Nicolson, G. L. (1976). Organ selectivity for implantation, survival and growth of B16 melanoma variant tumor lines. J. Natl. Cancer Inst. *57*, 1199–1202.

Fidler, I. J. and Nicolson, G. L. (1987). The process of cancer invasion and metastasis. Cancer Bull. *39*, 126–131.

Fidler, I. J. and Radinsky, R. (1990). Genetic control of cancer metastasis. J. Natl. Cancer Inst. *82*, 166–168.

Fiebig, H. H., Schmid, J. R., Bieser, W., Henss, H. and Lohr, G. W. (1987). Colony assay with human tumor xenografts, murine tumors and human bone marrow. Potential for anticancer drug development. Eur. J. Clin. Oncol. *23*, 937–948.

Finne, J., Tao, T. and Burger, M. M. (1980). Carbohydrate changes in glycoproteins of a poorly metastasising wheat germ agglutinin-resistant melanoma clone. Cancer Res. *40*, 2580–2587.

Finne, J., Burger, M. M. and Prieels, J. P. (1982). Enzymatic basis for a lectin-resistant phenotype: increase in a fucosyltransferase in mouse melanoma cells. J. Cell Biol. *92*, 277–282.

Finne, J., Castori, S., Feizi, T. and Burger, M. M. (1989). Lectin-resistant variants and revertants of mouse melanoma cells: differential expression of a fucosylated cell-surface antigen and altered metastasizing capacity. Int. J. Cancer *43*, 300–304.

Fisher B. and Fisher E. R. (1967a). The organ distribution of disseminated 51Cr-labeled tumor cells. Cancer Res. *27*, 412–420.

Fisher B. and Fisher E. R. (1967b). Metastasis of cancer cells. Meth. Cancer Res. *1*, 243–286.

Fisher, B., Gunduz, N., Coyle, J., Rudock, C. and Saffer, E. (1989). Presence of a growth-stimulating factor in serum following primary tumor removal in mice. Cancer Res. *49*, 1996–2001.

Fogler, W. E. and Fidler, I. J. (1985). Nonselective destruction of murine neoplastic cells by syngeneic tumoricidal macrophages. Cancer Res. *45*, 14–18.

Folkman, J. (1984). Angiogenesis: Initiation and modulation. Cancer Invasion and Metastasis: Biologic and Therapeutic Aspects. Raven Press, New York, pp. 201–209.

Folkman, J. (1986). How is blood vessel growth regulated in normal and neoplastic tissue? Cancer Res. *46*, 467–473.

Folkman, J. (1992). The role of angiogenesis in tumor growth. Semin. Cancer Biol. *3*, 65–71.

Folkman, J. (1995a) Angiogenesis inhibitors generated by tumors. Mol, Med. *1*, 120–122.

Folkman, J. (1995b) Angiogenesis in cancer, vascular, rheumatoid and other disease. Nat. Med. *1*, 27–31.

Folkman, J. and Cotran, R. (1976). Relation of vascular proliferation to tumor growth. Int. Rev. Exp. Pathol. *16*, 207–248.

Folkman, J., Haudenschild, C. C. and Zetter, B. R. (1979). Long term culture of capillary endothelial cells. Proc. Natl. Acad. Sci. USA *76*, 5217–5221

Folkman, J., Watson, K., Ingber, D. and Hanahan, D. (1989). Induction of angiogenesis during the transition from hyperplasia to neoplasia. Nature *339*, 58–61.

Foon, K. A. (1989). Biological response modifier: The new immunotherapy. Cancer Res. *49*, 1621–1639.

Foster, B. A., Gingrich, J. R., Kwon, E. D., Madias, C. and Greenberg, N. M. (1997). Characterization of prostatic epithelial cell lines derived from transgenic adenocarcinoma of the mouse prostate (TRAMP) model. Cancer Res. *57*, 3325–3330.

Foty, R. A. and Steinberg, M. S. (1997). Measurement of tumor cell cohesion and suppression of invasion by E- or P-cadherin. Cancer Res. *57*, 5033–5036.

Foty, R. A., Corbett, S. A., Schwarzbauer, J. E. and Steinberg, M. S. (1998). Dexamethasone up-regulates cadherin expression and cohesion of HT–1080 human fibrosarcoma cells. Cancer Res. *58*, 3586–3589.

Freije, J. M., MacDonald, N. J. and Steeg, P. S. (1996). Differential gene expression in tumor metastasis: Nm23. Curr. Top. Microbiol. Immunol. *213*, 215–232.

Fridman, R., Kibbey, M. C., Royce, L. S., Zain, M., Sweeney M, Jicha, D. L., Yannelli, J. R., Martin, G. R. and Kleinman, H. K. (1991). Enhanced tumor growth of both primary and established human and murine tumor cells in athymic mice after coinjection with matrigel. J. Natl. Cancer Inst. *83*, 769–774.

Fridman, R., Sweeney, T. M., Zain, M., Martin, G. R. and Kleinman, H. K. (1992). Malignant transformation of NIH 3T3 cells after subcutaneous co-injection with a reconstituted basement membrane (matrigel). Int. J. Cancer *51*, 740–744.

Frisch, S. M. and Ruoslahti, E. (1997). Integrins and anoikis. Curr. Opin. Cell Biol. *9*, 701–706.

Frixen, U. W., Behrens, J., Sachs, M., Eberle, G., Voss, B., Warda, A., Loechner, D. and Birchmeier, W. (1991). E-cadherin-mediated cell-cell adhesion prevents invasiveness of human carcinoma cells. J. Cell Biol. *113*, 173–185.

Frost, P., Kerbel, R. S., Hunt, B., Man, S. and Pathak, S. (1987). Selection of metastatic variants with identifiable karyotypic changes from a nonmetastatic murine tumor after treatment with 2′-deoxy–5-azacytidine or hydroxyurea: implications for the mechanisms of tumor progression. Cancer Res. *47*, 2690–2695.

Fu, X., Besterman, J. M., Monosov, A. and Hoffman, R. M. (1991). Models of human metastatic colon cancer in nude mice orthotopically constructed by using histologically intact patient specimens. Proc. Natl. Acad. Sci. USA *88*, 9345–9349.

Fujihara, T., Sawada, T., Chung, K. H. Y. S., Yashiro, M., Inoue, T. and Sowa, M. (1998). Establishment of lymph node metastatic model for human gastric cancer in nude mice and analysis of factors associated with metastasis. Clin. Exptl. Metastasis *16*, 389–398.

Fujimaki, T., Ellis, L. M., Bucana, C. D., Radinsky, R., Price, J. E. and Fidler, I. J. (1993). Simultaneous radiolabel, genetic tagging and proliferation assays to study the organ distribution and fate of metastatic cells. Intl. J. Oncol. *2*, 895–901.

Fulton, A. B. (1984). The Cytoskeleton: Cellular Architecture and Choreography. Chapman and Hall, New York, pp. 55–58.

Furlong, R. A., Takehara, T., Taylor, W. G., Nakamura, T. and Rubin, J. R. (1991). Comparison of biological and immunochemical properties indicates that scatter factor and hepatocyte growth factor are indistinguishable. J. Cell Sci. *100*, 173–177.

Gaiano, N., Amsterdam, A., Kawakami, K., Allende, M., Becker, T. and Hopkins, N. (1996). Insertional mutagenesis and rapid cloning of essential genes in zebrafish. Nature *383*, 829–832.

Gallo, O., Masini, E., Morbidelli, L., Franchi, A., Fini-Storchi, I., Vergari, W. and Ziche, M. (1998). Role of nitric oxide in angiogenesis and tumor progression in head and neck cancer. J. Natl. Cancer Inst. *90*, 587–596.

Gallo-Hendrikx E., Copps J., Percy D., Croy B. A. and Wildeman A. G. (1994). Enhancement of pancreatic tumor metastasis in transgenic immunodeficient mice. Oncogene *9*, 2983–2990.

Garabedian E. M., Humphrey P. A. and Gordon J. I. (1998). A transgenic mouse model of metastatic prostate cancer originating from neuroendocrine cells. Proc. Natl. Acad. Sci. USA 95, 15382–15387.

Garbisa, S., Kniska, K., Tryggvason, K., Foltz, C. M. and Liotta, L. A. (1980a) Quantitation of basement membrane collagen degradation by living tumor cells in vitro. Cancer Lett. 9, 359–366.

Garbisa, S., Tryggvason, K., Foidart, J. M. and Liotta, L. A. (1980b) Assay for radiolabeled type IV collagen in the presence of other proteins using a specific collagenase. Anal. Biochem. 107, 187–192.

Garbisa, S., Scagliotti G., Masiero L., Di Francesco C., Caenazzo C., Onisto M., Micela M., Stetler-Stevenson, W. G. and Liotta, L. A. (1992). Correlation of serum metalloproteinase levels with lung cancer metastasis and response to therapy; Cancer Res. 52, 4548–4549.

Garofalo A., Chriivi R. G. S., Scanziani E., Mayo J. G., Vecchi A. and Giavazzi R. (1993). Comparative study on the metastatic behavior of human tumors in nude, beige/nude/Xid and severe combined immunodeficient mice. Invasion Metastasis 13, 82–91.

Garrido, F. and Ruiz-Cabello, F. (1991). MHC expression on human tumors: its relevance for local tumor growth and metastasis. Semin. Cancer Biol. 2, 3–10.

Gasic, G. J. (1984). Role of plasma, platelets and endothelial cells in tumor metastasis. Cancer Metastasis Rev. 3, 99–116.

Gasic, G. J. and Gasic, T. B. (1962). Removal of sialic acid from the cell coat in tumor cells and vascular endothelium and its effects on metastasis. Proc. Natl. Acad. Sci. USA 48, 1172–1177.

Gasic, G. J., Gasic, T. B. and Stewart, C. C. (1968). Antimetastatic effects associated with platelet reduction. Proc. Natl. Acad. Sci. USA 61, 46–52.

Gasic, G. J., Gasic, T. B., Galanti, N., Johnson, T. and Murphy, S. (1973). Platelet-tumor cell interaction in mice: the role of platelets in the spread of malignant disease. Int. J. Cancer 11, 704–718.

Geimer, P. and Bade, E. G. (1991). The EGF-induced migration of rat liver epithelial cells is associated with a transient inhibition of DNA synthesis. J. Cell Sci. 100, 349–351.

Gerlier, D. and Thomasset, N. (1986). Use of MTT colorimetric assay to measure cell activation. J. Immunol. Meth. 94, 57–63.

Gherardi, E., Gray, J., Stoker, M., Perryman, M. and Furlong, R. (1989). Purification of scatter factor, a fibroblast-derived basic protein that modulates epithelial interactions and movement. Proc. Natl. Acad. Sci. USA 86, 5844–5848.

Giancotti, F. G. and Ruoslahti, E. (1990). Elevated levels of the alpha 5 beta 1 fibronectin receptor suppress the transformed phenotype of Chinese hamster ovary cells. 60, 849–859.

Giavazzi, R., Campbell, D. E., Jessup, J. M., Cleary, K. and Fidler, I. J. (1986). Metastatic behavior of tumor cells isolated from primary and metastatic human colorectal carcinomas implanted into different sites in nude mice. Cancer Res. *46*, 1928–1933.

Giavazzi, R., Jessup J. M., Campbell D. E., Walker S. M. and Fidler I. J. (1986). Experimental nude mouse model of human colorectal cancer liver metastases. J. Natl. Cancer Inst. *77*, 1303–1308.

Giavazzi, R., Foppolo, M., Dossi, R. and Remuzzi, A. (1993). Rolling and adhesion of human tumor cells on vascular endothelium under physiological flow conditions. J. Clin. Invest. *92*, 3038–3044.

Gill, G. N. and Lazar, C. S. (1981). Increased phosphotyrosine content and inhibition of proliferation in EGF-treated A431 cells. Nature *293*, 305–307.

Gillies, R. J., Didier, N. and Denton, M. (1986). Determination of cell number in monolayer cultures. Anal. Biochem. *159*, 109–113.

Gimbrone, M. A., Cotran, R. S., Leapman, S. B. and Folkman, J. (1974a) Tumor growth and neovascularization: an experimental model using the rabbit cornea. J. Natl. Cancer Inst. *52*, 413–427.

Gimbrone, M. A. Jr., Cotran, R. S. and Folkman, J. (1974b) Human vascular endothelial cells in culture. Growth and DNA synthesis. J. Cell Biol. *60*, 673–684.

Gingrich, J. R., Barrios, R. J., Morton, R. A., Boyce, B. F., DeMayo, F. J., Finegold, M. J., Angelopoulou, R., Rosen, J. M. and Greenberg, N. M. (1996). Metastatic prostate cancer in a transgenic mouse. Cancer Res. *56*, 4096–4102.

Gingrich, J. R., Barrios, R. J., Kattan, M. W., Nahm, H. S., Finegold, M. J. and Greenberg, N. M. (1997). Androgen-independent prostate cancer progression in the TRAMP model. Cancer Res. *57*, 4687–4691.

Giraldi T., Nisi C. and Sava G. (1977). Antimetastatic effects of N-diazoacetyl-glycine derivatives in C57BL mice. J. Natl. Cancer Inst. *58*, 1129–1130.

Giraldi, T., Perissin, L., Zorzet, S., Piccini, P. and Rapozzi, V. (1989). Effects of stress on tumor growth and metastasis in mice bearing Lewis lung carcinoma. Eur. J. Cancer Clin. Oncol. *25*, 1583–1588.

Giraldi, T., Perissin, L., Zorzet, S and Rapozzi, V. (1994a) Rotational stress reduces the effectiveness of antitumor drugs in mice. Ann. NY Acad. Sci. *741*, 234–243.

Giraldi, T., Perissin, L., Zorzet, S., Rapozzi, V. and Rodani, M. G. (1994b) Metastasis and neuroendocrine system in stressed mice. Intern. J. Neurosci. *74*, 265–278.

Gitlin, J. and D'Amore, P. (1983). Culture of retinal capillary cells using selective growth media. Microvasc. Res. *26*, 74–80.

Glaves D. (1983). Correlation between circulating cancer cells and incidence of metastasis. Br. J. Cancer *48*, 665–673.

Gleave, M., Hsieh, J. T., Gao, C. A., von Eschenbach, A. C. and Chung, L. W. (1991). Acceleration of human prostate cancer growth in vivo by factors produced by prostate and bone fibroblasts. Cancer Res. *51*, 3753– 3761.

Gleave, M., Hsieh, J. T., von Eschenbach, A. C. and Chung, L. W. (1992). Prostate and bone fibroblasts induce human prostate cancer growth in vivo: implications for bidirectional tumor-stromal cell interaction in prostate carcinoma growth and metastasis. J. Urol. *147*, 1151–1159.

Glinsky, G. V. (1993). Cell adhesion and metastasis: is the site specificity of cancer metastasis determined by leukocyte-endothelial cell recognition and adhesion? Crit. Rev. Oncol/Hematol. *14*, 229–278.

Goff, S. P. (1987). Gene isolation by retroviral tagging. Met. Enzimol., *152*, 469–481.

Goldfarb R. H., Albertsson P., Nannmark U. and Kitson R. P. (1998). Cytolytic activities of IL-2 activated NK cells from MMTV/v-Ha-ras transgenic oncomice during tumor progression. In Vivo *12*, 589–592.

Goldman, R. and Bar-Shavit, Z. (1979). Dual effect of normal and stimulated macrophages and their conditioned media on target cell proliferation. J. Natl. Cancer Inst. *63*, 1009–1016.

Goodall, H. and Johnson, M. H. (1982). Use of carboxyfluorescein diacetate to study formation of permeable channels between mouse blastomeres. Nature *295*, 524–526.

Goodall, H. and Johnson, M. H. (1984). The nature of intercellular coupling within the preinplantation mouse embryo. J. Embryol. Exp. Morphol. *79*, 53–76.

Goodall, H. and Maro, B. (1986). Major loss of junctional coupling during mitosis in early mouse embryos. J. Cell Biol. *102*, 568–575.

Gopas, J., Rager-Zisman, B., Bar-Eli, M., Hammerling, G. J. and Segal, S. (1989). The relationship between MHC antigen expression and metastasis. Adv. Cancer Res. *53*, 89–115.

Gordon, P. B., Sussman, H. and Hatcher, V. B. (1983). Long term culture of human endothelial cells. In Vitro *19*, 661–671.

Gordon, E. L., Danielsson, P. E., Nguyen, T. and Winn, H. R. (1991). A comparison of primary cultures of rat cerebral microvascular endothelial cells to rat aortic endothelial cells. In Vitro Cell. Dev. Biol. *27A*, 312–326.

Gorelik, E., Segal, S. and Feldman, M. (1980). Control of lung metastasis progression in mice: role of growth kinetics of 3LL Lewis lung carcinoma and host immune reactivity. J. Natl. Cancer Inst. *65*, 1257–1264.

Gorelik, E., Segal, S., Shapiro, J., Katzav, S., Ron, Y. and Feldman, M. (1982). Interactions between the local tumor and its metastases. Cancer Metastasis Rev. *1*, 83–94.

Gorelik, E., Duty, L., Anaraki, F. and Galili, U. (1995). Alterations of cell surface carbohydrates and inhibition of metastatic property of murine melanomas

by alpha 1,3 galactosyltransferase gene transfection. Cancer Res. *55*, 4168–4173.
Gorelik, E., Xu, F., Henion, T., Anaraki, F. and Galili, U. (1997). Reduction of metastatic properties of BL6 melanoma cells expressing terminal fucose(alpha)1–2-galactose after alpha1,2-fucosyltransferase cDNA transfection. Cancer Res. *57*, 332–336.
Goren, D., Grob, M., Lorenzoni, P. and Burger, M. M. (1997). Human bone cells stimulate the growth of human breast carcinoma cells. Tumor Biol. *18*, 341–349.
Gospodarowicz, D., Moran, J., Braun, D. and Birdwell, C. (1976). Clonal growth of bovine vascular endothelial cells: FGF as a survival agent. Proc. Natl. Acad. Sci. USA *73*, 4120–4124.
Gospodarowicz, D., Greenburg, G., Vlodavsky, I., Alvarado, J. and Johnson, L. K. (1979). The identification and localization of fibronectin in cultured corneal endothelial cells: cell surface polarity and physiological implications. Exp. Eye Res. *29*, 485–509.
Gospodarowicz, D., Gonzalez, R. and Fujii, D. K. (1983). Are factors originating form serum, plasma or cultured cells involved in the growth promoting effect of the ECM produced by cultured bovine corneal EC? J. Cell Physiol. *114*, 191–202.
Gottlieb, C., Baenziger, J. and Kornfeld, S. (1975). Deficient uridine diphosphate-N-acetylglucosamine:glycoprotein N- acetylglucosaminyltransferase activity in a clone of Chinese hamster ovary cells with altered surface glycoproteins. J. Biol. Chem. *250*, 3303–3309.
Graessman, A., Graessman, M. and Mueller, C. (1980). Microinjection of early SV40 DNA fragments and T antigen. Meth. Enzymol. *65*, 816–825.
Graf, J., Ogle, R. C., Robey, F. A., Sasaki, M., Martin, G. R., Yamada, Y. and Kleinman, H. K. (1987a) A pentapeptide from the laminin B1 chain mediates cell adhesion and binds the 67,000 laminin receptor. Biochemistry *26*, 6896–6900.
Graf, J., Iwamoto, Y., Sasaki, M., Martin, G. R., Kleinman, H. K., Robey, F. A. and Yamada, Y. (1987b) Identification of an amino acid sequence in laminin mediating cell attachment, chemotaxis, and receptor binding. Cell *48*, 989–996.
Greller, L. D., Tobin, F. L. and Poste, G. (1996). Tumor heterogeneity and progression: Conceptual foundations for modeling. Invasion Metastasis *16*, 177–208.
Grignani, G., Pacchiarini, L., Almasio, P., Pagliarino, M., Gamba, G., Rizzo, S. C. and Ascari, A. (1986a). Characterization of the platelet-aggregating activity of cancer cells with different metastatic potential. Int. J. Cancer *38*, 237–244.

Grignani, G., Pacchiarini, L. and Pagliarino, M. (1986b). The possible role of blood platelets in tumor growth and dissemination. Haematologica *71*, 245–255.

Grinnell, F. (1986). Focal adhesion sites and the removal of substratum-bound fibronectin. J. Cell Biol. *103*, 2697–2706.

Gualandris, A., Rusnati, M., Belleri, M., Nelli, E. E., Bastaki, M., Molinari-Tosatti, M. P., Bonardi, F., Parolini, S., Albini, A., Morbidelli, L., Ziche, M., Corallini, A., Possati, L., Vacca, A., Ribatti, D. and Presta., M. (1996). Basic fibroblast growth factor overexpression in endothelial cells: an autocrine mechanism for angiogenesis and angioproliferative diseases. Cell Growth Differentiation *7*, 147–160.

Gualberto, A., Aldape, K., Kozakiewicz, K. and Tlsty, T. D. (1998). An oncogenic form of p53 confers a dominant, gain of function phenotype that disrupts spindle checkpoint control. Proc. Natl. Acad. Sci. USA *95*, 5166–5171.

Guise T. A. and Mundy G. R. (1998). Cancer and bone. Endocr. Rev. *19*, 18–54.

Gumkowski, F., Kaminska, G., Kaminski, M., Morissey, L. and Auerbach, R. (1987). Heterogeneity of mouse vascular endothelium: in vitro studies of lymphatic, large blood vessel and microvascular endothelial cells. Blood Vessels *24*, 11–23.

Gunthert, U., Hofmann, M., Rudy, W., Reber, S., Zoller, M., Haussmann, I., Matzku, S., Wenzel, A., Ponta, H. and Herrlich, P. (1991). A new variant of glycoprotein CD44 confers metastatic potential to rat carcinoma cells. Cell *65*, 13–24.

Habets, G. G., Scholtes, E. H., Zuydgeest, D., van der Kammen, R. A., Stam, J. C., Berns, A. and Collard, J. G. (1994). Identification of an invasion-inducing gene, Tiam–1, that encodes a protein with homology to GDP-GTP exchangers for Rho-like proteins. Cell *77*, 537–549.

Hagmar, B. and Ryd, W. (1977). Tumor cell locomotion: A factor in metastasis formation? Influence of cytochalasin B on a tumor dissemination pattern. Int. J. Cancer *19*, 576–580.

Hahn, K., DeBiasio, R. and Taylor, D. L. (1992). Patterns of elevated free calcium and calmodulin activation in living cells. Nature *359*, 736–738.

Hakomori, S. (1994). Novel endothelial cell activation factor(s) released from activated platelets which induce E-selectin expression and tumor cell adhesion to endothelial cells: a preliminary note. Biochem. Biophys. Res. Comm. *203*, 1605–1613.

Hakomori, S. (1996). Tumor malignancy defined by aberrant glycosilation and sphingo(glyco)lipid metabolism. Cancer Res. *56*, 5309–5318.

Hall, S. J. and Thompson, T. C. (1997). Spontaneous but not experimental metastatic activities differentiate primary tumor-derived vs metastasis-derived mouse prostate cancer cell lines. Clin. Exptl. Metastasis *15*, 630–638.

Hamada, J., Takeichi, N. and Kobayashi, H. (1988). Metastatic capacity and intercellular communication between normal cells and metastatic cell clones derived from a rat mammary carcinoma. Cancer Res. *48*, 5129–5132.

Hamada, J., Cavanaugh, P. G., Lotan, O. and Nicolson, G. L. (1992). Separable growth and migration factors for large-cell lymphoma cells secreted by microvascular endothelial cells derived from target organs for metastasis. Br. J. Cancer *66*, 349–354.

Hamburger, A. W., Jones, S. E. and Salmon, S. E. (1980). Soft-agar cloning of cells from patients with lymphoma. Prog. Clin. Biol. Res. *48*, 43–52.

Hammer, D. A. and Apte, S. M. (1992). Simulation of cell rolling and adhesion on surfaces in shear flow: general results and analysis of selectin-mediated neutrophil adhesion. Biophys. J. *63*, 35–57.

Hanahan, D. and Folkman, J. (1996). Patterns and emerging mechanisms of the angiogenic switch during tumorigenesis. Cell *86*, 353–364.

Hanna N. (1982). Role of natural killer cells in control of cancer metastasis. Cancer Metastasis Rev. *1*, 45–65.

Hanna, N. (1985). The role of natural killer cells in the control of tumor growth and metastasis. Biochim. Biophys. Acta *780*, 213–226.

Hanna, N. and Fidler, I. J. (1980). The role of natural killer cells in the destruction of circulating tumor emboli. J. Natl. Cancer Inst. *65*, 801–809.

Hanna, N. and Schneider, M. (1983). Enhancement of tumor metastasis and suppression of natural killer cell activity by β-estradiol treatment. J. Immunol. *130*, 974–980.

Hanna, N., Davis, T. W. and Fidler, I. J. (1982). Environmental and genetic factors determine the level of NK activity of nude mice and affect their suitability as models for experimental metastasis. Intl. J. Cancer *30*, 371–376.

Hansen, M. B., Nielsen, S. E. and Berg, K. (1989). Re-examination and further development of a precise and rapid dye method for measuring cell growth/cell kill. J. Immunol. Meth. *119*, 203–210.

Hanss, M. (1983). Erythrocyte filtrability measurement by the initial flow rate method. Biorheology *20*, 199–211.

Haq, M., Goltzman, D., Tremblay, G. and Brodt, P. (1992). Rat prostate adenocarcinoma cells disseminate to bone and adhere preferentially to bone-marrow-derived endothelial cells. Cancer Res. *52*, 4613–4619.

Haranaka, K., Satomi, N. and Sakurai, A. (1984). Antitumor activity of murine tumor necrosis factor (TNF). against transplanted murine tumors and heterotransplanted human tumors in nude mice. Int. J. Cancer *34*, 263–267.

Hart, I. R. and Fidler, I. J. (1980). Role of organ selectivity in the determination of metastatic patterns of B16 melanoma. Cancer Res. *40*, 2281–2287.

Hart, I. R. and Fidler, I. J. (1981). The implications of tumor heterogeneity for studies on the biology and therapy of cancer metastasis. Biochim. Biophys. Acta *651*, 37–50.

Hart, I. R., Raz, A. and Fidler, I. J. (1980). Effect of cytoskeleton-disrupting agents on the metastatic behavior of melanoma cells. J. Natl. Cancer Inst. *64*, 891–900.

Hart, I. R., Talmadge, J. E. and Fidler, I. J. (1981). Metastatic behavior of a murine reticulum cell sarcoma exhibiting organ-specific growth. Cancer Res. *41*, 1281–1287.

Hart, I. R., Birch, M. and Marshall, J. F. (1991). Cell adhesion receptor expression during melanoma progression and metastasis. Cancer Metastasis Rev. *10*, 115–128.

Hasday, J. D., Shah, E. M. and Lieberman, A. P. (1990). Macrophage tumor necrosis factor-alpha release is induced by contact with some tumors. J. Immunol. *145*, 371–379.

Hay, E. D. (1982). Cell Biology of Extracellular Matrix. Plenum Press, New York.

Hedrick, L., Cho, K. R. and Vogelstein, B. (1993). Cell adhesion molecules as tumor suppressors. Trends Cell Biol. *3*, 36–39.

Heider, K. H., Dammrich, J., Skroch-Angel, P., Muller-Hermelink, H. K., Vollmers, H. P., Herrlich, P. and Ponta, H. (1993a) Differential expression of CD44 splice variants in intestinal- and diffuse-type human gastric carcinomas and normal gastric mucosa. Cancer Res. *53*, 4197–4203.

Heider, K. H., Hofmann, M., Hors, E., van den Berg, F., Ponta, H., Herrlich, P. and Pals, S. T. (1993b) A human homologue of the rat metastasis-associated variant of CD44 is expressed in colorectal carcinomas and adenomatous polyps. J. Cell Biol. *120*, 227–233.

Heim, S., Teixeira, M. R., Dietrich, C. U. and Pandis, N. (1997). Cytogenetic polyclonality in tumors of the breast. Cancer Genet. Cytogenet. *95*, 16–19.

Heitjan, D. F., Manni, A. and Santen, R. J. (1993). Statistical analysis of in vivo tumor growth experiments. Cancer Res. *53*, 6042–6050.

Hejna, M., Raderer, M. and Zielinski, C. C. (1999). Inhibition of metastases by anticoagulants. J. Natl. Cancer Inst. *91*, 22–36.

Hendrix, M. J. C., Seftor, E. A., Seftor, R. E. B., Misiorowski, R. L., Saba, P. Z., Sundareshan, P. and Welch, D. R. (1989). Comparison of tumor cell invasion assays: human amnion versus reconstituted basement membrane barriers. Invasion and Metastasis *9*, 278–297.

Heppner, G. (1984). Tumor heterogeneity. Cancer Res. *44*, 2259–2265.

Heppner G. H. and Miller F. R. (1997). The cellular basis of tumor progression. Int. Rev. Cytol. *177*, 1–56.

Heppner G. H., Dexter D. L., DeNucci T., Miller F. R. and Calabresi P. (1978). Heterogeneity in drug sensitivity among tumor cell subpopulations of a single mammary tumor. Cancer Res. *38*, 3758–3763.

Herlyn, M., Rodeck, U., Mancianti, M., Cardillo, F. M., Lang, A., Ross, A. H., Jambrosic, J. and Koprowski, H. (1987). Expression of melanoma- associated antigens in rapidly dividing human melanocytes in culture. Cancer Res. *47*, 3057–3061.

Herlyn, M., Kath, R., Williams, N., Valyi-Nagy, I. and Rodeck, U. (1990). Growth-regulatory factors for normal, premalignant, and malignant human cells in vitro. Adv. Cancer Res. *54*, 213–234.

Herman, M. A., Ch'ng, Q., Hettenbach, S. M., Ratliff, T. M., Kenyon, C. and Herman, R. K. (1999). EGL–27 is similar to a metastasis-associated factor and controls cell polarity and cell migration in C. elegans. Development *126*, 1055–1064.

Herrlich, P., Sleeman, J., Wainwright, D., Konig, H., Sherman, L., Hilberg, F. and Ponta, H. (1998). How tumor cells make use of CD44. Cell Adhes. Commun. *6*, 141–147.

Herrmann, F., Schmidt, R. E., Ritz, J. and Griffin, J. D. (1987). In vitro regulation of human hematopoiesis by natural killer cells: analysis at a clonal level. Blood *69*, 246–254.

Hertl, W., Ramsey, W. S. and Nowlan, E. D. (1984). Assessment of cell substrate adhesion by a centrifugal method. In Vitro *20*, 796–801.

Heussen C. and Dowdle, E. B. (1980). Electrophoretic analysis of plasminogen activators in polyacrylamide gels containing sodium dodecyl sulfate and co-polymerized substrates, Anal. Biochem. *102*, 196–202

Hilgard, P., Heller, H. and Schmidt, C. G. (1976). The influence of platelet aggregation inhibitors on metastasis formation in mice (3LL). Z. Krebsforsch. *86*, 243–250.

Hill, R. P., Chambers, A. F., Ling, V. and Harris, J. F. (1984). Dynamic heterogeneity: rapid generation of metastatic variants in mouse B16 melanoma cells. Science *224*, 998–1001.

Hocking, A. M., Shinomura, T. and McQuillan, D. J. (1998). Leucine-rich repeat glycoproteins of the extracellular matrix. Matrix Biol. *17*, 1–19.

Hoffman R. M. (1994). Orthotopic is orthodox: why are orthotopic-transplant metastatic models different from all other models? J. Cell Biol. *56*, 1–3.

Hoffmann, J., Schirner, M., Menrad, A. and Schneider, M. R. (1997). A highly sensitive model for quantification of in vivo tumor angiogenesis induced by alginate-encapsulated tumor cells. Cancer Res. *57*, 3847–3851.

Hofmann, M., Rudy, W., Zoller, M., Tolg, C., Ponta, H., Herrlich, P. and Gunthert, U. (1991). CD44 splice variants confer metastatic behavior in rats: homol-

ogous sequences are expressed in human tumor cell lines. Cancer Res. *51*, 5292–5297.
Holzmann, B., Gosslar, U. and Bittner, M. (1998). Alpha 4 integrins and tumor metastasis. Curr. Top. Microbiol. Immunol. *231*, 125–141.
Honn, K. V. and Tang, D. G. (1992). Adhesion molecules and tumor cell interaction with endothelium and subendothelial matrix. Cancer Metastasis Rev. *11*, 353–375.
Honn, K. V., Cicone, B. and Skoff, A. (1981). Prostacyclin: a potent antimetastatic agent. Science *212*, 1270–1272.
Honn, K. V., Busse, W. D. and Sloane, B. F. (1983). Prostacyclins and thromboxanes: implications for their role in tumor cell metastasis. Biochem. Pharmacol. *32*, 1–11.
Honn, K. V., Tang, D. G. and Chen, Y. Q. (1992a) Platelets and cancer metastasis: more than an epiphenomenon. Semin. Thromb. Haemost. *18*, 392–415.
Honn, K. V., Tang, D. G. and Crissman, J. D. (1992b). Platelets and cancer metastasis: a causal relationship? Cancer Metastasis Rev. *11*, 325–351.
Honn, K. V., Tang, D. G., Gao, X., Butovich, I., Liu, B., Timar, J. and Hagmann, W. (1994). 12-lypoxigenase and 12(S)-HETE: role in cancer metastasis. Cancer Metastasis Rev. *13*, 365–396.
Horak, E., Darling, D. L. and Tarin, D. (1986). Analysis of organ- specific effects on metastatic tumor formation by studies in vitro. J. Natl. Cancer Inst. *75*, 913–922.
Horwitz, D. A. and Garrett, M. A. (1971). Use of leukocyte chemotaxis in vitro to assay mediators generated by immune reactions. I. Quantitation of mononuclear and polymorphonuclear leukocyte chemotaxis with polycarbonate (nuclepore) filters. J. Immunol. *106*, 649–655.
House, S. D. and Lipowsky, H. H. (1987). Leukocyte-endothelium adhesion: microhemodynamics in mesentery of the cat. Microvasc. Res. *34*, 363–379.
Howard, J. M. (1963). Studies of autotransplantation of incurable cancer. Surg. Gynecol. Obstet. *117*, 567–572.
Howard, R. B., Chu, H., Zeligman, B. E., Marcell, T., Bunn, P. A., McLemore, T. L., Mulvin, D. W., Cowen, M. E. and Johnston, M. R. (1991). Irradiated nude rat model for orthotopic human lung cancers. Cancer Res. *51*, 3274–3280.
Howe, A., Aplin, A. E., Alahari, S. K. and Juliano, R. L. (1998). Integrin signaling and cell growth control. Curr. Opin. Cell Biol. *10*, 220–231.
Humphries, M. J. and Olden, K. (1989). Asparagine-linked oligosaccharides and tumor metastasis. Pharmac. Ther. *44*, 85–105.
Humphries, M. J., Matsumoto, K., White, S. L. and Olden, K. (1986a). Oligosaccharide modification by swainsonine treatment inhibits pulmonary colonization by B16-F10 murine melanoma cells. Proc. Natl. Acad. Sci. USA *83*, 1752–1756.

Humphries, M. J., Olden, K. and Yamada, K. M. (1986b) A synthetic peptide from fibronectin inhibits experimental metastasis of murine melanoma cells. Science *233*, 467–470.

Hutchinson, R., Fligiel, S., Appleyard, J., Varani, J. and Wicha, M. (1989). Attachment of neuroblastoma cells to ECM: correlation with metastatic capacity. J. Lab. Clin. Med. *113*, 561–568.

Hynes, R. O. (1976). Cell surface proteins and malignant transformation. Biochim. Biophys. Acta *458*, 73–107.

Hynes, N. E. and Stern, D. F. (1994). The biology of erbB–2/neu/HER- -2 and its role in cancer. Biochim. Biophys. Acta *1198*, 165–184.

Ibrahiem, E. H. I., Nigam, V. N., Brailovsky, C. A., Madarnas, P. and Elhilali, M. (1983). Orthotopic implantation of primary (N-[4-(5-nitro-furyl)-2-thiazolyl]formamide-induced bladder cancer in bladder submucosa: an animal model for bladder cancer study. Cancer Res. *43*, 617–622.

Ilic, D., Almeida, E. A. C., Schlaepfer, D. D., Dazin, P., Aizawa, S., Damsky, C. H. (1998). Extracellular matrix survival signals transduced by focal adhesion kinase suppress p53-mediated apoptosis. J. Cell Biol. *143*, 547–560.

Imai, Y., Inoue, T. and Ishikawa, T. (1998). Mutations of the human MUT S homologue 6 gene in ampullary carcinoma and gastric cancer. Int. J. Cancer *78*, 576–580.

Imhof, B. A. and Dunon, D. (1995). Leukocyte migration and adhesion. Adv. Immunol. *58*, 345–416.

Inoue, T., Cavanaugh, P. G., Steck, P. A., Brunner, N. and Nicolson, G. L. (1993). Differences in transferrin response and numbers of transferrin receptors in rat and human mammary carcinoma lines of different metastatic potentials. J. Cell Physiol. *56*, 212–217.

Irimura, T., Gonzales, R. and Nicolson, G. L. (1981). Effects of tunicamycin on metastatic B16 melanoma cell surface glycoproteins and blood-borne arrest and survival properties. Cancer Res. *41*, 3411–3418.

Irving, M., Roll, J., Huang, S. and Bissel, D. (1984). Characterization and culture of sinusoidal endothelium from normal rat liver: lipoprotein uptake and collagen phenotype. Gastroenterology *87*, 1233–1247.

Isaacs, J. T., Heston, W. D. W., Weissman, R. M. and Coffey, D. S. (1978). Animal models of the hormone-sensitive and -insensitive prostatic adenocarcinomas, Dunning R-3327-H, R-3327-HI, and R-3327-AT. Cancer Res. *38*, 4353–4359.

Ishikawa, M., and Kerbel, R. S. (1989). Characterization of a metastasis-deficient lectin-resistant human melanoma mutant. Intl. J. Cancer *43*, 134–139.

Ishikawa, M., Fernandez, B. and Kerbel, R. S. (1988). Highly pigmented human melanoma variants which metastasizes widely in nude mice, including to skin and brain. Cancer Res. *48*, 4897–4903.

Isoai, A., Giga-Hama, Y., Sinkai, K., Mukai, M., Akedo, H. and Kumagai, H. (1990). Purification and characterization of tumor invasion-inhibiting factors. Jpn. J. Cancer Res. *81*, 909–914.

Iurlaro, M., Benelli, R., Masiello, L., Rosso, M. and Albini, A. (1998). Beta Interferon inhibits HIV–1 Tat induced angiogenesis; synergism with 13-cis Retinoic Acid. Eur. J. Cancer *34*, 570–576.

Iwamoto, Y., Robey, F. A., Graf, J., Sasaki, M., Kleinman, H. K., Yamada, Y. and Martin, G. R. (1987). YIGSR, a synthetic laminin pentapeptide, inhibits experimental metastasis formation. Science *238*, 1132–1134.

Iwamoto, Y., Nomizu, M., Yamada, Y., Ito, Y., Tanaka, K. and Sugioka, Y. (1996). Inhibition of angiogenesis, tumour growth and experimental metastasis of human fibrosarcoma cells HT1080 by a multimeric form of the laminin sequence Tyr-Ile-Gly-Ser-Arg (YIGSR). Br. J. Cancer *73*, 589–595.

Jaffe, E. A., Nachman, R. L., Becker, C. G. and Minick, C. R. (1973). Culture of human endothelial cells derived from umbilical veins. Identification by morphologic and immunologic criteria. J. Clin. Invest. *52*, 2745–2756.

Jain, R. K., Schlenger, K., Hockel, M. and Yuan, F. (1997). Quantitative angiogenesis assays: Progress and problems. Nat. Med. *3*, 1203–1208.

Jakob, W. and Voss, K. (1984). Utilization of image analysis for the quantification of vascular responses in the chick chorioallantoic membrane. Exp. Pathol. *26*, 93–99.

Jakob, W., Jentzsch, K. D., Manersberger, B. and Heider, G. (1978). The chick chorioallantoic membrane as bioassay for angiogenesis factors: reactions induced by carrier materials. Exp. Pathol. *15*, 241–249.

Janik, P., Briand, P. and Hartmann, N. R. (1975). The effect of estrone-progesterone treatment on cell proliferation kinetics of hormone- dependent GR mouse mammary tumors. Cancer Res. *35*, 3698–3704.

Jeffers, M., Rong, S., Woude, G. F. (1996). Hepatocyte growth factor/scatter factor-Met signaling in tumorigenicity and invasion/metastasis. J. Mol. Med. *74*, 505–513.

Jiang, D., Yang, H., Willson, J. K. V., Liang, J., Humphrey, L. E., Zborowska, E., Wang, D., Foster, J., Fan, R., Brattain, M. G. (1998). Autocrine transforming growth factor alpha provides a growth advantage to malignant cells by facilitating re-entry into the cell cycle from suboptimal growth states. J. Biol. Chem. *47*, 31471–31479.

Jiang, W. G. and Hiscox, S. (1997). Hepatocyte growth factor/scatter factor, a cytokine playing multiple and converse roles. Histol. Histopathol. *12*, 537–555.

Jiang, W. G., Lloyds, D., Puntis, M. and Hallet, M. (1993). Monocyte-conditioned media possess a novel factor which increases motility of cancer cells. Int. J. Cancer *53*, 426–431.

Johnson, J. P. (1991). Cell adhesion molecules of the immunoglobulin supergene family and their role in malignant transformation and progression to metastatic disease. Cancer Metastasis Rev. *10*, 11–22.

Johnson, J. P., Stade, B. G., Holzmann, B., Schwaeble, W. and Riethmueller, G. (1989). De novo expression of intercellular adhesion molecule 1 in melanoma correlates with increased risk of metastasis. Proc. Natl. Acad. Sci. USA *86*, 641–644.

Jouanneau, J., Gavrilovich, J., Caruelle, D., Jaye, M., Moens, G., Caruelle, J. P. and Thiery, J. P. (1991). Secreted or non secreted forms of aFGF produced by transfected epithelial cells influence cell morphology, motility, and invasive potential. Proc. Natl. Acad. Sci. USA *88*, 2893–2897.

Juacaba, S. F., Horak, E., Price, J. E. and Tarin, D. (1989). Tumor cell dissemination patterns and metastasis of murine mammary carcinoma. Cancer Res. *49*, 570–575.

Juhasz, I., Albelda, S. M., Elder, D. E., Murphy, G. F., Adachi, K., D., Valyi-Nagy, I. T. and Herlyn, M. (1993). Growth and invasion of human melanomas in human skin grafted into immunodeficient mice. Am. J. Pathol. *143*, 528–537.

Juliano, R. (1994). Signal transduction by integrins and its role in the regulation of tumor growth. Cancer Metastasis Rev. *13*, 25–30.

Justice, A. (1985). Review of the effects of stress on cancer in laboratory animals: importance of the time of stress application and tipe of tumor. Psychol. Bull. *98*, 108–138.

Kaido, T., Bandu, M. T., Maury, C., Ferrantini, M., Belardelli, F. and Gresser, I. (1995). IFN-alpha 1 gene transfection completely abolishes the tumorigenicity of murine B16 melanoma cells in allogeneic DBA/2 mice and decreases their tumorigenicity in syngeneic C57BL/6 mice. Int. J. Cancer *60*, 221–229.

Kamibayashi, Y., Oyamada, Y., Mori, M. and Oyamada, M. (1995). Aberrant expression of gap junction proteins (connexins) is associated with tumor progression during multistage mouse skin carcinogenesis in vivo. Carcinogenesis *16*, 1287–1297.

Kanemoto, T., Reich, R., Royce, L., Greatorex, D., Adler, S. H., Shiraishi, N., Martin, G. R., Yamada, Y. and Kleinman, H. K. (1990). Identification of an amino acid sequence from the laminin A chain that stimulates metastasis and collagenase IV production. Proc. Natl. Acad. Sci. USA *87*, 2279–2283.

Kannagi, R. (1997). Carbohydrate-mediated cell adhesion involved in hematogenous metastasis of cancer. Glycoconj. J. *14*, 577–584.

Kapoun, A. M. and Shackleford, G. M. (1997). Preferential activation of Fgf8 by proviral insertion in mammary tumors of Wnt1 transgenic mice. Oncogene *14*, 2985–2989.

Karpatkin, S., Pearlstein, E., Ambrogio, C. and Coller, B. S. (1988). Role of adhesive proteins in platelet tumor interaction in vitro and metastasis formation in vivo. J. Clin. Invest. *81*, 1012–1019.

Kasper, S., Sheppard, P. C., Yan, Y., Pettigrew, N., Borowsky, A. D., Prins, G. S., Dodd, J. G., Duckworth, M. L. and Matusik, R. J. (1998). Development, progression, and androgen-dependence of prostate tumors in probasin-large T antigen transgenic mice: a model for prostate cancer. Lab. Invest. *78*, 1–15.

Kath, R., Rodeck, U., Parmiter, A., Jambrosic, J. and Herlyn, M. (1990). Growth factor independence in vitro of primary melanoma cells from advanced but not early or intermediate lesions. Cancer Ther. Control *1*, 179–191.

Kaufman, R. J., Davies, M. V., Wasley, L. C. and Michnick, D. (1991). Improved vectors for stable expression of foreign genes in mammalian cells by use of the untranslated leader sequence from EMC virus. Nucleic Acids Res. *19*, 4485–4490.

Kaufmann, A. M., Lichtner, R. B., Schirrmacher, V. and Khazaie, K. (1996). Induction of apoptosis by EGF receptor in rat mammary adenocarcinoma cells coincides with enhanced spontaneous tumour metastasis. Oncogene *13*, 2349–2358.

Kawaguchi, T., Kawaguchi, M., Miner, K. M., Lembo, T. M. and Nicolson, G. L. (1983). Brain meninges tumor formation by in vivo-selected metastatic B16 melanoma variants in mice. Clin. Exp. Metastasis *1*, 247–259.

Kawamata, H., Kameyama, S., Kawai, K., Tanaka, Y., Nan, L., Barch, D. H., Stetler-Stevenson, W. G. and Oyasu, R. (1995a) Marked acceleration of the metastatic phenotype of a rat bladder carcinoma cell line by the expression of human gelatinase A. Intl. J. Cancer *63*, 568–575.

Kawamata, H., Kawai, K., Kameyama, S., Johnson, M. D., Stetler-Stevenson, W. G. and Oyasu, R. (1995b) Over-expression of tissue inhibitor of matrix metalloproteinases (TIMP1 and TIMP2). suppresses extravasation of pulmonary metastasis of a rat bladder carcinoma. Intl. J. Cancer *63*, 680–687.

Keely, P., Parise, L. and Juliano, R. (1998). Integrins and GTPases in tumour cell growth, motility and invasion. Trends Cell Biol. *8*, 101–106.

Kelsall, S. R. and Mintz, B. (1998). Metastatic cutaneous melanoma promoted by ultraviolet radiation in mice with transgene-initiated low melanoma susceptibility. Cancer Res. *58*, 4061–4065.

Kerbel, R. S. and Man, M. S. (1984). Single-step selection of unique human melanoma variants displaying unusually aggressive metastatic behavior in nude athymic mice. Invasion Metastasis *4*(Suppl. 1), 31–43.

Kerbel, R. S., Twiddy, R. R. and Robertson, D. M. (1978). Induction of a tumor with greatly increased metastatic growth potential by injection of cells from a low-metastatic H-2 heterozygous tumor cell line into an H-2 incompatible parental strain. Intl. J. Cancer *22*, 583–594.

Kerbel, R. S., Waghorne, C., Man, M. S., Elliott, B. and Breitman, M. L. (1987). Alteration of the tumorigenic and metastatic properties of neoplastic cells is associated with the process of calcium phosphate-mediated DNA transfection. Proc. Natl. Acad. Sci. USA *84*, 1263–1267.

Kerbel, R. S., Cornil, I. and Korczak, B. (1989). New insights into the evolutionary growth of tumors revealed by Southern gel analysis of tumors genetically tagged with plasmid or proviral DNA insertions. J. Cell Sci. *94*, 381–387.

Kerbel, R. S., Cornil, I. and Theodorescu, D. (1991). Importance of orthotopic transplantation procedures in assessing the effects of transfected genes on human tumor growth and metastasis. Cancer Metastasis Rev. *10*, 201–215.

Kern, P., Knedler, A. and Eckel, R. (1983). Isolation and culture of microvascular endothelium from human adipose tissue. J. Clin. Invest. *71*, 1822–1829.

Key, M. E. (1983). Macrophages in cancer metastases and their relevance to metastatic growth. Cancer Metastasis Rev. *2*, 75–88.

Kibbey, M. C., Grant, D. S. and Kleinman, H. K. (1992). Role of the SIKVAV site of laminin in promotion of angiogenesis and tumor growth: an in vivo Matrigel model. J. Natl. Cancer Inst. *84*, 1633–1638.

Kidder, G. M., Rains, J. and McKeon J. (1987). Gap junction assembly in the preimplantation mouse conceptus is independent of microtubules, microfilaments, cell flattening and cytokines. Proc. Natl. Acad. Sci. USA *84*, 3718–3722.

Kieda, C. (1998). Role of lectin-glycoconjugate recognition in cell- cell interactions leading to tissue invasion. Adv. Exp. Med. Biol. *435*, 75–82.

Kikuchi, Y. (1991). Deformability of mammalian and fish erythrocytes: comparison of mean pore transit times of cells and estimation of cellular viscosity and elasticity. Jap. J. Physiol. *41*, 907–922.

Killion, J. J. and Fidler, I. J. (1998). Therapy of cancer metastasis by tumoricidal activation of tissue macrophages using liposome-encapsulated immunomodulators. Pharmacol. Ther. *78*, 141–154.

Kim, J., Yu, W., Kovalski, K. and Ossowski, L. (1998). Requirement for specific proteases in cancer cell intravasation as revealed by a novel semiquantitative PCR-based assay. Cell *94*, 353–362.

Kim, W. H., Jun, S. H., Kibbey, M. C., Thompson, E. W. and Kleinman, H. K. (1994). Expression of beta 1 integrin in laminin-adhesion-selected human colon cancer cell lines of varying tumorigenicity. Invasion Metastasis *14*, 147–155.

Kinsey, D. L. (1960). An experimental study of preferential metastasis. Cancer *13*, 674–676.

Kleinman, H. K., McGarvey, M. L., Hassel, J. R., Star, V. L., Cannon, F. B., Laurie, G. W. and Martin, G. R. (1986). Basement membrane complexes with biological activity. Biochemistry *25*, 312–318.

Klemke, R. L., Yebra, M., Bayna, E. M. and Cheresh, D. A. (1994). Receptor tyrosine kinase signaling required for integrin avb5-directed cell motility but not adhesion on vitronectin. J. Cell Biol. *127*, 859–866.

Klepfish, A., Greco, M. A. and Karpatkin, S. (1993). Thrombin stimulates melanoma tumor cell binding to endothelial cells and subendothelial matrix. Int. J Cancer *53*, 978–982.

Knauer, D. J. and Cunningham, D. D. (1983). A reevaluation of the response of human umbilical vein endothelial cells to certain growth factors. J. Cell Physiol. *117*, 397–406.

Knighton, D. R., Ausprunk, D. H., Tapper, D. and Folkman, J. (1977). Avascular and vascular phases of tumor growth in the chick embryo. Br. J. Cancer *35*, 347–356.

Knighton, D. R., Fiegel, V. D. and Phillips, G. D. (1991). The assays for angiogenesis. Progr. Clin. Biol. Res. *365*, 291–299.

Knox, J. D., Mack, C. F., Powell, W. C., Bowden, G. T. and Nagle, R. B. (1993). Prostate tumor cell invasion: A comparison of orthotopic and ectopic models. Invasion Metastasis *13*, 325–331.

Kobayashi, K., Nakanishi, H., Inada, K., Fujimitsu, Y., Yamachika, T., Shirai, T. and Tatematsu, M. (1996). Growth characteristics in the initial stage of micrometastasis formation by bacterial LacZ gene-tagged rat prostatic adenocarcinoma cells. Jpn. J. Cancer Res. *87*, 1227–1234

Koch, F., Kaempgen, E., Schuler, G. and Romani, N. (1992). Effective enrichment of murine epidermal Langerhans cells by a modified (mismatched) panning technique. J. Invest. Dermatol. *99*, 803–807.

Kohn, E. C. and Liotta, L. A. (1995). Molecular insights into cancer invasion: Strategies for prevention and intervention. Cancer Res. *55*, 1856–1862.

Koike, A., Moore, G. E., Mendoza, C. B. and Watne, A. L. (1963). Heterologous, homologous and autologous transplantation of human tumors. Cancer *16*, 1065–1071.

Kojima, N., Shiota, M., Sadahira, Y., Handa, K. and Hakomori, S. (1992). Cell adhesion in a dynamic flow system as compared to static system. Glycosphingolipid-glycosphingolipid interaction in the dynamic system predominates over lectin- or integrin-based mechanisms in adhesion of B16 melanoma cells to non-activated endothelial cells. J. Biol. Chem. *267*, 17264–17270.

Kooistra, A., Romijn, J. C. and Schroder, F. H. (1997). Stromal inhibition of epithelial cell growth in the prostate; overview of an experimental study. Urol. Res. *25*, S97–S105.

Koop, S., MacDonald, I. C., Luzzi, K., Schmidt, E. E., Morris, V. L., Grattan, M., Khokha, R., Chambers, A. F. and Groom, A. C. (1995). Fate of melanoma cells entering the microcirculation: Over 80% survive and extravasate. Cancer Res. 55, 2520–2523.

Koop, S., Schmidt, E. E., MacDonald, I. C., Morris, V. L., Khokha, R., Grattan, M., Leone, J., Chambers, A. F. and Groom, A. C. (1996). Independence of metastatic ability and extravasation: Metastatic ras-transformed and control fibroblasts extravasate equally well. Proc. Natl. Acad. Sci. USA 93, 11080–11084.

Korczak, B., Robson, I. B., Lamarche, C., Bernstein, A. and Kerbel, R. S. (1988). Genetic tagging of tumor cells with retrovirus vectors: clonal analysis of tumor growth and metastasis in vivo. Mol. Cell. Biol. 8, 3143–3149.

Korzeniewski, C. and Callewaert, D. M. (1983). An enzyme-release assay for natural cytotoxicity. J. Immunol. Meth. 64, 313–320.

Kramer, R. H. and Nicolson, G. L. (1979). Interactions of tumor cells with vascular endothelial cell monolayers: a model for metastatic invasion. Proc. Natl. Acad. Sci. USA 76, 5704–5708.

Kripke, M. L., Gruys, E. and Fidler, I. J. (1978). Metastatic heterogeneity of cells from an ultraviolet light induced murine fibrosarcoma of recent origin. Cancer Res. 38, 2962–2967.

Kruger, A., Sanchez-Sweatman, O. H., Martin, D. C., Fata, J. E., Ho, A. T., Orr, F. W., Ruther, U. and Khokha, R. (1998). Host TIMP–1 overexpression confers resistance to experimental brain metastasis of a fibrosarcoma cell line. Oncogene 16, 2419–2423.

Krutovskikh, V. A., Yamasaki, H., Tsuda, H. and Asamoto, M. (1998). Inhibition of intrinsic gap-junction intercellular communication and enhancement of tumorigenicity of the rat bladder carcinoma cell line BC31 by a dominant-negative connexin 43 mutant. Mol. Carcinog. 23, 254–261.

Kubota, T. (1994). Metastatic models of human cancer xenografted in the nude mouse: The importance of orthotopic transplantation. J. Cell. Biochem. 56, 4–8.

Kueng, W., Silber, E. and Eppenberger, U. (1989). Quantification of cells cultured on 96-well plates. Anal. Biochem. 182, 16–19.

Kuhnen, C., Tolnay, E., Steinau, H. U., Voss, B. and Muller, K. M. (1998). Expression of c-met receptor and hepatocyte growth factor/scatter factor in synovial sarcoma and epithelioid sarcoma. Virchows Arch. 432, 337–342.

Kurachi, H., Morishige, K., Amemiya, K., Adachi, H., Hirota, K., Miyake, A. and Tanizawa, O. (1991). Importance of TGFα/EGFr autocrine growth mechanism in an ovarian cancer cell line in vivo. Cancer Res. 51, 5956–5959.

Kurebayashi, J., McLeskey, S. W., Johnson, M. D., Lippman, M. E., Dickson, R. B. and Kern, F. G. (1993). Quantitative demonstration of spontaneous metastais by MCF7 human breast cancer cells cotransfected with fibroblast growth factor 4 and LacZ. Cancer Res. 53, 2178–2187.

Kuzu, I., Bicknell, R., Harris, A. L., Jones, M., Gatter, K. C. and Mason, D. Y. (1992). Heterogeneity of vascular endothelial cells with relevance to diagnosis of vascular tumors. J. Clin. Pathol. *45*, 143–148.

Kyriazis, A. A. and Kyriazis, A. P. (1980). Preferential sites of growth of human tumors in nude mice following subcutaneous transplantation. Cancer Res. *40*, 4509–4511.

LaBarba, R. C. (1970). Experiential and environmental factors in cancer. Psychosom. Med. *32*, 258–276.

LaBiche, R. A., Tressler, R. J. and Nicolson, G. L. (1993). Selection for enhanced adhesion to microvessel endothelial cells or resistance to interferon-gamma modulates the metastatic potential of murine RAW117 large-cell lymphoma cells. Clin. Exp. Metastasis *11*, 472–481.

Lacy, E. R., Kuwayama, H., Cowart, K. S., King, J. S., Deutz, A. H. and Sistrunk, S. (1991). A rapid, accurate immunohistochemical method to label proliferating cells in the digestive tract. A comparison with tritiated thymidine. Gastroenterology *100*, 259–262.

Ladanyi, A., Timar, J., Paku, S., Molnar, G. and Lapis, K. (1990). Selection and characterization of human melanoma lines with different liver-colonizing capacity. Int. J. Cancer *46*, 456–461.

Lafrenie, R., Shaughnessy, S. G. and Orr, F. W. (1992a) Cancer cell interactions with injured or activated endothelium. Cancer Metastasis Rev. *11*, 377–388.

Lafrenie, R. M., Podor, T. J., Buchanan, M. R. and Orr, F. W. (1992b). Up-regulated biosynthesis and expression of endothelial cell vitronectin receptor enhances cancer cell adhesion. Cancer Res. *52*, 2202–2208.

Lafrenie, R. M., Gallo, S., Podor, T. J., Buchanan, M. R. and Orr, F. W. (1994). The relative roles of vitronectin receptor, E-selectin and $\alpha_4\beta_1$ in cancer cell adhesion to IL-1-treated endothelial cells. Eur. J. Cancer *30A*, 2151–2158

Lafreniere, R. and Rosenberg, S. A. (1986). A novel approach to the generation and identification of experimental hepatic metastases in a murine model. J. Natl. Cancer Inst. *76*, 309–322.

Lam, W. C., Delikatny, E. J., Orr, F. W., Wass, J., Varani, J. and Ward, P. A. (1981). The chemotactic response of tumor cells. Am. J. Pathol. *104*, 69–76.

Lane, T. A., Lamkin, G. E. and Wancewicz, E. (1989). Modulation of endothelial cell expression of intercellular adhesion molecule 1 by protein kinase C activation. Biochem. Biophys. Res. Comm. *161*, 945–952.

Lang, R. A. and Burgess, A. W. (1990). Autocrine growth factors and tumorigenic transformation. Immunol. Today *11*, 244–249.

Langer, R. and Folkman, J. (1976). Polymers for the sustained release of proteins and other macromolecules. Nature *363*, 475–482.

Lapis, K., Paku, S. and Liotta, L. A. (1988). Endothelialization of embolized tumor cells during metastasis formation. Clin. Exp. Metastasis *6*, 73–89.

Latif, F., Tory, K., Gnarra, J., Yao, M., Duh, F. M., Orcutt, M. L., Stackhouse, T., Kuzmin, I., Modi, W., Geil, L. et al. (1993). Identification of the von Hippel-Landau disease tumor suppressor gene. Science 260, 1317–1320.

Lauffenburger, D., Rothman, C. and Zigmond, S. H. (1983). Measurement of leukocyte motility and chemotaxis parameters with a linear under-agarose migration assay. J. Immunol. 131, 940–947.

Lauri, D., Needham, L., Martin-Padura, I. and Dejana, E. (1991). Tumor cell adhesion to endothelial cells: ELAM-1 as an inducible adhesive receptor specific for colon carcinoma cells. J. Natl. Cancer Inst. 83, 1321–1324.

Lawrence, M. B. and Springer, T. A. (1991). Leukocytes roll on a selectin at physiologic flow rates: distinction from and prerequisite for adhesion through integrins. Cell 65, 859–873.

Lawrence, M. B. and Springer, T. A. (1993). Neutrophils roll on E-selectin. J. Immunol. 151, 6338–6346.

Lawrence, M. B., McIntire, L. V. and Eskin, S. G. (1987). Effect of flow on PMN leukocytes/endothelial cell adhesion. Blood 70, 1284–1290.

Lawrence, M. B., Smith, C. W., Eskin, S. G. and McIntire, L. V. (1990). Effect of venous shear stress on CD18-mediated neutrophil adhesion to cultured endothelium. Blood 75, 227–237.

Layton, M. G. and Franks, L. M. (1986). Selective suppression of metastasis but not tumorigenicity of a mouse lung carcinoma by cell hybridization. Int. J. Cancer 37, 723–730.

Leber, T. M. and Balkwill, F. R. (1998). Regulation of monocyte MMP-9 production by TNF-alpha and a tumour-derived soluble factor (MMPSF). Br. J. Cancer 78, 724–732.

Lee, K., Tanaka, M., Hatanaka, M. and Kuze, F. (1987). Reciprocal effects of epidermal growth factor and transforming growth factor β on the anchorage-dependent and -independent growth of A431 epidermoid carcinoma cells. Exp. Cell Res. 173, 156–162.

Lee, F. S., Lane, T. F., Kuo, A., Shackleford, G. M. and Leder, P. (1995). Insertional mutagenesis identifies a member of the Wnt gene family as a candidate oncogene in the mammary epithelium of int-2/Fgf-3 transgenic mice. Proc. Natl. Acad. Sci. USA 92, 2268–2272.

Leene, W., Duyzings, M. J. M. and VonSteeg, C. (1973). Lymphoid stem cell identification in the developing thymus and bursa of Fabricius of the chick. Z. Zellforsh 136, 521–533.

Lester, B. R., Weinstein, L. S., McCarthy, J. B., Sun, Z., Smith, R. S. and Furcht, L. T. (1991). The role of G-protein in matrix-mediated motility of highly and poorly invasive melanoma cells. Int. J. Cancer 48, 113–120.

Leung-Tack, J., Capo, C., de Lapeyriere, O., Benoliel, A. M., Arnaud, D. and Bongrand, P. (1988). Relationship between cellular adhesiveness and metasta-

tic activity in polyomavirus-transformed FR3T3 rat cell lines. Int. J. Cancer 42, 946–951.
Levesque, J. P., Hatzfeld, A. and Hatzfeld, J. (1991). Mitogenic properties of major extracellular proteins. Immunol. Today 12, 258–262.
Levy, J. A., White, A. C. and McGrath, C. M. (1982). Growth and histology of a human mammary carcinoma cell line at different sites in the athymic mouse. Br. J. Cancer 45, 375–383.
Levy, J. P., Muldoon, R. R., Zolotukhin, S. and Link, C. J. Jr. (1996). Retroviral transfer and expression of a humanized, red-shifted green fluorescent protein gene into human tumor cells. Nat. Biotechnol. 14, 610–614.
Ley, K. and Tedder, T. F. (1995). Leukocyte interactions with vascular endothelium. New insights into selectin-mediated attachment and rolling. J. Immunol. 155, 525–528.
Li, L., Shin, D. M. and Fidler, I. J. (1990). Ectopic organ environment favors tumorigenicity and production of metastasis by Lewis Lung tumor cells. Invasion Metastasis 10, 129–141.
Li, L., Nicolson, G. L. and Fidler, I. J. (1991). Direct in vitro lysis of metastatic tumor cells by cytokine-activated murine vascular endothelial cells. Cancer Res. 51, 245–254.
Lichtenberg, J., Hansen, C. A., Skak, N. T., Bay, C., Mortensen, J. T. and Binderup, L. (1997). The rat subcutaneous air sac model: a new and simple method for in vivo screening of antiangiogenesis. Pharmacol. Toxicol. 81, 280–284.
Lichtner, R. B., Belloni, P. N. and Nicolson, G. L. (1989). Differential adhesion of metastatic rat mammary carcinoma cells to organ-derived microvessel endothelial cells and subendothelial matrix. Expl. Cell Biol. 57, 146–152.
Liebrich, W., Schlag, P., Manasterski, M., Lehner, B., Stohr, M., Moller, P. and Schirrmacher, V. (1991). In vitro and clinical characterisation of a Newcastle disease virus-modified autologous tumor cell vaccine for treatment of colorectal cancer patients. Eur. J. Cancer 27, 703–710.
Lieubeau-Teillet, B., Rak, J., Jothy, S., Iliopoulos, O., Kaelin, W. and Kerbel, R. S. (1998). von Hippel-Lindau gene-mediated growth suppression and induction of differentiation in renal cell carcinoma cells grown as multicellular tumor spheroids. Cancer Res. 58, 4957–4962.
Lifsted, T., Le Voyer, T., Williams, M., Muller, W., Klein-Szanto, A., Buetow, K. H. and Hunter, K. W. (1998). Identification of inbred mouse strains harboring genetic modifiers of mammary tumor age of onset and metastatic progression. Int. J. Cancer 77, 640–644.
Lin, S., Rusciano, D., Lorenzoni, P., Hartmann, G., Birchmeier, W., Giordano, S., Comoglio, P. and Burger, M. M. (1998). C-met activation is necessary but not sufficient for liver colonization by B16 murine melanoma cells. Clin. Exp. Metastasis 16, 253–265.

Lin, W. C., Pretlow, T. P., Pretlow, T. G. 2d and Culp, L. A. (1990a) Development of micrometastases: earliest events detected with bacterial lacZ gene-tagged tumor cells. J. Natl. Cancer Inst. *82*, 1497–1503.

Lin, W. C., Pretlow, T. P., Pretlow, T. G. 2d and Culp, L. A. (1990b) Bacterial lacZ gene as a highly sensitive marker to detect micrometastasis formation during tumor progression. Cancer Res. *50*, 2808–2817.

Lin, W. C., Pretlow, T. P., Pretlow, T. G. and Culp, L. A. (1992). High resolution analyses of two different classes of tumor cells in situ tagged with alternative histochemical marker genes. Am. J. Pathol. *141*, 1331–1342.

Lindahl, T. (1994). DNA surveillance defect in cancer cells. Current Biol. *4*, 249–251.

Liotta, L. A. (1986). Tumor invasion and metastasis–role of the extracellular matrix. Cancer Res. *46*, 1–7.

Liotta, L. A., Kleinerman, J. and Saidel, G. M. (1974). Quantitative relationships of intravascular tumor cells, tumor vessels, and pulmonary metastases following tumor implantation. Cancer Res. *34*, 997–1004.

Liotta, L. A., Kleinerman, E. J. and Saidel, G. M. (1976). The significance of hematogenous tumor cell clumps in the metastatic process. Cancer Res. *36*, 889–894.

Liotta, L. A., Vembu D., Saini R. K. and Boone C. (1978). In vivo monitoring of the death rate of artificial murine pulmonary micrometastases. Cancer Res. *38*, 1231–1236.

Liotta, L. A., Tryggvason, K., Garbisa, S., Hart, I., Foltz, C. M. and Shafie, S. (1980a) Metastatic potential correlates with enzymatic degradation of basement membrane collagen. Nature *284*, 67–68.

Liotta, L. A., Lee, C. W. and Morakis, D. J. (1980b) New method for preparing large surfaces of intact human basement membrane for tumor invasion studies. Cancer Lett. *11*, 141–152.

Liotta, L., Rao, C. N. and Wewer, U. (1986a) Biochemical interactions of tumor cells with the basement membrane. Ann. Rev. Biochem. *55*, 1037–1057.

Liotta, L. A., Mandler, R., Murano, G., Katz, D. A., Gordon, R. K., Chiang, P. K. and Schiffmann, E. (1986b) Tumor cell autocrine motility factor. Proc. Natl. Acad. Sci. USA *83*, 3302–3306.

Liotta, L., Steeg, P. and Stetler-Stevenson, W. (1991). Cancer metastasis and angiogenesis: An imbalance of positive and negative regulation. Cell *64*, 327–336.

Lipton, A., Klinger, I., Paul, D. and Holley, R. W. (1971). Migration of mouse 3T3 fibroblasts in response to a serum factor. Proc. Natl. Acad. Sci. USA *68*, 2799–2801.

Lloyd, B. H., Platt-Higgins, A., Rudland, P. S. and Barraclough, R. (1998). Human S100A4 (p9Ka) induces the metastatic phenotype on benign tumour cells. Oncogene 17, 465–473.

Lollini, P. L., Bosco, M. C., Cavallo, F., de Giovanni, C., Giovarelli, M., Landuzzi, L., Musiani, P., Modesti, A., Nicoletti, G., Palmieri, G., et al. (1993). Inhibition of tumor growth and enhancement of metastasis after transfection of the gamma-interferon gene. Int. J. Cancer 55, 320–329.

Lotz, M. M., Burdsal, C. A., Erickson, H. P. and McClay, D. R. (1989). Cell adhesion to fibronectin and tenascin: quantitative measurements of initial binding and subsequent stregthening J. Cell Biol. 109, 1795–1805.

Lu, C. and Kerbel, R. S. (1993). Interleukin 6 undergoes transition from paracrine growth inhibitor to autocrine stimulator during human melanoma progression. J. Cell Biol. 120, 1281–1288.

Lu, S. J., Man, S., Bani, M. R., Adachi, D., Hawley, R. G., Kerbel, R. S. and Ben-David, Y. (1995). Retroviral insertional mutagenesis as a strategy for the identification of genes associated with cis-diamminedichloroplatinum(II) resistance. Cancer Res. 55, 1139–1145.

Lukanidin, E. M. and Georgiev, G. P. (1996). Metastasis-related mts1 gene. Curr. Top. Microbiol. Immunol. 213, 171–195.

Luzzi, K. J., MacDonald, I. C., Schmidt, E. E., Kerkvliet, N., Morris, V. L., Chambers, A. F. and Groom, A. C. (1998). Multistep nature of metastatic inefficiency—Dormancy of solitary cells after successful extravasation and limited survival of early micrometastases. Am. J. Pathol. 153, 865–873.

Mabry, M., Speak, J. A., Griffin, J. D., Stahel, R. A. and Bernal, S. D. (1985). Use of SM-1 monoclonal antibody and human complement in selective killing of small cell carcinoma of the lung. J. Clin. Invest. 75, 1690–1695.

Maciag, T., Hoover, G. A., Stemerman, M. B. and Weinstein, R. (1981). Serial propagation of human endothelial cells in vitro. J. Cell Biol. 91, 420–426.

Maciag, T., Kadish, J., Wilkins, L., Stemerman, M. B. and Weinstein, R. (1982). Organizational behavior of human umbelical vein endothelial cells. J. Cell Biol. 94, 511–520.

Madri, J. A. and Williams, S. K. (1983). Capillary endothelial cell cultures: phenotypic modulation by matrix components. J. Cell Biol. 97, 153–165.

Mahalingam, M., Ugen, K. E., Kao, K. J. and Klein, P. A. (1988). Functional role of platelets in experimental metastasis studied with cloned murine fibrosarcoma cell variants. Cancer Res. 48, 1460–1464.

Mahoney, P. A., Weber, U., Onofrechuk, P., Biessmann, H., Bryant, P. J. and Goodman, C. S. (1991). The fat tumor suppressor gene in Drosophila encodes a novel member of the cadherin gene superfamily. Cell 67, 853–868.

Majuri, M. L., Mattila, P. and Renkonen, R. (1992). Recombinant E-selectin protein mediates tumor cell adhesion via Slea and Slex. Biochem. Biophys. Res. Commun. *182*, 1376–1382.

Manishen, W. J., Sivananthan, K. and Orr, F. W. (1986). Resorbing bone stimulates tumor cell growth; a role for the host microenvironment in bone metastasis. Am. J. Pathol. *123*, 39–45.

Marchetti, D., Menter, D., Jin, L., Nakajima, M. and Nicolson, G. L. (1993). Nerve growth factor effects on human and mouse melanoma cell invasion and heparanase production. Int. J. Cancer *55*, 692–699.

Marengo, S. R. and Chung, L. K. (1994). An orthotopic model for the study of growth factors in the ventral prostate of the rat: effects of epidermal growth factor and basic fibroblast growth factor. J. Androl. *15*, 277–286.

Marincola, F. M., Drucker, B. J., Siao, D., Hough, K. L. and Holder, W. D. (1989). The nude mouse as a model for the study of human pancreatic cancer. J. Surg. Res. *47*, 520–529.

Markowitz, S. D., Myeroff, L., Cooper, M. J., Traicoff, J., Kochera, M., Lutterbaugh, J., Swiriduk, M. and Willson, J. K. (1994). A benign cultured colon adenoma bears three genetically altered colon cancer oncogenes, but progresses to tumorigenicity and transforming growth factor-beta independence without inactivating the p53 tumor suppressor gene. J. Clin. Invest. *93*, 1005–1013.

Maroulakou, I. G., Shibata, M. A., Jorcyk, C. L., Chen, X. X. and Green, J. E. (1997). Reduced p53 dosage associated with mammary tumor metastases in C3(1)/TAG transgenic mice. Mol. Carcinog. *20*, 168–174.

Martin-Padura, I., Mortarini, R., Lauri, D., Bernasconi, S., Sanchez-Madrid, F., Parmiani, G., Mantovani, A., Anichini, A. and Dejana, E. (1991). Heterogeneity in human melanoma cell adhesion to cytokine activated endothelial cells correlates with VLA-4 expression. Cancer Res. *51*, 2239–2241.

Martinez-Zaguilan, R., Martinez, G. M., Gomez, A., Hendrix, M. J. C. and Gillies, R. J. (1998). Distinct regulation of pHin and [Ca2+]in in human melanoma cells with different metastatic potential. J. Cell Physiol. *176*, 196–205.

Marvin, M. R., Southall, J. C., Trokhan, S., DeRosa, C. and Chabot, J. (1998). Liver metastases are enhanced in homozygous deletionally mutant ICAM-1 or LFA-1 mice. J. Surg. Res. *80*, 143–148.

Marzullo, A., Vacca, A., Roncali, L., Pollice, L. and Ribatti, D. (1998). Angiogenesis in hepatocellular carcinoma: an experimental study in the chick embryo chorioallantoic membrane. Int. J. Oncol. *13*, 17–21.

Massagué, J. and Pandiella, A. (1993). Membrane-anchored growth factors. Ann. Rev. Biochem. *62*, 515–541.

Mattei, S., Colombo, M. P., Melani, C., Silvani, A., Parmiani, G. and Herlyn, M. (1994). Expression of cytokine/growth factors and their receptors in human melanoma and melanocytes. Int. J. Cancer 56, 853–857.
Mayadas, T. N., Johnson, R. C., Rayburn, H., Hynes, R. O. and Wagner, D. D. (1993). Leukocyte rolling and extravasation are severely compromised in P selectin-deficient mice. Cell 74, 541–554.
Mayhew, E. and Glaves, D. (1984). Quantitation of tumorigenic disseminating and arrested cancer cells. Br. J. Cancer 50, 159–166.
Mayrovitz, H. N. (1992). Leukocyte rolling: a prominent feature of venules in intact skin of anesthetized hairless mice. Am. J. Physiol. 262, H157–161.
McCarthy, J. B., Palm, S. L. and Furcht, L. T (1983). Migration by haptotaxis of a schwann cell tumor line to the basement membrane glycoprotein laminin. J. Cell Biol. 97, 772–777.
McCarthy, S. A., Kuzu, I., Gatter, K. C. and Bicknell, R. (1991). Heterogeneity of the endothelial cell and its role in organ preference of tumor metastasis. Trends in Physiol. Sci. 12, 462–467.
McClatchey, A. I., Saotome, I., Mercer, K., Crowley, D., Gusella, J. F., Bronson, R. T. and Jacks, T. (1998). Mice heterozygous for a mutation at the Nf2 tumor suppressor locus develop a range of highly metastatic tumors. Genes Dev. 12, 1121–1133.
McClay, D. R., Wessel, G. M. and Marchase, R. B. (1981). Intercellular recognition: quantitation of initial binding events. Proc. Natl. Acad. Sci. USA 78, 4975–4979.
McCutcheon, M., Coman, D. R. and Moore, M. B. (1948). Studies on invasion in cancer: adhesiveness of malignant cells in various human adenocarcinomas. Cancer 1, 460–467.
McDonald, J. A. (1988). Extracellular matrix assembly. Ann. Rev. Cell Biol. 4, 183–207.
McLemore, T. L., Liu, M. C., Blacker, P. C., Gregg, M., Alley, M. C., Abbott, B. J., Shoemaker, R. H., Bohlman, M. E., Litterst, C. C., Hubbard, W. C., Brennan, R. H., McMahon, J. B., Fine, D. L., Eggleston, J. C., Mayo, J. G. and Boyd, M. R. (1987). Novel intrapulmonary model for orthotopic propagation of human lung cancers in athymic nude mice. Cancer Res. 47, 5132–5140.
McLeskey, S. W., Zhang, L. R., Kharbanda, S., Kurebayashi, J., Lippman, M. E., Dickson, R. B. and Kern, F. G. (1996). Fibroblast growth factor overexpressing breast carcinoma cells as models of angiogenesis and metastasis. Breast Cancer Res. Treat. 39, 103–117.
McNeil, P. L., Murphy, R. F., Lanni, F. and Taylor, D. L. (1984). A method for incorporating macromolecules into adherent cells. J. Cell Biol. 98, 1556–1564.
Mege, J. L., Capo, C., Benoliel, A. M. and Bongrand, P. (1986). Determination of binding strength and kinetics of binding initiation. A model study made on

the adhesive properties of P388D1 macrophage-like cells. Cell. Biophys. *8*, 141–159.
Mehta, P. P., Bertram, J. S. and Loewenstein, W. R. (1986). Growth inhibition of transformed cells correlates with their junctional communication with normal cells. Cell *44*, 187–196.
Mehta, P., Lawson, D., Ward, M. B., Kimura, A. and Gee, A. (1987). Effect of human tumor cells on platelet aggregation: potential relevance to pattern of metastasis. Cancer Res. *47*, 3115–3117.
Menter, D. G., Steinert, B. W., Sloane, B. F., Gundlach, N., O'Gara, C. Y., Marnett, L. J., Diglio, C., Walz, D., Taylor, J. D. and Honn, K. V. (1987). Role of platelet membrane in enhancement of tumor cell adhesion to endothelial cell ECM. Cancer Res. *47*, 6751–6762.
Meredith Jr., J. E., Fazeli, B. and Schwartz, M. A. (1993). The extracellular matrix as a cell survival factor. Mol. Biol. Cell *4*, 953–961.
Meschter, C. L., Connolly, J. M. and Rose, D. P. (1992). Influence of regional location of the inoculation site and dietary fat on the pathology of MDA-MB-435 human breast cancer cell-derived tumors grown in nude mice. Clin. Exptl. Metastasis *10*, 167–173.
Meyvisch, C. (1983). Influence of implantation site on formation of metastases. Cancer Metastasis Rev. *2*, 295–306.
Meyvisch, C., Storme, G., Bruyneel, E. and Mareel, M. (1983). Invasiveness and tumorigenicity of MO_4 mouse fibrosarcoma cells pretreated with microtubule inhibitors. Clin. Exp. Metabol. *1*, 17–28.
Michie, S. A., Streeter, P. R., Bolt, P. A., Butcher, E. C. and Picker, L. J. (1993). The human peripheral lymph node vascular addressin. An inducible endothelial antigen involved in lymphocyte homing. Am. J. Pathol. *143*, 1688–1698.
Miele, M. E., Robertson, G., Lee, J. H., Coleman, A., McGary, C. T., Fisher, P. B., Lugo, T. G. and Welch, D. R. (1996). Metastasis suppressed, but tumorigenicity and local invasiveness unaffected, in the human melanoma cell line MelJuSo after introduction of human chromosomes 1 or 6. Mol. Carcinog. *15*, 284–299.
Mignatti, P., Morimoto, T. and Rifkin, D. B. (1991). Basic FGF released by single, isolated cells stimulates their migration in an autocrine manner. Proc. Natl. Acad. Sci. USA *88*, 11007–11011.
Miller, B. E., Aslakson, C. J. and Miller, F. R. (1990). Efficient recovery of clonogenic stem cells from solid tumors and occult metastatic deposits. Invasion Metastasis *10*, 101–112.
Miller, F. R. (1981). Comparison of metastasis of mammary tumors growing in the mammary fat pad versus the subcutis. Invasion Metastasis *1*, 220–226.

Miller, F. R. and McInerney, D. (1988). Epithelial component of host-tumor interactions in the orthotopic site preference of a mouse mammary tumor. Cancer Res. *48*, 3698–3701.

Miller, T. M. and Goodenough, D. A. (1985). Gap junction structures after experimental alteration of junctional channel conductance. J. Cell Biol. *101*, 1741–1748.

Miner, K. M., Kawaguchi, T., Uba, G. W. and Nicolson, G. L. (1982). Clonal drift of cell surface, melanogenic and experimental metastatic properties of in vivo-selected brain meninges-colonizing murine B16 melanoma. Cancer Res. *42*, 4631–4638.

Miossec, P., Cavender, D. and Ziff, M. (1986). Production of interleukin 1 by human endothelial cells. J. Immunol. *136*, 2486–2491.

Mittelman, L., Levin, S., Verschueren, H., de Baetselier, P. and Korenstein, R. (1994). Direct correlation between cell membrane fluctuations, cell filterability and the metastatic potential of lymphoid cell lines. Biochem. Biophys. Res. Commun. *203*, 899–906.

Miyake, M., Nakano, K., Itoi, S., Koh, T. and Taki, T. (1996). Motility-related Protein-1 (MRP-1/CD9). reduction as a factor of poor prognosis in breast cancer. Cancer Res. *56*, 1244–1249.

Mizuno, K. and Nakamura, T. (1993). Molecular characteristics of HGF and the gene, and its biochemical aspects. EXS *65*, 1–29.

Mogi, Y., Kogawa, K., Takayama, T., Yoshizaki, N., Bannai, K., Muramatsu, H., Koike, K., Kohgo, Y., Watanabe, N. and Niitsu, Y. (1991). Platelet aggregation induced by ADP release from cloned murine fibrosarcoma cells is positively correlated with the experimental metastatic potential of the cells. Jpn. J. Cancer Res. *82*, 192–198.

Mohler, J. L. (1993). Cellular motility and prostatic carcinoma metastases. Cancer Metastasis Rev. *12*, 53–67.

Mohler, J. L., Partin, A. W., Isaacs, W. B. and Coffey, D. S. (1987a) Time lapse videomicroscopic identification of Dunning R–3327 adenocarcinoma and normal rat prostate cells. J. Urol. *137*, 544–547.

Mohler, J. L., Partin, and Coffey, D. S. (1987b) Prediction of metastatic potential by a new grading system of cell motility: validation in the Dunning R–3327 prostatic adenocarcinoma model. J. Urol. *138*, 168–170.

Mohler, J. L., Levy, F. and Sharief, Y. (1991). Metastatic potential and substrate dependence of cell motility and attachment in the Dunning R–3327 rat prostatic adenocarcinoma model. Cancer Res. *51*, 6580–6585.

Montesano, R., Orci, L. and Vassalli, P. (1983). In vitro rapid organization of endothelial cells into capillary-like networks is promoted by collagen matrices. J. Cell Biol. *97*, 1648–1652.

Montesano, R., Pepper, M. S., Vassalli, J. D. and Orci, L. (1992). Proteolityc Balance and Capillary Morphogenesis in Vitro. Angiogenesis: Key Principles-Science-Technology-Medicine. Birkauser Verlag, Basel.

Moretta, A., Sivori, S., Ponte, M., Mingari, M. C. and Moretta, L. (1998). Stimulatory receptors in NK and T cells. Curr. Top. Microbiol. Immunol. *230*, 15–23.

Mori, M., Mimori, K., Shiraishi, T., Haraguchi, M., Ueo, H., Barnard, G. F. and Akiyoshi, T. (1998). Motility related protein 1 (MRP1/CD9). Expression in colon cancer. Clin. Cancer Res. *4*, 1507–1510.

Morikawa, K., Walker, S. M., Nakajima, M., Pathak, S., Jessup, J. M. and Fidler, I. J. (1988). Influence of organ microenvironment on the growth, selection and metastasis of human colon carcinoma cells in nude mice. Cancer Res. *48*, 6863–6871.

Morris, V. L., Schmidt, E. E., MacDonald, I. C., Groom, A. C. and Chambers, A. F. (1997). Sequential steps in hematogenous metastasis of cancer cells studied by in vivo videomicroscopy. Invasion Metastasis *17*, 281–296.

Mosmann, T. (1983). Rapid colorimetric assay for cellular growth and survival: application to proliferation and cytotoxicity assays. J. Immunol. Meth. *65*, 55–63.

Mueller, B. M., Romerdahl, C. A., Trent, J. M. and Reisfeld, R. A. (1991). Suppression of spontaneous metastasis in Scid mice with an antibody to the epidermal growth factor receptor. Cancer Res. *51*, 2193–2198.

Mukaida, H., Hirabayashi, N., Hirai, T., Iwata, T., Saeiki, S. and Toge, T. (1991). Significance of freshly cultured fibroblasts from different tissues in promoting cancer cell growth. Int. J. Cancer *48*, 423–427.

Murphy, L. C. (1998). Mechanisms of hormone independence in human breast cancer. In Vivo *12*, 95–106.

Murray, J. C., Liotta, L. A., Rennard, S. I. and Martin, G. R. (1980). Adhesion characteristics of murine metastatic and non metastatic tumor cells in vitro. Cancer Res. *40*, 347–351.

Muthukkaruppan, V. and Auerbach, R. (1979). Angiogenesis in the mouse cornea. Science *206*, 1416–1418.

Muto, M. G., Welch, W. R., Mok, S. C., Bandera, C. A., Fishbaugh, P. M., Tsao, S. W., Lau, C. C., Goodman, H. M., Knapp, R. C. and Berkowitz, R. S. (1995). Evidence for a multifocal origin of papillary serous carcinoma of the peritoneum. Cancer Res. *55*, 490–492.

Nabeshima, K., Kataska, H. and Koona, M. (1986). Enhanced migration of tumor cells in response to collagen degradation products and tumor cell collagenolytic activity. Invas. Metast. *6*, 270–286.

Nagamachi, Y., Tani, M., Shimizu, K., Tsuda, H., Niitsu, Y. and Yokota, J. (1998). Orthotopic growth and metastasis of human non-small cell lung carcinoma cells injected into the pleural cavity of nude mice. Cancer Lett. *127*, 203–209.

Naito, S., Kanamori, T. and Hisano, S. (1982). Human renal cell carcinoma: establishment and characterization of two new cell lines. J. Urol. *128*, 1117–1121.

Naito, S., von Eschenbach, A. C., Giavazzi, R. and Fidler, I. J. (1986). Growth and metastasis of tumor cells isolated from a human renal cell carcinoma implanted into different organs of nude mice. Cancer Res. *46*, 4109–4115.

Nakagima, M., Irimura, T. and Nicolson, G. L. (1986). A solid-phase substrate of heparanase: its application to assay of human melanoma for heparan sulfate degradative activity. Anal. Biochem. *157*, 162–171.

Nakamura, T., Nawa, K. and Ichihara, A. (1984). Partial purification and characterization of hepatocyte growth factor from serum of hepatectomized rats. Biochem. Biophys. Res. Comm. *122*, 1450–1459.

Nathan, C. and Sporn, M. (1991). Cytokines in context. J. Cell Biol. *113*, 981–986.

Negro A., Onisto, M., Pellati D. and Garbisa, S. (1997). CNTF up-regulation of TIMP-2 in neuroblastoma cells. Mol. Brain Res. *48*, 30–36.

Nelson, R. D., Quie, P. G. and Simmons, R. L. (1975). Chemotaxis under agarose: a new and simple method for measuring chemotaxis and spontaneous migration of human polymorphonuclear leukocytes and monocytes. J. Immunol. *115*, 1650–1656.

Neri, A., Welch, D. R., Kawaguchi, T. and Nicolson, G. L. (1982). Development and biologic properties of malignant cell sublines and clones of a spontaneously metastasizing rat mammary adenocarcinoma. J. Natl. Cancer Inst. *68*, 507–517.

Netland, P. A. and Zetter, B. R. (1984). Organ-specific adhesion of metastatic tumor cells in vitro. Science *224*, 1113–1115.

Netland, P. A. and Zetter, B. R. (1985). Metastatic potential of B16 melanoma cells after in vitro selection for organ-specific adherence. J. Cell Biol. *101*, 720–724.

Nguyen, M., Shing, Y. and Folkman, J. (1994). Quantitation of angiogenesis and antiangiogenesis in the chick embryo chorioallantoic membrane. Microvasc Res. *7*, 31–40.

Nicolson, G. L. (1973). Neuraminidase 'unmasking' and failure of trypsin to 'unmask'-D-galactose-like sites on erythrocyte, lymphoma, and normal and virus-transformed fibroblast cell membranes. J. Natl. Cancer Inst. *50*, 1443–1451.

Nicolson, G. L. (1987). Differential growth properties of metastatic large-cell lymphoma cells in target organ- conditioned medium. Exp. Cell Res. *168*, 572–577.

Nicolson, G. L. (1988a) Organ specificity of tumor metastasis: role of preferential adhesion, invasion and growth of malignant cells at specific secondary sites. Cancer Metastasis Rev. *7*, 143–188.

Nicolson, G. L. (1988b). Cancer metastasis: tumor cell and host organ properties important in colonization of specific secondary sites. Biochim. Biophys. Acta *948*, 175–224.

Nicolson, G. L. (1993). Paracrine/autocrine growth mechanisms in tumor metastasis. Oncol. Res. *4*, 389–399.

Nicolson, G. L. (1994). Tumor microenvironment: Paracrine and autocrine growth mechanisms and metastasis to specific sites. Front. Radiat. Ther. Oncol. *28*, 11–24.

Nicolson, G. L. and Custead, S. E. (1982). Tumor metastasis is not due to adaptation of cells to a new organ environment. Science *215*, 176–178.

Nicolson, G. L. and Dulski, K. M. (1986). Organ specificity of metastatic tumor colonization is related to organ- selective growth properties of malignant cells. Int. J. Cancer *38*, 289–294.

Nicolson, G. L. and Moustafa, A. S. (1998). Metastasis-associated genes and metastatic tumor progression. In Vivo *12*, 579–588.

Nicolson, G. L. and Winkelhake, J. L. (1975). Organ-specificity of blood-borne tumor metastasis determined by cell adhesion? Nature *255*, 230–232.

Nicolson, G. L., Mascali, J. J. and McGuire, E. J. (1982). Metastatic RAW117 lymphosarcoma as a model for malignant-normal cell interactions. Possible roles for cell surface antigens in determining the quantity and location of secondary tumors. Oncodev. Biol. Med. *4*, 149–159.

Nicolson, G. L., Irimura, T., Nakajima, M., Updyke, T. V. and Poste, G. (1984). The cellular interactions of tumor cells with special reference to endothelial cells and their basal-lamina like matrix. In: Hemostatic Mechanisms and Metastasis (Honn, K. V. and Sloane, B. F., eds.). Martinus Nijhoff, Dordrecht, pp. 295–318.

Nicolson, G. L., Dulski, K., Basson, C. and Welch, D. R. (1985). Preferential organ attachment and invasion in vitro by B16 melanoma cells selected for differing metastatic colonization and invasive properties. Invasion Metastasis *5*, 144–158.

Nicolson, G. L., Dulski, K. M. and Trosko, J. E. (1988). Loss of intercellular junctional communication correlates with metastatic potential in mammary adenocarcinoma cells. Proc. Natl. Acad. Sci. USA *85*, 473–476.

Nicolson, G. L., Belloni, P. N., Tressler, R. J., Dulski, K. M., Inoue, T. and Cavanaugh, P. G. (1989). Adhesive, invasive and growth properties of selected metastatic variants of a murine large-cell lymphoma. Invasion Metastasis *9*, 102–116.

Nicolson, G. L., Inoue, T., van Pelt, C. S. and Cavanaugh, P. G. (1990). Differential expression of a Mr approximately 90,000 cell surface transferrin receptor-related glycoprotein on murine B16 metastatic melanoma sublines selected for enhanced brain or ovary colonization. Cancer Res. *50*, 515–520.

Niederkorn, J., Streilein, J. W. and Shadduck, J. A. (1981). Deviant immune responses to allogeneic tumors injected intracameraly and subcutaneously in mice. Invest. Ophthal. Vis. Sci. *20*, 355–363.

Nierodzik, M. L., Plotkin, A., Kajumo, F. and Karpatkin, S. (1991). Thrombin stimulates tumor-platelet adhesion in vitro and metastasis in vivo. J. Clin. Invest. *87*, 229–236.

Nierodzik, M. L., Kajumo, F. and Karpatkin, S. (1992). Effect of thrombin treatment of tumor cells on adhesion of tumor cells to platelets in vitro and tumor metastasis in vivo. Cancer Res. *52*, 3267–3272.

Nimgaonkar, M., Kemp, A., Lancia, J. and Ball, E. D. (1996). A combination of CD34 selection and complement-mediated immunopurging (anti-CD15 monoclonal antibody) eliminates tumor cells while sparing normal progenitor cells. J. Hematother. *5*, 39–48.

Nip, J., Shibata, H., Loskutoff, D. J., Cheresh, D. A. and Brodt, P. (1992). Human melanoma cells derived from lymphatic metastases use integrin alfa-V beta-3 to adhere to lymphnode vitronectin. J. Clin. Invest. *90*, 1406–1413.

Noel, A., de Pauw-Gillet, M. C., Purnell, G., Nusgens, B., Lapiere, C. M. and Foidart, J. M. (1993). Enhancement of tumorigenicity of human breast adenocarcinoma cells in nude mice by matrigel and fibroblasts. Br. J. Cancer *68*, 909–915.

Noel, A., Gilles, C., Bajou, K., Devy, L., Kebers, F., Lewalle, J. M., Maquoi, E., Munaut, C., Remacle, A. and Foidart, J. M. (1997). Emerging roles for proteinases in cancer. Invasion Metastasis *17*, 221–239.

Noel, A., Hajitou, A., L'Hoir, C., Maquoi, E., Baramova, E., Lewalle, J. M., Remacle, A., Kebers, F., Brown, P., Calberg-Bacq, C. M. and Foidart, J. M. (1998). Inhibition of stromal matrix metalloproteases: effects on breast- tumor promotion by fibroblats. Int. J. Cancer *76*, 267–273.

Noonan, D. and Hassel, J. R. (1993). Perlecan, the large low-density proteoglycan of basement membranes: structure and variant forms. Kidney Int. *43*, 53–60.

Noonan, D. M., Fulle, A., Valente, P., Cai, S., Horigan, E., Sasaki, M., Yamada, Y. and Hassell, J. R. (1991). The complete sequence of perlecan, a basement membrane heparan sulfate proteoglycan, reveals extensive similarity with laminin A chain, low density lipoprotein-receptor and the neural cell adhesion molecule. J. Biol. Chem. *266*, 22939–22947.

Nordt, F. J. (1983). Hemorheology in cerebrovascular diseases: approaches to drug development. Ann. N. Y. Acad. Sci. *416*, 651–661.

Nowell, P. C. (1976). The clonal evolution of tumor cell populations. Science *194*, 23–28.
O'Reilly, M. S., Holmgren, L., Shing, Y., Chen, C., Rosenthal, R. A., Moses, M., Lane, W. S., Cao, Y., Sage, E. H. and Folkman, J. (1994). Angiostatin: a novel angiogenesis inhibitor that mediates the suppression of metastases by a Lewis lung carcinoma. Cell *79*, 315–328.
O'Reilly, M. S., Holmgren, L., Chen, C. and Folkman, J. (1996). Angiostatin induces and sustains dormancy of human primary tumors in mice. Nat. Med. *2*, 689–692.
O'Reilly, M. S., Boehm, T., Shing, Y., Fukai, N., Vasios, G., Lane, W. S., Flynn, E., Birkhead, J. R., Olsen, B. R. and Folkman, J. (1997). Endostatin: an endogenous inhibitor of angiogenesis and tumor growth. Cell *88*, 277–285.
Ochalek, T., Nordt, F. J., Tullberg, K. and Burger, M. M. (1988). Correlation between cell deformability and metastatic potential in B16-F1 melanoma cell variants. Cancer Res. *48*, 5124–5128.
Okumura, Y., Hamada, J., Cavanaugh, P. G. and Nicolson, G. L. (1992). Preferential growth stimulation of metastataic rat mammary adenocarcinoma cells by organ-derived syngeneic fibroblasts in vitro. Invasion Metastasis *12*, 275–283.
Olumi, A. F., Dazin, P. and Tlsty, T. D. (1998). A novel coculture technique demonstrates that normal human prostatic fibroblasts contribute to tumor formation of LNCaP cells by retarding cell death. Cancer Res. *58*, 4525–4530.
Orr, C. W. and Roseman, S. (1969). Intercellular adhesion. I. A quantitative assay for measuring the rate of adhesion. J. Memb. Biol. *1*, 109–124.
Osbakken, M. D., Kreider, J. W. and Taczanowsky, P. (1986). Nuclear magnetic resonance imaging characterization of a rat mammary tumor. Magnet. Reson. Med. *3*, 1–9.
Otsuka, T., Takayama, H., Sharp, R., Celli, G., LaRochelle, W. J., Bottaro, D. P., Ellmore, N., Vieira, W., Owens, J. W., Anver, M. and Merlino, G. (1998). c-met autocrine activation induces development of malignant melanoma and acquisition of the metastatic phenotype. Cancer Res. *58*, 5157–5167.
Ozawa, S., Ueda, M., Ando, N., Abe, O., Hirai, M. and Shimizu, N. (1987). Stimulation by EGF of the growth of EGF receptor-hyperproducing tumor cells in athymic mice. Int. J. Cancer *40*, 706–710 .
Pacchiarini, L., Zucchella, M., Milanesi, G., Tacconi, F., Bonomi, E., Carnevari, A. and Grignani, G. (1991). Thromboxane production by platelets during tumor cell-induced platelet aggregation. Invasion Metastasis *11*, 102–109.
Paget, S. (1889). Distribution of secondary growths in cancer of the breast. Lancet *1*, 571–573.
Paku, S., Rot, A., Ladànyi, A. and Lapis, K. (1989). Demonstration of the organ preference of liver selected high metastatic Lewis lung tumor cell line. Clin. Exp. Metastasis *7*, 599–607.

Pàl, K., Kopper, L. and Lapis, K. (1983). Increased metastatic capacity of Lewis lung tumor cells by in vivo selection procedure. Invasion Metastasis *3*, 174–182.

Parent, C. A., Devreotes, P. N. (1999). A Cell's sense of direction. Science *284*, 765–770.

Parker, G., Roseman, B. J., Bucana, C. D., Tsan, R. and Radinsky, R. (1998). Preferential activation of the epidermal growth factor receptor in human colon carcinoma liver metastases in nude mice. J. Histochem. Cytochem. *46*, 595–602.

Partin, A. W., Schoeniger, J. S., Mohler, J. L. and Coffey, D. S. (1989). Fourier analysis of cell motility: correlation of motility with metastatic potential. Proc. Natl. Acad. Sci. USA *86*, 1254–1258.

Partin, A. W., Mohler, J. L. and Coffey, D. S. (1992). Cell motility as an index of metastatic ability in prostate cancers: results with an animal model and with human cancer cells. In: Therapy for Genitourinary Cancer (Lepor, H. and Lawson, R. K., eds.). Kluwer Academic Publishers, Dordrecht, pp. 121–130.

Passaniti, A. and Hart, G. W. (1988). Cell surface sialylation and tumor metastasis. J. Biol. Chem. *263*, 7591–7603.

Passaniti, A., Taylor, R. M., Pili, R., Guo, Y., Long, P. V. and Haney, J. A. (1992). A simple, quantitative method for assessing angiogenesis and antiangiogenic agents using reconstituted basement membrane, heparin, and fibroblast growth factor. Lab. Invest. *67*, 519–528.

Pauli, B. U. and Lee, C. L. (1988). Organ preference of metastasis. The role of organ-specifically modulated endothelial cells. Lab. Invest. *58*, 379–387.

Pauli, B. U., Augustin-Voss, H. G., El-Sabban, M. E., Johnson, R. C. and Hammer, D. A. (1990). Organ preference of metastasis. The role of endothelial cell adhesion molecules. Cancer and Metastasis Rev. *9*, 175–189.

Penna, D., Schmidt, A. and Beermann, F. (1998). Tumors of the retinal pigment epithelium metastasize to inguinal lymph nodes and spleen in tyrosinase-related protein 1 SV40 T antigen transgenic mice. Oncogene *17*, 2601–2607.

Perez-Stable, C., Altman, N. H., Mehta, P. P., Deftos, L. J. and Roos, B. A. (1997). Prostate cancer progression, metastasis, and gene expression in transgenic mice. Cancer Res. *57*, 900–906.

Perissin, L., Zorzet, S., Piccini, P., Rapozzi, V. and Giraldi, T. (1989). Corticosterone does not mediate the effects of stress on tumour metastasis. Pharmacol. Res. *21*, 461–462.

Perissin, L., Zorzet, S., Piccini, P., Rapozzi, V. and Giraldi, T. (1991). Effects of rotational stress on the effectiveness of cyclophosphamide and razoxane in mice bearing Lewis lung carcinoma. Clin. Exptl. Metastasis *9*, 541–549.

Perissin, L., Rapozzi, V., Zorzet, S. and Giraldi, T. (1997). Survival time in mice bearing TLX5 lymphoma subjected to rotational stress and chemotherapy with CCNU. Anticancer Res. *17*, 4355–4358.

Perissin, L., Zorzet, S., Rapozzi, V., Carignola, R., Angeli, A. and Giraldi, T. (1998). Seasonal effects of rotational stress on Lewis lung carcinoma metastasis and T-lymphocyte subsets in mice. Life Sciences *63*, 711–719.

Perl, A. K., Wilgenbus, P., Dahl, U., Semb, H. and Christofori, G. (1998). A causal role for E-cadherin in the transition from adenoma to carcinoma. Nature (London) *392*, 190–193.

Perper, R. J., Zee, T. W. and Mickelson, M. M. (1968). Purification of lymphocytes and platelets by gradient centrifugation. J. Lab. Clin. Med. *72*, 842–848.

Perucho, M. (1996). Microsatellite instability: the mutator that mutates the other mutator. Nature Med. *2*, 630–631.

Phillips, K. K., Welch, D. R., Miele, M. E., Lee, J.-H., Wei, L. L. and Weissman, B. E. (1996). Suppression of MDA-MB-435 breast carcinoma cell metastasis following the introduction of human chromosome 11. Cancer Res. *56*, 1222–1226.

Phillips, R. A., Jewett, M. A. S. and Gallie, B. L. (1989). Growth of human tumors in immune-deficient SCID mice and nude mice. Curr. Topics Microbiol. Immunol. *152*, 259–263.

Phillips, S. M., Bendall, A. J. and Ramshaw, I. A. (1990). Isolation of gene associated with high metastatic potential in rat mammary adenocarcinomas. J. Natl. Cancer Inst. *82*, 199–203.

Plaksin, D., Gelber, C., Feldman, M. and Eisenbach, L. (1988). Reversal of the metastatic phenotype in Lewis lung carcinoma cells after transfection with syngeneic H-2Kb gene. Proc. Natl. Acad. Sci. USA *85*, 4463–4467.

Pober, J. S., Gimbrone, M. A., Cotran, R. S., Reiss, C. S., Burakoff, S. J., Fiers, W. and Ault, K. A. (1983). Ia expression by vascular endothelium is inducible by activated T cells and by human gamma interferon. J. Exp. Med. *157*, 1339–1353.

Polissar, M. J. and Shimkin, M. B. (1954). A quantitative interpretation of the distribution of induced pulmonary tumors in mice. J. Natl. Cancer Inst. *15*, 377–403.

Pollack, V. A. and Fidler, I. J. (1982). Use of young nude mice to select subpopulations of tumor cells with increased metastatic potential from nonsyngeneic neoplasms. J. Natl. Cancer Inst. *69*, 137–141.

Poste, G, Doll, J., Hart, I. R. and Fidler I. J. (1980). In vitro selection of murine B-16 melanoma variants with enhanced tissue-invasive properties. Cancer Res. *40*, 1636–1644.

Poste, G., Tzeng, J., Doll, J., Greig, R., Rieman, D. and Zeidman, I. (1982). Evolution of tumor cell heterogeneity during progressive growth of individual lung metastases. Proc. Natl. Acad. Sci. USA *79*, 6574–6578.

Potgens, A. J., Westphal, H. R., de Waal, R. M. and Ruiter, D. J. (1995). The role of vascular permeability factor and basic fibroblast growth factor in tumor angiogenesis. Biol. Chem. Hoppe Seyler *376*, 57–70.

Potter, K. M., Juacaba, S. F., Price, J. E. and Tarin, D. (1983). Observations on organ distribution of fluoroscein-labelled tumour cells released intravascularly. Invasion Metastasis *3*, 221–233.

Powell, D. J., Jr., Russell, J., Nibu, K., Li, G. Q., Rhee, E., Liao, M., Goldstein, M., Keane, W. M., Santoro, M., Fusco, A. and Rothstein, J. L. (1998). The RET/PTC3 oncogene: Metastatic solid-type papillary carcinomas in murine thyroids. Cancer Res. *58*, 5523–5528.

Prat, M., Pennacchietti, S., Crepaldi, T., Chiara, F. and Comoglio, P. M. (1995). The HGF receptor (Met): transduction of signals for invasive cell growth. Antibody, Immunoconjugates and Pharmaceuticals *8*, 341–361.

Preissner, K. T. (1991). Structure and biological role of vitronectin. Ann. Rev. Cell Biol. *7*, 275–310.

Prezioso, J. A., Wang, N., Duty, L., Bloomer, W. D. and Gorelik, E. (1993). Enhancement of pulmonary metastasis formation and gamma-glutamyltranspeptidase activity in B16 melanoma induced by differentiation in vitro. Clin. Exptl. Metastasis *11*, 263–274.

Price, J. (1987). Retroviruses and the study of cell lineage. Development *01*, 409–419.

Price, J. E. (1996). Metastasis from human breast cancer cell lines. Breast Cancer Res. Treat. *39*, 93–102.

Price, J. E. and Zhang, R. D. (1990). Studies of human breast cancer metastasis using nude mice. Cancer Metast. Rev. *8*, 285–297.

Price, J. E., Carr, D. and Tarin, D. (1984). Spontaneous and induced metastasis of naturally occurring tumors in mice: analysis of cell shedding into the blood. J. Natl. Cancer Inst. *73*, 1319–1326.

Price, J. E., Aukerman, S. L. and Fidler, I. J. (1986). Evidence that the process of murine melanoma metastasis is sequential and selective and contains stochastic elements. Cancer Res. *46*, 5172–5178.

Price, J. E., Naito, S. and Fidler, I. J. (1988). Growth in an organ microenvironment as a selective process in metastasis. Clin. Exp. Metastasis *6*, 91–102.

Price, J. E., Polyzos, A., Zhang, R. D. and Daniels, L. M. (1990). Tumorigenicity and metastasis of human breast carcinoma cell lines in nude mice. Cancer Res. *50*, 717–721.

Price, E. A., Coombe, D. R. and Murray, J. C. A simple fluorometric assay for quantifying the adhesion of tumor cells to endothelial monolayers. (1995). Clin. Exp. Metastasis *13*, 155–164.

Prieto, A. L., Edelman, G. M. and Crossin, K. L. (1993). Multiple integrins mediate cell attachment to cytotactin/tenascin. Proc. Natl. Acad. Sci. USA *90*, 10154–10158.

Puck, T. T., Marcus, P. I. and Cieciura, S. J. (1956). Clonal growth of mammalian cells in vitro. J. Exp. Med. *103*, 273–289.

Qian, F., Vaux, D. L. and Weissman, I. L. (1994). Expression of the integrin $\alpha_4\beta_1$ on melanoma cells can inhibit the invasive stage of metastasis formation. Cell *77*, 335–347.

Quinn, K. A., Treston, A. M., Unsworth, E. J., Miller, M. J., Vos, M., Grimley, C., Battey, J., Mulshine, J. L. and Cuttitta, F. (1996). Insulin- like growth factor expression in human cancer cell lines. J. Biol. Chem. *271*, 1477–11483.

Rabbani, S. A., Gladu, J., Harakidas, P., Jamison, B. and Goltzman, D. (1999). Over-production of parathyroid hormone-related peptide results in increased osteolytic skeletal metastasis by prostate cancer cells in vivo. Intl. J. Cancer *80*, 257–264.

Radinsky, R. (1995a) Modulation of tumor cell gene expression and phenotype by the organ specific metastatic environment. Cancer Metastasis Rev. *14*, 323–338.

Radinsky, R. (1995b) Molecular mechanisms for organ-specific colon carcinoma metastasis. Eur. J. Cancer, [A] *31A*, 1091–1095.

Radinsky, R. and Fidler, I. J. (1992). Regulation of tumor cell growth at organ-specific metastases. In Vivo *6*, 325–332.

Radinsky, R., Risin, S., Fan, D., Dong, Z., Bielenberg, D., Bucana, C. and Fidler, I. J. (1995). Level and function of epidermal growth factor receptor predict the metastatic potential of human colon carcinoma cells. Clin. Cancer Res. *1*, 19–31.

Rak, J. and Kerbel, R. S. (1996). Treating cancer by inhibiting angiogenesis: New hopes and potential pitfalls. Cancer Metastasis Rev. *15*, 231–236.

Ramesh, S., McMillan, S., Gallagher, G., Greenhalgh, D. and McCulloch, P. (1998). Tpr-met in neoplastic and pre-neoplastic gastric carcinogenesis. Proc. AACR *39*, 592.

Ramshaw, I. A., Carlsen, S., Wang, H. C. and Badenoch-Jones, P. (1983). The use of cell fusion to analyse factors involved in tumour cell metastasis. Int. J. Cancer *32*, 471–478.

Raz, A. and Ben-Ze'ev, A. (1987). Cell-contact and -architecture of malignant cells and their relationship to metastasis. Cancer Metast. Rev. *6*, 3–21.

Raz, A., Hanna, N. and Fidler, I. J. (1981). In vivo isolation of a metastatic tumor cell variant involving selective and nonadaptive processes. J. Natl. Cancer Inst. *66*, 183–189.

Reid, L. M. (1989). Defining hormone and matrix requirements for differentitaed epithelia. Meth. Mol. Biol. *5*, 237–276.

Rembrink, K., Romijn, J. C., van der Kwast, T. H., Rübben, H. and Schröder, F. H. (1997). Orthotopic implantation of human prostate cancer cell lines: a clinically relevant animal model for metastatic prostate cancer. Prostate *31*, 168–174.

Remels, L. M. and de Baetselier, P. C. (1987). Characterization of 3LL-tumor variants generated by in vitro macrophage-mediated selection. Int. J. Cancer *39*, 343–352.

Remels, L., Neirynck, A., Brys, L., Vercauteren, E. and de Baetselier, P. (1989). TNF-alpha mediated selection of macrophage-resistant gene-regulatory tumor variants. Clin. Exp. Metastasis *7*, 493–506.

Ren, J., Hamada, J., Takeichi, N., Fujikawa, S. and Kobayashi, H. (1990). Ultrastructural differences in junctional intercellular communication between highly and weakly metastatic clones derived from rat mammary carcinoma. Cancer Res. *50*, 358–362.

Repesh, L. A., Drake, S. R., Warner, M. C., Downing, S. W., Jyring, R., Seftor, E. A., Hendrix, M. J. and McCarthy, J. B. (1993). Adriamycin-induced inhibition of melanoma cell invasion is correlated with decreases in tumor cell motility and increases in focal contact formation. Clin. Exp. Metastasis *11*, 91–102.

Revesz, L. (1956). Effect of tumour cells killed by X-rays on the growth of admixed viable cells. Nature (London) *178*, 1391–1392.

Revesz, L. (1958). Effect of lethally damaged tumor cells on the development of admixed viable cells. J. Natl. Cancer Inst. *20*, 1157–1186.

Reyes, G., Villanueva, A., Garcìa, C., Sancho, F. J., Piulats, J., Lluìs, F. and Capellá, G. (1996). Orthotopic xenografts of human pancreatic carcinomas acquire genetic aberrations during dissemination in nude mice. Cancer Res. *56*, 5713–5719.

Ribatti, D., Urbinati, C., Nico, B., Rusnati, M., Roncali, L. and Presta, M. (1995). Endogenous basic fibroblast growth factor is implicated in the vascularization of the chick embryo chorioallantoic membrane. Dev. Biol. *170*, 39–49.

Ribatti, D., Vacca, A., Roncali, L. and Dammacco, F. (1996). The chick embryo chorioallantoic membrane as a model for in vivo research on angiogenesis. Int. J. Dev. Biol. *40*, 1189–1197.

Ribatti, D., Gualandris, A., Bastaki, M., Vacca, A., Iurlaro, M., Roncali, L. and Presta, M. (1997). New model for the study of angiogenesis and antiangiogenesis in the chick embryo chorioallantoic membrane: the gelatin sponge/chorioallantoic membrane assay. J. Vasc. Res. *34*, 455–463.

Ribatti, D., Alessandri, G., Vacca, A., Iurlaro, M. and Ponzoni, M. (1998). Human neuroblastoma cells produce extracellular matrix-degrading enzymes, induce endothelial cell proliferation and are angiogenic in vivo. Int. J. Cancer 77, 449–454.

Rice, E. G. and Bevilacqua, M. P. (1989). An inducible endothelial cell surface glycoprotein mediates melanoma adhesion. Science 246, 1303–1306.

Ridley, A. J. and Hall, A. (1992). The small GTP-binding protein rho regulates the assembly of focal adhesions and actin stress fibers in response to growth factors. Cell 70, 389–399.

Ridley, A. J., Paterson, H. F., Johnston, C. L., Diekmann, D. and Hall, A. (1992). The small GTP-binding protein rac regulates growth factor-induced membrane ruffling. Cell 70, 401–410.

Riley, V. (1981). Bio-behavioral factors in animal work on tumorgenesis. In: Perspectives on Behavioral Medicine (Weiss, S. M. Herd, J. A. and Fox, B. H., eds.). Academic Press, New York, pp. 183–214

Riley, V., Fitzmaurice, M. A. and Spackman, D. H. (1981). Animal models in bio-behavioral research: effects of anxiety stress on immunocompetence and neoplasia. In: Perspectives on Behavioral Medicine (Weiss, S. M. Herd, J. A. and Fox, B. H., eds.). Academic Press, New York, pp. 371–400.

Risau, W. (1997). Mechanisms of angiogenesis. Nature 386, 671–674.

Risio, M., Coverlizza, S., Poccardi, G., Candelaresi, G. L. and Gaiola, O. (1986). In vitro immunohistochemical localization of S-phase cells by a monoclonal antibody to bromodeoxyuridine. Basic Appl. Histochem. 30, 469–477.

Ritland, S. R., Rowse, G. J., Chang, Y. and Gendler, S. J. (1997). Loss of heterozygosity analysis in primary mammary tumors and lung metastases of MMTV-MTAg and MMTV-neu transgenic mice. Cancer Res. 57, 3520–3525.

Rockwell, S. C., Kallman, R. F. and Fajardo, L. F. (1972). Characteristics of a serially transplanted mouse mammary tumor and its tissue- culture-adapted derivative. J. Natl. Cancer Inst. 49, 735–749.

Rodier, J. M., Valles, A. M., Denoyelle, M., Thiery, J. P. and Boyer, B. (1995). pp60c-src is a positive regulator of growth factor-induced cell scattering in a rat bladder carcinoma cell line. J. Cell Biol. 131, 761–773.

Roehm, N. W., Rodgers, G. H., Hatfield, S. M. and Glasebrook, A. L. (1991). An improved colorimetric assay for cell proliferation and viability utilizing the tetrazolium salt XTT. J. Immunol. Methods, 142, 257–265.

Rogelj, S., Weinberg, R. A., Fanning, P. and Klagsbrun, M. (1988). Basic fibroblast growth factor fused to a signal peptide transform cells. Nature 331, 173–175.

Rogelj, S., Klagsbrun, M., Atzmon, R., Kurokawa, M., Haimovitz, A., Fuks, Z. and Vlodavski, I. (1989). Basic FGF is an ECM component required for supporting the proliferation of vascular EC and the differentiation of PC12 cells. J. Cell Biol. 109, 823–831.

Rong, S., Segal, S., Anver, M., Resau, J. H. and Vande Woude, J. F. (1993). Invasiveness and metastasis of NIH 3T3 cells induced by met-HGF/SF autocrine stimulation. Proc. Natl. Acad. Sci. USA *91*, 4731–4745.

Roos, E. (1991). Adhesion molecules in lymphoma metastasis. Cancer Metastasis Rev. *10*, 33–48.

Roos, E. and Dingemans, K. P. (1979). Mechanisms of metastasis. Biochim. Biophys. Acta *560*, 135–166.

Rosales, C., O'Brien, V., Kornberg, L. and Juliano, R. (1995). Signal transduction by cell adhesion receptors. Biochim. Biophys. Acta *1242*, 77–98.

Rosen, E., Goldberg, I., Liu, D., Setter, E., Donovan, M., Bhargava, M., Reiss, M. and Kacinski, B. (1991). TNF stimulates epithelial tumor cell motility. Cancer Res. *51*, 5315–5321.

Rossi, M. C. and Zetter, B. R. (1992). Selective stimulation of prostatic carcinoma cell proliferation by transferrin. Proc. Natl. Acad. Sci. USA *89*, 6197–6201.

Ruiz, P., Dunon, D., Sonnenberg, A. and Imhof, B. A. (1993). Suppression of mouse melanoma metastasis by EA–1, a monoclonal antibody specific for alfa–6 integrins. Cell Adh. Commun. *1*, 67–81.

Ruoslahti, E. (1994–95). Fibronectin and its alpha 5 beta 1 integrin receptor in malignancy. Invasion Metastasis *14*, 87–97.

Rusciano, D. and Burger, M. M. (1992). Why do cancer cells metastasize into particular organs? BioEssays, *14*, 185–194.

Ruoslahti, E. and Yamaguchi, Y. (1991). Proteoglycans as modulators of growth factor activities. Cell *64*, 867–869.

Rusciano, D. and Burger, M. M. (1993). Mechanisms of metastasis. In: Molecular Genetics of Nervous System Tumors Levine, A. J. and Schmidek, H. H., eds.). Wiley-Lyss, New York, pp. 357–369.

Rusciano, D., Lorenzoni, P. and Burger, M. M. (1991). The role of both specific cellular adhesion and growth promotion in liver colonization by F9 embryonal carcinoma cells. Int. J. Cancer *48*, 450–456.

Rusciano, D., Lorenzoni, P. and Burger, M. M. (1992). Specific growth stimulation in the absence of specific cellular adhesion in lung colonization by retinoic acid treated F9 teratocarcinoma cells. Int. J. Cancer *52*, 471–477.

Rusciano, D., Lorenzoni, P. and Burger, M. M. (1993). Paracrine growth response as a major determinant in liver-specific colonization by in vivo selected B16 murine melanoma cells. Invasion Metastasis *13*, 212–224.

Rusciano, D., Lorenzoni, P. and Burger, M. M. (1994). Murine models of liver metastasis. Invasion Metastasis *14*, 349–361.

Rusciano, D., Lorenzoni, P. and Burger, M. M. (1995). Expression of constitutively activated HGF/SF receptor (c-met) in B16 melanoma cells selected for enhanced liver colonization. Oncogene *11*, 1979–1987.

Rusciano, D., Lorenzoni, P. and Burger, M. M. (1996). Constitutive activation of c-met in liver metastatic B16 melanoma cells depends on both substrate adhesion and cell density, and is regulated by a cytosolic tyrosine phosphatase activity. J. Biol. Chem. *271*, 20763–20769.

Rusciano, D., Lin, S., Lorenzoni, P., Casella, N. and Burger, M. M. (1998a) The influence of HGF/SF on the metastatic phenotype of B16 melanoma cells. Tumor Biol. *19*, 335–345.

Rusciano, D., Lorenzoni, P., Lin, S. and Burger, M. M. (1998b) HGF/SF and hepatocytes are potent downregulators of tyrosinase expression in B16 melanoma cells. J. Cell. Biochem. *71*, 264–276.

Russell, P. J., Bennett, S. and Stricker, P. (1998). Growth factor involvement in progression of prostate cancer. Clin. Chem. *44*, 705–723.

Russo, R. (1986). Cell-substrate interactions in relation to metastasis. In Chadwick, C. M. (ed.). Receptors in Tumor Biology. Cambridge University Press, Cambridge, UK, pp. 131–168.

Ryan, U., White, L., Lopez, M. and Ryan, J. (1982). Use of microcarriers to isolate and culture pulmonary microvascular endothelium. Tissue Cell *14*, 597–606.

Sadano, H., Shimokawa-Kuroki, R. and Taniguchi, S. (1994). Intracellular localization and biochemical function of variant beta-actin, which inhibits metastasis of B16 melanoma. Jpn. J. Cancer Res. *85*, 735–743.

Safarians, S., Sternlicht, M. D., Freiman, C. J., Huaman, J. A. and Barsky, S. H. (1996). The primary tumor is the primary source of metastasis in a human melanoma scid model. Implications for the direct autocrine and paracrine epigenetic regulation of the metastatic process. Intl. J. Cancer *66*, 151–158.

Saiki, I., Murata, J., Iida, J., Sakurai, T., Nishi, N., Matsuno, K. and Azuma, I. (1989). Antimetastatic effects of synthetic polypeptides containing repeated structures of the cell adhesive Arg-Gly-Asp (RGD) and Tyr-Ile-Gly-Ser-Arg (YIGSR) sequences. Br. J. Cancer *60*, 722–728.

Saito, K., Uda, H., Tanaka, S., Kuwabara, H. and Sakamoto, H. (1992). A light and electron microscopic histochemical study on lectin binding to cells with high metastatic potential in lewis lung carcinoma. J. Exp. Pathol. *6*, 123–132.

Sakariassen, K. S., Aarts, P. A. M. M., de Groot, P. G., Houdijk, W. P. M. and Sixma, J. J. (1983). A perfusion chamber developed to investigate platelet interaction in flowing blood with human vessel wall cells, their ECM, and purified components. J. Lab. Clin. Med. *102*, 522–535.

Samiei, M. and Waghorne, C. G. (1991). Clonal selection within metastatic SP1 mouse mammary tumors is independent of metastatic potential. Int. J. Cancer *47*, 771–775.

Sargent, N. S. E., Oestreicher, M., Haidvogl, H., Madnick, H. M. and Burger, M. M. (1988). Growth regulation of cancer metastasis by their host organ. Proc. Natl. Acad. Sci. USA *85*, 7251–7255.

Sasaki, A., Boyce, B. F., Story, B., Wright, K. R., Chapman, M., Boyce, R., Mundy, G. R. and Yoneda, T. (1995). Bisphosphonate risedronate reduces metastatic human breast cancer burden in bone in nude mice. Cancer Res. 55, 3551–3557.

Sasaki, C. and Passaniti, A. (1998). Identification of anti-invasive but non-cytotoxic chemoterapeutic agents using the tetrazolium dye MTT to quantitate viable cells in Matrigel. Biotechniques 24, 1038–1043.

Sato, H. and Suzuki, M. (1976). Deformability and viability of tumor cells by transcapillary passage with reference to organ affinity in metastasis in cancer. In: Weiss, L. (ed.). Fundamental Aspects of Metastasis. North Holland, Amsterdam, pp. 311–318.

Sato, H., Khato, J., Sato, T. and Suzuki, M. (1977). Deformability and filterability of tumor cells through 'nucleopore' filter, with reference to viabiltiy and metastatic spread. GANN 20, 3–11.

Sato, H., Takino, T., Okada, Y., Cao, J., Shinagawa, A., Yamamoto, E. and Seiki, M. (1994). A matrix metallo-proteinase expressed on the surface of invasive tumor cells. Nature 370, 61–65.

Schackert, G. and Fidler, I. J. (1988a) Development of in vivo models for studies of brain metastasis. Int. J. Cancer 41, 589–594.

Schackert, G. and Fidler, I. J. (1988b) Site-specific metastasis of mouse melanomas and a fibrosarcoma in the brain or meninges of syngeneic animals. Cancer Res. 48, 3478–3484.

Schackert, G., Price, J. E., Bucana, C. D. and Fidler, I. J. (1989). Unique patterns of brain metastasis produced by different human carcinomas in athymic nude mice. Int. J. Cancer 44, 892–897.

Schackert, G., Price, J. E., Zhang, R. D., Bucana, C. D., Itoh, K. and Fidler, I. J. (1990). Regional growth of different human melanomas as metastases in the brain of nude mice. Am. J. Pathol. 136, 95–102.

Schmidt, C., Verschueren, H., Toussaint-Demylle, D., van den Berg, T., Kraal, G. and de Baetselier, P. (1994). The role of the spleen in the organ-specific metastasis of murine BW 5147 T lymphomas. Clin. Exp. Metastasis 12, 164–174.

Schmuke, J. J. and Welply, J. K. (1995). A method for measuring leukocyte rolling on the selectins. Anal. Biochem. 226, 197–201.

Schor, S. L., Schor, A. M., Grey, A. M. and Rushton, G. (1988). Foetal and cancer patients fibroblasts produce an autocrine migration stimulating factor not made by normal adult cells. J. Cell Sci. 90, 391–399.

Schor, S. L., Grey, A. M., Ellis, I., Schor, A. M., Coles, B. and Murphy, R. (1993). Migration stimulating factor (MSF): its structure, mode of action and possible function in health and disease. Symp. Soc. Exp. Biol. 47, 235–251.

Schuetze, S. M. and Goodenough, D. A. (1982). Dye transfer between cells of the embryonic chick lens becomes less sensitive to CO_2 treatment with development. J. Cell Biol. *92*, 694–705.

Scollay, R. and Shortman, K. (1985). Identification of early stages of T lymphocyte development in the thymus cortex and medulla. J. Immunol. *134*, 3632–3642.

Scudiero, D. A., Shoemaker, R. H., Paull, K. D., Monks, A., Tierney, S., Nofziger, T. H., Currens, M. J., Seniff, D. and Boyd, M. R. (1988). Evaluation of a soluble tetrazolium/formazan assay for cell growth and drug sensitivity in culture using human and other tumor cell lines. Cancer Res. *48*, 4827–4833.

Seebacher, T., Manske, M., Kornblihtt, A. R. and Bade, E. G. (1988). Cellular fibronectin is induced by EGF, but not by dexamethasone or cyclic AMP in rat liver epithelial cells. FEBS Lett. *239*, 113–116.

Seebacher, T., Manske, M., Zoller, J., Crabb, J. and Bade, E. G. (1992). The EGF-inducible protein EIP–1 of migrating normal and malignant rat liver epithelial cells is identical to plasminogen activator inhibitor 1, and is a component of the ECM migration tracks. Exp. Cell Res. *203*, 504–507.

Seftor, R. E., Seftor, E. A., Gehlsen, K. R., Stetler-Stevenson, W. G., Brown, P. D., Ruoslahti, E. and Hendrix, M. J. (1992). Role of the alpha-v-beta–3 integrin in human melanoma cell invasion. Proc. Natl. Acad. Sci. USA *89*, 1557–1561.

Sharma, B., Handler, M., Eichstetter, I., Whitelock, J. M., Nugent, M. A. and Iozzo, R. V. (1998). Antisense targeting of perlecan blocks tumor growth and angiogenesisIn vivo [In Process Citation]. J. Clin. Invest. *102*, 1599–608.

Sheikh, M. S., Garcia, M., Pujol, P., Fontana, J. A. and Rochefort, H. (1994–95). Why are estrogen receptor-negative breast cancers more aggressive than the estrogen receptor-positive breast cancers? Invasion Metastasis *14*, 329–336.

Sherer, G., Fitzharris, T., Faulk, W. and LeRoy, E. (1980). Cultivation of microvascular endothelial cells from human preputial skin. In Vitro *16*, 675–684.

Shibamoto, S., Hayakawa, M., Takeuchi, K., Hori, T., Oku, N., Miyazawa, K., Kitamura, K., Takeichi, M. and Ito, F. (1994). Tyrosine phosphorylation of beta-catenin and plakoglobin enhanced by HGF and EGF in human carcinoma cells. Cell Adh. Commun. *1*, 295–305.

Shimoyama, Y., Nagafuchi, A., Fujita, S., Gotoh, M., Takeichi, M., Tsukita, S. and Hirohashi, S. (1992). Cadherin dysfunction in a human cancer cell line: possible involvement of loss of alfa-catenin expression in reduced cell-cell adhesiveness. Cancer Res. *52*, 5770–5774.

Shoemaker, R. H., Dykes, D. J., Plowman, J., Harrison, S. D., Griswold, D. P., Abbott, B. J., Mayo, J. G., Fodstad, O. and Boyd, M. R. (1991). Practical spontaneous metastasis model for in vivo therapeutic studies using a human melanoma. Cancer Res. *51*, 2837–2841.

Shrestha, P., Sumitomo, S., Lee, C. H., Nagahara, K., Kamegai, A., Yamanaka, T., Takeuchi, H., Kusakabe, M. and Mori, M. (1996). Tenascin: growth and adhesion modulation—extracellular matrix degrading function: an in vitro study. Eur. J. Cancer B Oral Oncol. *32B*, 106–113.

Sidebottom, E. and Clark, S. R. (1983). Cell fusion segregates progressive growth from metastasis. Br. J. Cancer *47*, 399–405.

Silletti, S., Watanabe, H., Hogan, V., Nabi, I. R. and Raz, A. (1991). Purification of B16-F1 melanoma autocrine motility factor and its receptor. Cancer Res. *51*, 3507–3511.

Silletti, S., Yao, J. P., Pienta, K. J. and Raz, A. (1995). Loss of cell-contact regulation and altered responses to AMF correlate with increased malignancy. Int. J. Cancer *63*, 100–105.

Simon, C., Nemechek, A. J., Boyd, D., O'Malley, B. W., Jr., Goepfert, H., Flaitz, C. M. and Hicks, M. J. (1998). An orthotopic floor-of-mouth cancer model allows quantification of tumor invasion. Laryngoscope *108*, 1686–1691.

Singer, S. J. and Kupfer, A. (1986). The directed migration of eukaryotic cells. Ann. Rev. Cell Biol. *2*, 337–365.

Singh, R. K. and Fidler, I. J. (1996). Regulation of tumor angiogenesis by organ-specific cytokines. Curr. Top. Microbiol. Immunol. *213*, 1–11.

Singh, R. K., Bucana, C. D., Gutman, M., Fan, D., Wilson, M. R. and Fidler, I. J. (1994). Organ site-dependent expression of basic fibroblast growth factor in human renal cell carcinoma cells. Am. J. Pathol. *145*, 365–374.

Singh, R. K., Tsan, R. and Radinsky, R. (1997). Influence of the host microenvironment on the clonal selection of human colon carcinoma cells during primary tumor growth and metastasis. Clin. Exptl. Metastasis *15*, 140–150.

Sironi, M., Breviario, F., Proserpio, P., Biondi, A., Vecchi, A., Damme, J. V., Dejana, E. and Mantovani, A. (1989). IL-1 stimulates IL-6 production in endothelial cells. J. Immunol. *142*, 549–553.

Smail, E. H., Cronstein, B. N., Meshulam, T., Esposito, A. L., Ruggeri, R. W. and Diamond, R. D. (1992). In vitro, *Candida albicans* releases the immune modulator adenosine and a second, high-molecular weight agent that blocks neutrophil killing. J. Immunol. *148*, 3588–3595.

Smith, D. F., Larsen, R. D., Mattox, S., Lowe, J. B. and Cummings, R. D. (1990). Transfer and expression of a murine UDP-Gal-:β-D-Gal-α1,3-galactosyltransferase gene in transfected CHO cells. J. Biol. Chem. *265*, 6225–6234.

Smith, D. F., Larsen, R. D., Mattox, S., Lowe, J. B. and Cummings, R. D. (1990). Transfer and expression of a murine UDP-Gal-:β-D-Gal- α1,3-galactosyltransferase gene in transfected CHO cells. J. Biol. Chem. *265*, 6225–6234.

Smith, R. C., Litwin, M. S., Lu, Y. and Zetter, B. R. (1995). Identification of an endogenous inhibitor of prostatic carcinoma cell growth. Nature Med. *1*, 1040–1045.

Smith-Templeton, N., Brown, P. D., Levy, A. T., Margulies, I. M. K., Liotta, L. A. and Stetler-Stevenson, W. G. (1990). Cloning and characterization of human tumor cell interstitial collagenase. Cancer Res. *50*, 5431–5437.

Soininen, R., Huotari, M., Ganguly, A., Prockop, D. J. and Tryggvason. K. (1989). Structural organization of the gene for the alpha chain of human type IV collagen. J. Biol. Chem. *264(23)*, 13565–13571.

Southam, C. M. and Brunschwig, A. (1961). Quantitative studies of autotransplantation of human cancer. Cancer *14*, 971–978.

Southern, P. J. and Berg, P. (1982). Transformation of mammalian cells to antibiotic resistance with a bacterial gene under control of the SV40 early region promoter. J. Mol. Appl. Genet. *1*, 327–341.

Spanel-Burowski, K., Schnapper, U. and Heymer, B. (1988). The chick chorioallantoic membrane assay in the assessment of angiogenic factors. Biomed. Res. *9*, 253–260.

Spriggs, D. R., Imamura, K., Rodriguez, C., Sariban, E. and Kufe, D. W. (1988). Tumor necrosis factor expression in human epithelial tumor cell lines. J. Clin. Invest. *81*, 455–460.

St. John, J. J., Schroen, D. J. and Cheung, H. T. (1994). An adhesion assay using minimal shear force to remove non-adherent cells. J. Immunol. Meth. *170*, 159–166.

Stachowiak, M. K., Moffett, J., Maher, P., Tucholski, J. and Stachowiak, E. K. (1997). Mol. Neurobiol. *15*, 257–283.

Stallmach, A., von Lampe, B., Orzechowski, H. D., Matthes, H. and Riecken, E. O. (1994). Increased fibronectin-receptor expression in colon carcinoma-derived HT 29 cells decreases tumorigenicity in nude mice. Gastroenterology *106*, 19–27.

Stamper, H. B. and Woodruff, J. J. (1976). Lymphocyte homing into lymph nodes: in vitro demonstration of the selective affinity of recirculating lymphocytes for high-endothelial venules. J. Exp. Med. *144*, 828–833.

Stamper, H. B. and Woodruff, J. J. (1977). An in vitro model of lymphocyte homing. I. Characterization of the interaction between thoracic duct lymphocytes and specialized high-endotelial venules of lymph-nodes. J. Immunol. *119*, 772–780.

Stanley, P., Caillibot, V. and Siminovitch, L. (1975). Selection and characterization of eight phenotypically distinct lines of lectin-resistant Chinese hamster ovary cell. Cell *6*, 121–128.

Starkey, J. R., Davis, W. C. and Talmadge, J. E. (1982). Immunoselection of tumor variants resistant to antibody-mediated cytotoxicity. Their immunologic and metastatic characterization. Cancer Immunol. Immunother. *14*, 124–131.

Staroselsky, A. N., Radinsky, R., Fidler, I. J., Pathak, S., Chernajovsky, Y. and Frost, P. (1992). The use of molecular genetic markers to demonstrate the effect of organ environment on clonal dominance in a human renal-cell carcinoma grown in nude mice. Int. J. Cancer *51*, 130–138.

Steele, R. E. (1989). Membrane-anchored growth factors work! Trends Biochem. Sci. *14*, 201–202.

Steinberg, M. S. and Foty, R. A. (1997). Intercellular adhesions as determinants of tissue assembly and malignant invasion. J. Cell Physiol. *173*, 135–139.

Stephenson, R. A., Dinney, C. P., Gohji, K., Ordonez, N. G., Killion, J. J. and Fidler, I. J. (1992). Metastatic model for human prostate cancer using orthotopic implantation in nude mice. J. Natl. Cancer Inst. *84*, 951–957.

Steplewski, Z., Robinson Goldman, P. and Vogel, W. H. (1987). Effect of housing stress on the formation and development of tumors in rats. Cancer Lett. *34*, 257–261.

Stetler-Stevenson, W. G., Brown, P. D., Onisto, M., Levy, A. T. and Liotta, L. A. (1990). Tissue inhibitor of metalloproteinase–2 (TIMP-2). mRNA expression in tumor cell lines and human tumor tissues. J. Biol. Chem. *265*, 13933–13938.

Stoker, M. (1989). Effect of scatter factor on motility of epithelial cells and fibroblasts. J. Cell Physiol. *139*, 565–569.

Stoker, M. and Gherardi, E. (1991). Regulation of cell movement: the motogenic cytokines. Biochem. Biophys. Acta *1072*, 81–102.

Stoker, M. and Perryman, M. (1985). An epithelial scatter factor released by embryo fibroblasts. J. Cell Sci. *77*, 209–223.

Stoker, M., Gherardi, E., Perryman, M. and Gray, J. (1987). Scatter factor is a fibroblast-derived modulator of epithelial cell mobility. Nature *327*, 239–242.

Stokes, C. L., Rupnick, M. A., Williams S. K. and Lauffenburger, D. A. (1990). Chemotaxis of human microvessel endothelial cells in response to acidic FGF. Lab. Invest. *63*, 657–668.

Stossel, T. P. (1989). From signal to pseudopod. How cells control cytoplasmic actin assembly. J. Biol. Chem. *264*, 18261–18264.

Stossel, T. P. (1993). On the crawling of animal cells. Science *260*, 1086–1094.

Stracke, M. L., Krutzsch, H. C., Unsworth, E. J., Arestad, A., Cioce, V., Schiffmann, E. and Liotta, L. (1992). Identification, purification and partial sequence analysis of autotaxin, a novel motility-stimulating protein. J. Biol. Chem. *267*, 2524–2529.

Striker, G., Soderland, C., Schmer, G., Johnson, A., Ross, R. and Striker, L. (1984). Isolation, characterization and propagation in vitro of human glomerular endothelial cells. J. Exp. Med. *160*, 323–328.

Strobel, E. S., Strobel, H. G., Bross, K. J., Winterhalter, B., Fiebig, H. H., Shildge, J. U. and Loehr, J. W. (1989). Effects of human bone marrow stroma on the growth of human tumor cells. Cancer Res. *49*, 1001–1007.

Sugarbaker, E. V. (1952). The organ selectivity of experimentally induced metastases in rats. Cancer *5*, 606–612.

Sugarbaker, E. V. (1979). Cancer metastasis: a product of tumor-host interactions. Curr. Probl. Cancer *3*, 1–59.
Sugarbaker, E. V. (1981). Patterns of metastasis in human malignancies. Cancer Biol. Rev. *2*, 235–303.
Sugarbaker, E. V., Cohen, A. M. and Ketcham, A. S. (1971). Do metastases metastasize? Ann. Surg. *2174*, 161–166.
Sugarbaker E. V., Thornthwaite J. and Ketcham A. S. (1977). Inhibitory effect of a primary tumor on metastasis In: Progress in cancer research and therapy. (SB Day, WPL Myers, P Stansly, S Garratini and MG Lewis., eds.). Raven Press, New York, pp. 227–240.
Sugimoto, Y., Watanabe, M., Oh-hara, T., Sato, S., Isoe, T. and Tsuruo, T. (1991). Suppression of experimental lung colonization of a metastatic variant of murine colon adenocarcinoma 26 by MoAb 8F11 inhibiting TCIPA. Cancer Res. *51*, 921–925.
Sun, P. C., El-Mofty, S. K., Haughey, B. H. and Scholnick, S. B. (1995). Allelic loss in squamous cell carcinomas of the larynx: discordance between primary and metastatic tumors. Genes Chromosomes Cancer *14*, 145–148.
Sung, V., Stubbs, J. T. 3rd, Fisher, L., Aaron, A. D. and Thompson, E. W. (1998a) Bone sialoprotein supports breast cancer cell adhesion proliferation and migration through differential usage of the alpha(v)beta3 and alpha(v)beta5 integrins. J. Cell Physiol. *176*, 482–494.
Sung, V., Gilles, C., Murray, A., Clarke, R., Aaron, A. D., Azumi, N. and Thompson, E. W. (1998b) The LCC15-MB human breast cancer cell line expresses osteopontin and exhibits an invasive and metastatic phenotype. Exp. Cell Res. *241*, 273–284.
Swallow, C. J., Murray, M. P. and Guillem, J. G. (1996). Metastatic colorectal cancer cells induce matrix metalloproteinase release by human monocytes. Clin. Exp. Metastasis *14*, 3–11.
Swanson, J. A. and McNeil, P. L. (1987). Nuclear reassembly excludes large macromolecules. Science *238*, 548–550.
Sweeney, T. M., Kibbey, M. C., Zain, M., Fridman, R. and Kleinman, H. K. (1991). Basement membrane and the SIKVAV laminin-derived peptide promote tumor growth and metastases. Cancer Metastasis Rev. *10*, 245–254.
Sy, M. S., Liu, D., Schiavone, R., Ma, J., Mori, H. and Guo, Y. (1996). Interactions between CD44 and hyaluronic acid: thier role in tumor growth and metastasis. Curr. Top. Microbiol. Immunol. *213(3)*, 129–153.
Symmans, W. F., Liu, J., Knowles, D. M. and Inghirami, G. (1995). Breast cancer heterogeneity: evaluation of clonality in primary and metastatic lesions. Hum. Pathol. *26*, 210–216.
Tada, H., Shiho, O., Kuroshima, K., Koyama, M. and Tsukamoto, K. (1986). An improved colorimetric assay for interleukin 2. J. Immunol. Meth. *93*, 157–165.

Taipale, J. and Keski-Oja, J. (1997). Growth factors in the extracellular matrix. FASEB J. *11*, 51–59.

Tajima, H., Matsumoto, K. and Nakamura, T. (1992). Regulation of cell growth and motility by hepatocyte growth factor and receptor expression in various cell species. Exp. Cell Res. *202*, 423–431.

Takeichi, M. (1977). Functional correlation between cell adhesive properties and some cell surface proteins. J. Cell Biol. *75*, 464–474.

Takeichi, M. and Okada, T. S. (1972). Roles of magnesium and calcium ions in cell-to-substrate adhesion. Exp. Cell Res. *74*, 51–60.

Talmadge, J. E. and Zbar, B. (1987). Clonality of pulmonary metastases from the bladder 6 subline of the B16 melanoma studied by southern hybridization. J. Natl. Cancer Inst. *78*, 315–320.

Tan, M. H. and Chu, T. M. (1985). Characterization of the tumorigenic and metastatic properties of a human pancreatic tumor cell line (ASPC–1). implanted orthotopically into nude mice. Tumor Biol. *6*, 89–98.

Tang, C. K., Perez, C., Grunt, T., Waibel, C., Cho, C. and Lupu, R. (1996). Involvement of heregulin-beta2 in the acquisition of the hormone- independent phenotype of breast cancer cells. Cancer Res. *56*, 3350–3358.

Tang, D. G. and Honn, K. V. (1994). Adhesion molecules and tumor metastasis: An update. Invasion Metastasis *14*, 109–122.

Tang, D. G., Onoda, J. M., Steinert, B. W., Grossi, I. M., Nelson, K. K., Umbarger, L., Diglio, C. A., Taylor, J. D. and Honn, K. V. (1993). Phenotypic properties of cultured tumor cells: integrin aIIb3 expression, TCIPA, and tumor cell adhesion to endothelium as important parameters of experimental metastasis. Int. J. Cancer *54*, 338–347.

Taniguchi, S., Tatsuka, M., Nakamatsu, K., Inoue, M., Sadano, H., Okazaki, H., Iwamoto, H. and Baba, T. (1989). High invasiveness associated with augmentation of motility in a fos-transferred highly metastatic rat 3Y1 cell line. Cancer Res. *49*, 6738–6744.

Taniguchi, S., Miyamoto, S., Sadano, H. and Kobayashi, H. (1991). Rat elongation factor 1 alpha: sequence of cDNA from a highly metastatic fos-transferred cell line. Nucl. Acids Res. *19*, 6949.

Tao, T. and Burger, M. M. (1977). Non-metastasising variants selected from metastasising melanoma cells. Nature *270*, 437–438.

Tao, T-W. and Burger, M. M. (1982). Lectin-resistant variant of mouse melanoma cells. I. Altered metastasizing capacity and tumorigenicity. Int. J. Cancer *29*, 425–430.

Tao, T., Matter, A., Vogel, K. and Burger, M. M. (1979). Liver-colonizing melanoma cells selected from B16 melanoma. Int. J. Cancer *23*, 854–857.

Tao, T-W., Jenkins, J. M., Vosbeck, K., Matter, A., Miller, M., Jockusch, B. M., Shen, Z. and Burger, M. M. (1983). Lectin-resistant variant of mouse melanoma cells. II. In vitro characteristics. Int. J. Cancer *31*, 239–247.

Taraboletti, G., Roberts, D. D. and Liotta, L. A. (1987). Thrombospondin-induced tumor cell migration: haptotaxis and chemotaxis are mediated by different molecular domains. J. Cell Biol. *105*, 2409–2415.

Taraboletti, G., Belotti, D., Giavazzi, R., Sobel, M. E. and Castronovo, V. (1993). Enhancement of metastatic potential of murine and human melanoma cells by laminin receptor peptide G: attachment of cancer cells to subendothelial matrix as a pathway for hematogenous metastasis. J. Natl. Cancer Inst. *85*, 235–240.

Tarin, D. and Price, J. E. (1981). Influence of microenvironment and vascular anatomy on 'metastatic' colonization potential of mammary tumors. Cancer Res. *41*, 3604–3609.

Tarin, D., Price, J. E, Kettlewell, M. G., Souter, R. G., Vass, A. C. and Crossley, B. (1984a) Mechanisms of human tumor metastasis studied in patients with peritoneovenous shunts. Cancer Res. *44*, 3584–3592.

Tarin, D., Price, J. E, Kettlewell, M. G., Souter, R. G., Vass, A. C. and Crossley, B. (1984b) Clinicopathological observations on metastasis in man studied in patients treated with peritoneovenous shunts. Br. Med. J. (Clin. Res. Ed.) *288*, 749–751.

Tarin, D., Vass, A. C., Kettlewell, M. G. and Price, J. E. (1984c) Absence of metastatic sequelae during long-term treatment of malignant ascites by peritoneo-venous shunting. A clinico-pathological report. Invasion Metastasis *4*, 1–12.

Tashiro, K., Sephel, G. C., Weeks, B., Sasaki, M., Martin, G. R., Kleinman, H. K. and Yamada, Y. (1989). A synthetic peptide containing the IKVAV sequence from the A chain of laminin mediates cell attachment, migration, and neurite outgrowth. J. Biol. Chem. *264*, 16174–16182.

Tashiro, K., Hagiya, M., Nishizawa, T., Seki, T., Shimonishi, M., Shimizu, S. and Nakamura, T. (1990). Deduced primary structure of rat hepatocyte growth factor and expression of the mRNA in rat tissue. Proc. Natl. Acad. Sci. USA *87*, 3200–3204.

Tatsuka, M., Jinno, S. and Kakunaga, T. (1989). Quantitative measurement of cell motility associated with transformed phenotype. Jpn. J. Cancer Res. *80*, 408–412.

Taub, M., Wang, Y., Szczesny, T. M. and Kleinman, H. K. (1990). Epidermal growth factor or transforming growth factor a is required for kidney tubulogenesis in matrigel cultures in serum-free medium. Proc. Natl. Acad. Sci. USA *87*, 4002–4006.

Terrana, B., Rusciano, D. and Pacenti, L. (1987). Organ colonization pattern of retinoic acid-treated and -untreated mouse embryonal carcinoma F9 cells. Cancer Res. *47*, 3791–3797.

Terranova, V. P., Williams, J. E., Liotta, L. A. and Martin, G. R. (1984). Modulation of the metastatic activity of melanoma cells by laminin and fibronectin. Science *226*, 982–985.

Terranova, V. P., Maslow, D. and Markus, G. (1989). Directed migration of murine and human tumor cells to collagenases and other proteases. Cancer Res. *49*, 4835–4841.

Theodorescu, D., Cornil, I., Fernandez, B. J. and Kerbel, R. S. (1990). Overexpression of normal and mutated forms of hras induces orthotopic bladder invasion in a human transitional cell carcinoma. Proc. Natl. Acad. Sci. USA *87*, 9047–9051.

Theodorescu, D., Cornil, I., Sheehan, C., Man, M. S. and Kerbel, R. S. (1991a) Ha-ras induction of the invasive phenotype results in upregulation of epidermal growth factor receptors and altered responsiveness to epidermal growth factor in human papillary transitional cell carcinoma cell. Cancer Res. *51*, 4475–4491.

Theodorescu, D., Cornil, I., Sheehan, C., Man, S. and Kerbel, R. S. (1991b) Dominance of metastatically competent cells in primary murine breast neoplasms is necessary for distant metastatic spread. Int. J. Cancer *47*, 118–123.

Thiery, J. P. and Boyer, B. (1992). The junction between cytokines and cell adhesion. Curr. Opin. Cell Biol. *4*, 782–792.

Thorgeirsson, U. P., Turpeenniemi-Hujanen, T., Neckers, L. M., Johnson, D. W. and Liotta, L. A. (1984). Protein synthesis, but not DNA synthesis is required for tumor cell invasion in vitro. Invasion Metastasis *4*, 73–84.

Timar, J., Trikha, M., Szekeres, K., Bazaz, R., Tovari, J., Silletti, S. and Honn, K. V. (1996). Autocrine motility factor signals integrin-mediated metastatic melanoma cell adhesion and invasion. Cancer Res. *56*, 1902–1908.

Tjernberg, B. and Zajichek, J. (1965). Cannulation of lymphatics leaving cancerous nodes in studies of tumor spread. Acta Cytol. *9*, 197–202.

Tlsty, T. D. (1997). Genomic instability and its role in neoplasia. Curr. Top. Microbiol. Immunol. *221*, 37–46.

Toezeren, A. and Ley, K. (1992). How do selectins mediate leukocyte rolling in venules? Biophys. J. *63*, 700–709.

Toezeren, A., Kleinman, H. K., Grant, D. S., Morales, D., Mercurio, A. M. and Byers, S. W. (1995). E-selectin-mediated dynamic interactions of breast and colon cancer cells with endothelial cell monolayers. Int. J. Cancer *60*, 426–431.

Togo, S., Shimada, H., Kubota, T., Moossa, A. R. and Hoffman, R. M. (1995). Host organ specifically determines cancer progression. Cancer Res. *55*, 681–684.

Toh, Y., Pencil, S. D. and Nicolson, G. L. (1994). A novel candidate metastasis-associated gene, mta1, differentially expressed in highly metastatic mammary adenocarcinoma cell lines. cDNA cloning, expression, and protein analyses. J. Biol. Chem. *269*, 22958–22963.

Toh, Y., Oki, E., Oda, S., Tokunaga, E., Ohno, S., Maehara, Y., Nicolson, G. L. and Sugimachi, K. (1997). Overexpression of the MTA1 gene in gastrointestinal carcinomas: correlation with invasion and metastasis. Int. J. Cancer *74*, 459–463.

Tokuyama, S., Moriya, S., Taniguchi, S., Yasui, A., Miyazaki, J., Orikasa, S. and Miyagi, T. (1997). Suppression of pulmonary metastasis in murine B16 melanoma cells by transfection of a sialidase cDNA. Int. J. Cancer *73*, 410–415.

Tollefsen, D. M., Feagler, J. R. and Majerus, P. W. (1974). The binding of thrombin to human platelets. J. Biol. Chem. *249*, 2646–2651.

Topley, P., Jenkins, D. C., Jessup, E. A. and Stables, J. N. (1993). Effect of reconstituted basement membrane components on the growth of a panel of human tumor cell lines in nude mice. Br. J. Cancer *67*, 953–958.

Torosian, M. H. and Bartlett, D. L. (1993). Inhibition of tumor metastasis by a circulating suppressor factor. J. Surg. Res. *55*, 74–79.

Tsuruoka, T., Tsuji, T., Nojiri, H., Holmes, E. H. and Hakomori, S. (1993). Selection of a mutant cell line based on differential expression of glycosphingolipid, utilizing anti-lactosylceramide antibody and complement. J. Biol. Chem. *268*, 2211–2216.

Tucker, R. F., Shipley, G. D., Moses, H. L. and Holley, R. W. (1986). Growth inhibitor from BSC–1 cells closely related to platelet type b transforming growth factor. Science *226*, 705–707.

Tullberg, K. F. and Burger, M. M. (1985). Selection of B16 melanoma cells with increased metastatic potential and low intercellular cohesion using nucleopore filters. Invasion Metastasis *5*, 1–15.

Tullberg, K. F., Haidvogl, H., Obrist, R., Burger, M. M. and Obrecht, J. P. (1989). Selection of highly malignant tumor cells using reconstituted basement membrane matrix. Invasion Metastasis *9*, 15–26.

Turpeenniemi-Hujanen, T., Thorgeirsson, U. P., Rao, C. N. and Liotta, L. A. (1986). Laminin increases the release of type IV collagenase from malignant cells. J. Biol. Chem. *261*, 1883–1889.

Uchida, S., Shimada, Y., Watanabe, G., Li, Z. G., Hong, T., Miyake, M. and Imamura, M. (1999). Motility-related protein (MRP–1/CD9). and KAI1/CD82 expression inversely correlate with lymph node metastasis in oesophageal squamous cell carcinoma. Br. J. Cancer *79*, 1168–1173.

Uchino, S., Tsuda, H., Noguchi, M., Yokota, J., Terada, M., Saito, T., Koboyashi, M., Sugimura, T. and Hirohashi, S. (1992). Frequent loss of heterozygosity at the DCC locus in gastric cancer. Cancer Res. 52, 3099–3102.

Uchiyama, A., Essner, R., Doi, F., Nguyen, T., Ramming, K. P., Nakamura, T., Morton, D. L. and Hoon, D. S. (1996). Interleukin 4 inhibits hepatocyte growth factor-induced invasion and migration of colon carcinomas. J. Cell. Biochem. 62, 443–453.

Ugen, K. E., Mahalingam, M., Klein, P. A. and Kao, K. J. (1988). Inhibition of TCIPA and experimental tumor metastasis by the synthetic Gly-Arg-Gly-Asp-Ser peptide. J. Natl. Cancer Inst. 80, 1461–1466.

Updyke, T. V. and Nicolson, G. L. (1986). Malignant melanoma cell lines selected in vitro for increased homotypic adhesion properties have increased experimental metastasis potential. Clin. Exp. Metastasis 4, 273–284.

Urdal, D. L., Kawase, I. and Henney, C. S. (1982). NK cell-target interactions: approaches toward definition of recognition structures. Cancer Metastasis Rev. 1, 66–83.

Urushihara, H., Takeichi, M., Hakura, A. and Okada, T. S. (1976). Different cation requirements for aggregation of BHK cells and their transformed derivatives. J. Cell Sci. 22, 685–695.

Vacca, A., Ribatti, D., Iurlaro, M., Albini, A., Minischetti, M., Bussolino, F., Pellegrino, A., Ria, R., Rusnati, M., Presta, M., Vincenti, V., Persico, M. G. and Dammacco, F. (1998). Human lymphoblastoid cells produce extracellular matrix-degrading enzymes and induce endothelial cell proliferation, migration, morphogenesis, and angiogenesis. Int. J. Clin. Lab. Res. 28, 55–68.

Vachula, M. and van Epps, D. E. (1992). In vitro models of lymphocyte transendothelial migration. Invasion Metastasis 12, 66–81.

Valles, A. M., Boyer, B., Badet, J., Tucker, G. C., Barritault, D., Thiery, J. P. (1990). Acidic FGF is a modulator of epithelial plasticity in a rat bladder carcinoma line. Proc. Natl. Acad. Sci. USA 87, 1124–1128.

Valle, E. F., Zalka, A. D., Groszek, L. and Stackpole, C. W. (1992). Patterning of B16 melanoma metastasis and colonization generally relates to tumor cell growth-stimulating or growth-inhibiting effects of organs and tissues. Clin. Exp. Metastasis 10, 419–429.

Van den Brenk, H., Burch, W. M., Kelly, H. and Orton, C. (1974). Venous diversion, trapping and growth of blood-borne cancer cells en route to the lungs. Br. J. Cancer 32, 46–61.

Van Leeuwen, F. N., van der Kammen, R. A., Habets, G. G. and Collard, J. G. (1995). Oncogenic activity of Tiam1 and Rac1 in NIH3T3 cells. Oncogene 11, 2215–2221

Van Lohuizen, M. and Berns, A. (1990). Tumorigenesis by slow-transforming retroviruses—an update. Biochim. Biophys. Acta 1032, 213–235.

Van Lohuizen, M., Verbeek, S., Scheijen, B., Wientjens, E., van der Gulden, H. and Berns, A. (1991). Identification of cooperating oncogenes in E mu-myc transgenic mice by provirus tagging. Cell 65, 737–752.

Vande Woude, G. F., Jeffers, M., Cortner, J., Alvord, G., Tsarfaty, I. and Resau, J. (1997). Met-HGF/SF: tumorigenesis, invasion and metastasis. Ciba Found. Symp. 212, 119–130.

Varani, J., Orr, W. and Ward, P. A. (1978). A comparison of the migration patterns of normal and malignant cells in two assay systems. Am. J. Pathol. 90, 159–172.

Vidal-Vanaclocha, F., Rocha, M., Assumendi, A. and Barbera-Guillem, E. (1993). Isolation and enrichment of two sublobular compartment-specific endothelial cell subpopulations from liver sinusoids. Hepatology 18, 328–339.

Vidal-Vanaclocha, F., Amezaga, C., Asumendi, A., Kaplanski, G. and Dinarello, C. A. (1994). IL-1 receptor blockade reduces the number and size of murine B16 melanoma hepatic metastases. Cancer Res. 54, 2667–2672.

Vink, J., Thomas, L., Etoh, T., Bruijn, J. A., Mihm, M. C., Gattoni-Celli, S. and Byers, H. R. (1993). Role of beta-1 integrins in organ-specific adhesion of melanoma cells in vitro. Lab. Invest. 68, 192–203.

Vistica, D. T., Skehan, P., Scudiero, D., Monks, A., Pittman, A. and Boyd, M. R. (1991). Tetrazolium-based assays for cellular viability: a critical examination of selected parameters affecting formazan production. Cancer Res. 51, 2515–2520 (published erratum appears in Cancer Res. 51, 4501).

Vlaeminck, M. N., Adenis, L., Mouton, Y. and Demaille, A. (1972). Etude expérimentale de la diffusion métastatique chez l'oeuf de poule embryonné. Repartition, microscopie et ultrastructure des foyers tumoraux. Int. J. Cancer 10, 619–631.

Vleminckx, K., Vakaet Jr., L., Mareel, M., Fiers, W. and van Roy, F. (1991). Genetic manipulation of E-cadherin expression by epithelial tumor cells reveals an invasion suppressor role. Cell 66, 107–119.

Vlodavski, I., Ariav, Y., Atzmon, R. and Fuks, Z. (1982). Tumor cell attachment to the vascular endothelium and subsequent degradation of the subendothelial ECM. Expl.Cell Res. 140, 149–159.

Vlodavsky, I., Fuks, Z. and Schirrmacher, V. (1983). In vitro studies on tumor cell interaction with the vascular endothelium and the subendothelial basal lamina: relationship to tumor cell metastasis. In: The Endothelial Cell-a Pluripotent Control Cell of the Vessel Wall (Thilo-Koemer, D. G. S. and Freshney, R. I., eds.). Karger, Basel, pp. 126–157.

Vlodavsky, I., Korner, G., Ishai-Michaeli, R., Bashkin, P., Bar-Shavit, R. and Fuks, Z. (1990). Extracellular matrix-resident growth factors and enzymes: possible involvement in tumor metastasis and angiogenesis. Cancer Metastas. Rev. 9, 203–226.

Vlodavsky, I., Bar-Shavit, R., Ishai-Michaeli, R., Bashkin, P. and Fuks, Z. (1991). Extracellular sequestration and release of fibroblast growth factor: a regulatory mechanism? Trends Biochem. Sci. *16*, 268–271.

Voest, E., Kenyon, B., O'Really, M., Truitt, G., D'Amato, R. and Folkman, J. (1995). Inhibition of angiogenesis in vivo by interleukin 12. J. Natl. Cancer Inst. *87*, 581–586.

Vogel, W. H. (1987). Stress—the neglected variable in experimental pharmacology and toxicology. Trends Pharmacol. Sciences *8*, 35–37.

Volk, T., Geiger, B. and Raz, A. (1984). Motility and adhesive properties of high and low metastatic murine neoplastic cells. Cancer Res. *44*, 811–824.

Voyta, J. C., Via, D. P., Butterfield, C. E. and Zetter, B. R. (1984). Identification and isolation of endothelial cells based on their increased uptake of acetylated low density lipoprotein. J. Cell Biol. *99*, 2034–2040.

Vukicevic, S., Kleinman, H. K., Luyten, F. P., Roberts, A. B., Roche, N. S. and Reddy, A. H. (1992). Identification of multiple active growth factors in basement membrane matrigel suggests caution in interpretation of cellular activity related to extracellular matrix components. Exp. Cell Res. *202*, 1–8.

Wacher, M. P., Krutzsch, H. C., Liotta, L. A. and Stetler-Stevenson, W. G. (1990). Development of a novel substrate capture immunoassay for the detection of a neutral metalloproteinase capable of degrading basement membrane (type IV) collagen. J. Immunolog. Metho. *126*, 239–245.

Wadsworth, P. (1999). Microinjection of mitotic cells. Methods Cell Biol. *61*, 219–231.

Waghorne, C., Thomas, M., Lagarde, A., Kerbel, R. S. and Breitman, M. L. (1988). Genetic evidence for progressive selection and overgrowth of primary tumors by metastatic cell subpopulations. Cancer Res. *48*, 6109–6114.

Wagner, C. R., Vetto, R. M. and Burger, D. R. (1983). The mechanism of antigen presentation by endothelial cells. Immunobiology *168*, 453–469.

Walther, B. T., Oehman, R. and Roseman, L. (1973). A quantitative assay for intercellular adhesion. Proc. Natl. Acad. Sci. USA *70*, 1569–1573.

Walz, G., Aruffo, A., Kolanus, W., Bevilacqua, M. P. ans Seed, B. (1990). Recognition by ELAM–1 of the Slex determinant on myeloid and tumor cells. Science *250*, 1132–1135.

Wang, X. J., Greenhalgh, D. A., Jiang, A., He, D. C., Zhong, L., Brinkley, B. R. and Roop, D. R. (1998). Analysis of centrosome abnormalities and angiogenesis in epidermal-targeted p53172H mutant and p53-knockout mice after chemical carcinogenesis: Evidence for a gain of function. Mol. Carcinog. *23*, 185–192.

Warn, R. (1994). A scattering of factors. Curr. Biol. *4*, 1043–1045.

Warren B. A., Chauvin W. J. and Philips J. (1977). Blood-borne tumor emboli and their adherence to vessel walls. In. Progress in Cancer Research and Therapy

(Day, S. B., Myers, W. P. L., Stansly, P., Garratini, S. and Lewis, M. G., eds.). Raven Press, New York, pp. 185–197.

Watanabe, M., Okochi, E., Sugimoto, Y. and Tsuruo, T. (1988). Identification of a platelet aggregating factor of murine colon adenocarcinoma 26, Mw 44,000 membrane protein as determined by MoAbs. Cancer Res. *48*, 6411–6416.

Watanabe, H., Carmi, P., Hogan, V., Raz, T., Silletti, S., Nabi, I.R. and Raz, A. (1991a) Purification of human tumor cell autocrine motility factor and molecular cloning of its receptor. J. Biol. Chem. *266*, 13442–13448.

Watanabe, H., Nabi, I. R. and Raz, A. (1991b) The relationship between motility factor receptor internalization and the lung colonization capacity of murine melanoma cells. Cancer Res. *51*, 2699–2705.

Watanabe, H., Shinozaki, T., Raz, A. and Chigira, M. (1993). Expression of AMF receptor in serum- and protein-independent fibrosarcoma cells: implications for autonomy in tumor-cell motility and metastasis. Int. J. Cancer *53*, 689–695.

Webster, M. A. and Muller, W. J. (1994). Mammary tumorigenesis and metastasis in transgenic mice. Semin. Cancer Biol. *5*, 69–76.

Weidner, K. M., Arakaki, N., Hartmann, G., Vandekerckhove, J., Weingart, S., Rieder, H., Fonatsch, C., Tsubouchi, H., Hishida, T., Daikuhara, Y. and Birchmeier, W. (1991). Evidence for the identity of human scatter factor and human hepatocyte growth factor. Proc. Natl. Acad. Sci. USA *88*, 7001–7005.

Weidner, K. M., Sachs, M. and Birchmeier, W. (1993a) The met receptor tyrosine kinase transduces motility, proliferation, and morphogenic signals of HGF/SF in epithelial cells. J. Cell Biol. *121*, 145–154.

Weidner, N., Carroll, P. R., Flax, J., Blumenfeld, W. and Folkman, J. (1993b) Tumor angiogenesis correlates with metastasis in invasive prostate carcinoma. Am. J. Path. *143*, 401–409.

Weinstat-Saslow, D. L. and Steeg, P. S. (1994). Angiogenesis and colonization in the tumor metastatic process: Basic and applied advances. FASEB J. *8*, 401–407.

Weinstat-Saslow, D. L., Zabrenetzky, V. S., van Houtte, K., Frazier, W.A., Roberts, D.D. and Steeg, P. S. (1994). Transfection of thrombospondin 1 complementary DNA into a human breast carcinoma cell line reduces primary tumor growth, metastatic potential, and angiogenesis. Cancer Res. *54*, 6504–6511.

Weinstein, J. N., Yoshikami, S., Henkart, P., Blumenthal, R. and Hagins, W. A. (1977). Liposome-cell interaction: transfer and intracellular release of a trapped fluorescent marker. Science *195*, 489–492.

Weislow, O. S., Kiser, R., Fine, D. L., Bader, J., Shoemaker, R. H. and Boyd, M. R. (1989). New soluble-formazan assay for HIV-1 cytopathic effects: application to high-flux screening of synthetic and natural products for AIDS-antiviral activity J. Natl. Cancer Inst. *81*, 577–586 (published erratum appears in J. Natl. Cancer Inst. *81*, 963).

Weiss, L. (1976). Cell deformability: some general considerations. In: Fundamental Aspects of Metastasis (Weiss, L., ed.). North-Holland, Amsterdam/American Elsevier, New York, pp. 305–310.

Weiss, L. (1985). Principles of Metastasis. Academic Press, San Diego.

Weiss, L. (1986). Metastatic inefficiency: causes and consequences. Cancer Rev. 3, 1–24.

Weiss, L. (1990). Metastatic inefficiency. Adv. Cancer Res. 54, 159–211.

Weiss, L. (1991). The biomechanics of cancer cell traffic, arrest, and intravascular destruction. In: Microcirculation in Cancer Metastasis (Orr, F. W., Buchanan, M. R. and Weiss, L., eds.). CRC Press, Boca Raton, pp. 131–144.

Weiss, L. (1994). Cell adhesion molecules: A critical examination of their role in metastasis. Invasion Metastasis 14, 192–197.

Weiss, L. and Dimitrov, D. S. (1984). A fluid mechanical analysis of the velocity, adhesion and destruction of cancer cells in capillaries during metastasis. Cell Biophys.6, 9–22

Weiss, L., Orr, F. W. and Honn, K. V. (1988). Interaction of cancer cells with the microvasculature during metastasis. FASEB J. 2, 12–21.

Welch, D. R. (1986). Discussion of the suitability, availability and requirements for in vivo and in vitro models of metastasis In: Cancer Metastasis: Experimental and Clinical Strategies. (Welch, D. R., Bhuyan, B. K. and Liotta, L. A., eds.). Alan R. Liss, New York, pp. 123–150.

Welch, D. R. (1997). Technical considerations for studying cancer metastasis in vivo. Clin. Exptl. Metastasis 15, 272–306.

Welch, D. R. and Tomasovic, S. P. (1985). Implications of tumor progression on clinical oncology. Clin. Exptl. Metastasis 3, 151–188.

Welch, D. R., Neri, A. and Nicolson, G. L. (1983). Comparison of 'spontaneous' and 'experimental' metastasis using rat 13762 mammary adenocarcinoma cell clones. Invasion Metastasis 3, 65–80.

Welch, D. R., Bhuyan, B. K. and Liotta, L. A. (1986), In: Cancer Metastasis: Experimental and Clinical Strategies (Welch, D. R., Bhuyan, B. K. and Liotta, L. A., eds.). Alan R. Liss, New York.

Welch, D. R., Bisi, J. E., Miller, B. E., Conaway, D., Seftor, E. A., Yohem, K. H., Gilmore, L. B., Seftor, R. E. B., Nakajima, M. and Hendrix, M. J. C. (1991). Characterization of a highly invasive and spontaneously metastatic human malignant melanoma cell line. Intl. J. Cancer 47, 227–237.

Wexler, H. (1966). Accurate identification of experimental pulmonary metastases. J. Natl. Cancer Inst. 36, 641–645.

Whalen, G. F. and Sharif, S. F. (1992). Locally increased metastatic efficiency as a reason for preferential metastasis of solid tumors to lymph nodes. Ann. Surg. 215, 166–171.

Whiteside, T. L. and Herberman, R. B. (1989). The role of natural killer cells in human disease. Clin. Immunol. Immunopathol. *53*, 1–23.

Whiteside, T. L., Vujanovic, N. L. and Herberman, R. B. (1998). Natural killer cells and tumor therapy. Curr. Top. Microbiol. Immunol. *230*, 221–244.

Wicha, M. S., Lowrie, G., Kohn, E., Bagavandoss, P. and Mahn, T. (1982). ECM promotes mammary epithelial growth and differentiation in vitro. Proc. Natl. Acad. Sci. USA *79*, 3213–3217.

Wilding, G. (1995). Endocrine control of prostate cancer. Cancer Surv. *23*, 43–62.

Wilhelm, S. M., Collier, I. E. Marmer, B. L., Eisen, A. Z., Grant, G. A. and Goldberg, G. I. (1989). SV-40-transformed human lung fibroblasts secrete a 92-kDa type IV collagenase which is identical to that secreted by normal human macrophages. J. Biol. Chem. *264(29)*, 17213–17221.

Willis, R. A. (1973). The Spread of Tumors in the Human Body, 3rd edn. Butterworth, London.

Wiltrout, R. H., Herberman, R. B., Zhang, S. R., Chirigos, M. A., Ortaldo, J. R., Green, K. M. Jr. and Talmadge, J. E. (1985). Role of organ-associated NK cells in decreased formation of experimental metastases in lung and liver. J. Immunol. *134*, 4267–4275.

Wojtukievicz, M. Z., Tang, D. G., Ciarelli, J. J., Nelson, K. K., Walz, D. A., Diglio, C. A., Mammen, E. F. and Honn, K. V. (1993). Thrombin increases the metastatic potential of tumor cells. Int. J. Cancer *54*, 793–806.

Wood, J. S., Holyoke, E. D., Clason, W. P. C., Sommers, S. C. and Warren, S. (1954). An experimental study of the relationship between tumor size and number of lung metastases. Cancer *7*, 437–443.

Woodruff, M. (1980). The Interactions of Cancer and the Host. Grune and Stratton, New York.

Woodruff, M. F. (1988). Tumor clonality and its biological significance. Adv. Cancer Res. *50*, 197–229.

Woodruff, M. (1990). Cellular Variation and Adaptation in Cancer. Oxford University Press, New York.

Wright, J. A. and Huang, A. (1996). Growth factors in mechanisms of malignancy: roles for TGF-beta and FGF. Histol. Histopathol. *11*, 521–536.

Wright, J. A., Egan, S. E. and Greenberg, A. H. (1990). Genetic regulation of metastatic progression. Anticancer Res. *10*, 1247–1255.

Wu, H. C., Hsieh, J. T., Gleave, M. E., Brown, N. M., Pathak, S. and Chung, L. W. (1994). Derivation of androgen-independent human LNCaP prostatic cancer cell sublines: role of bone stromal cells. Int. J. Cancer *57*, 406–412.

Wu, Z., O'Reilly, M. S., Folkman, J. and Shing, Y. (1997). Suppression of tumor growth with recombinant murine angiostatin. Biochem. Biophys. Res. Commun. *236*, 651–654.

Wysocki, L. J. and Sato, V. L. (1978). 'Panning' for lymphocytes: a method for cell selection. Proc. Natl. Acad. Sci. USA 75, 2844–2848.

Xie, X., Brunner, N., Jensen, G., Albrectsen, J., Gotthardsen, B. and Rygaard, J. (1992). Comparative studies between nude and scid mice on the growth and metastatic behavior of xenografted human tumors. Clin. Exptl. Metastasis 10, 201–210.

Yagel, S., Khokha, R., Denhardt, D. T., Kerbel, R. S., Parhar, R. S. and Lala, P. K. (1989). Mechanisms of cellular invasiveness: a comparison of amnion invasion in vitro and metastatic behavior in vivo. J. Natl. Cancer Inst. 81,768–775.

Yamada, K. M., Akiyama, S. K., Hasegawa, T., Hasegawa, E., Humphries, M. J., Kennedy, D. W., Nagata, K., Urushihara, H., Olden K. and Chen, W. T. (1985). Recent advances in research on fibronectin and other cell attachment proteins. J. Cell. Biochem. 28, 79–97.

Yamada, K. M., Kennedy, D. W., Yamada, S. S., Gralnick, H., Chen, W. T. and Akiyama, S. K. (1990). Monoclonal antibody and synthetic peptide inhibitors of human tumor cell migration. Cancer Res. 50, 4485–4496.

Yamamura, K., Kibbey, M. C. and Kleinman, H. K. (1993). Melanoma cells selected for adhesion to laminin peptides have different malignant properties. Cancer Res. 53, 423–428.

Yamamura, Y., Fischer, B. C., Harnaha, J. B. and Proctor, J. W. (1984). Heterogeneity of murine mammary adenocarcinoma cell subpopulations. In vitro and in vivo resistance to macrophage cytotoxicity and its association with metastatic capacity. Int. J. Cancer 33, 67–72.

Yang, F., Demma, M., Warren, V., Dharmawardhane, S. and Condeelis, J. (1990). Identification of an actin-binding protein from Dictyostelium as elongation factor 1a. Nature 347, 494–496.

Yang, M., Hasegawa, S., Jiang, P., Wang, X. O., Tan, Y. Y., Chishima, T., Shimada, H., Moossa, A. R. and Hoffman, R. M. (1998). Widespread skeletal metastatic potential of human lung cancer revealed by green fluorescent protein expression. Cancer Res. 58, 4217–4221.

Yang, M., Jiang, P., Sun, F. X., Hasegawa, S., Baranov, E., Chishima, T., Shimada, H., Moossa, A. R. and Hoffman, R. M. (1999). A fluorescent orthotopic bone metastasis model of human prostate cancer. Cancer Res. 59, 781–786.

Yano, S., Nokihara, H., Hanibuchi, M., Parajuli, P., Shinohara, T., Kawano, T. and Sone, S. (1997). Model of malignant pleural effusion of human lung adenocarcinoma in SCID mice. Oncol. Res. 9, 573–579.

Yee, D., Cullen, K. J., Paik, S., Perdue, J. F., Hampton, B., Schwartz, A., Lippman, M. E. and Rosen, N. (1988). Insulin-like growth factor II mRNA expression in human breast cancer. Cancer Res. 48, 6691–6696.

Yin, J. J., Selander, K., Chirgwin. J. M., Dallas, M., Grubbs, B. G., Wieser, R., Massagué, J., Mundy, G. R. and Guise, T. A. (1999). TGF-β signaling blockade inhibits PTHrP secretion by breast cancer cells and bone metastases development. J. Clin. Invest. *103*, 197–206.

Yogeeswaran, G. and Salk, P. L. (1981). Metastatic potential is positively correlated with cell surface sialylation of cultured murine tumor cell lines. Science *212*, 1514–1516.

Yoneda, T., Sasaki, A. and Mundy, G. R. (1994). Osteolytic bone metastasis in breast cancer. Breast Cancer Res. Treat. *32*, 73–84.

Yoshimura, M., Nishikawa, A., Ihara, Y., Taniguchi, S. and Taniguchi, N. (1995). Suppression of lung metastasis of B16 mouse melanoma by N-acetylglucosaminyltransferase III gene transfection. Proc. Natl. Acad. Sci. USA *92*, 8754–8758.

Young, M. R. I., Duffie, G. P., Lozano, Y., Young, M. E. and Wright, M. A. (1990). Association of a funtional prostaglandin-E2-protein kinase A coupling with responsiveness of metastatic Lewis lung carcinoma variants to prostaglandin E2 and to prostaglandin E2-producing nonmetastatic Lewis lung carcinoma variants. Cancer Res. *50*, 2973–2978.

Yu, S. F., von Ruden, T., Kantoff, P. W., Garber, C., Seiberg, M., Ruther, U., Anderson, W. F., Wagner, E. F. and Gilboa, E. (1986). Self-inactivating retroviral vectors designed for transfer of whole genes into mammalian cells. Proc. Natl. Acad. Sci. USA *83*, 3194–3198.

Yurchenko, P. D. and Schittny, J. C. (1990). Molecular architecture of basement membranes. FASEB J. *4*, 1577–1590.

Zarnegar, R. and Michalopoulos, G. K. (1995). The many faces of HGF: from hepatopoiesis to hematopoiesis. J. Cell Biol. *129*, 1177–1180.

Zeidman, I. (1961). The fate of circulating tumor cells. I. Passage of cells through capillaries. Cancer Res. *21*, 38–39.

Zeidman, I. and Buss, J. M. (1952). Transpulmonary passage of tumor cell emboli. Cancer Res. *12*, 731–733.

Zetter, B. R. (1988). In: Endothelial Cells, Vol. 2 (Ryan, U., ed.). CRC Press, Boca Raton, pp. 63–80.

Zetter, B. R. (1990). The cellular basis of site-specific tumor metastasis. N. Engl. J. Med. *322*, 605–612.

Zetter, B. R. and Brightman, S. E. (1990). Cell motility and the extracellular matrix. Curr. Opin. Cell Biol. *2*, 850–856.

Zhu, D., Cheng, C. and Pauli, B. U. (1991). Mediation of lung metastasis of murine melanomas by a lung-specific endothelial cell adhesion molecule. Proc. Natl. Acad. Sci. USA *88*, 9568–9572.

Ziche, M., Alessandri, G. and Gullino, P. M. (1989). Gangliosides promote the angiogenic response. Lab. Invest. *61*, 629–634.

Ziche, M., Maglione, D., Ribatti, D., Morbidelli, L., Lago, C., Battisti, M., Paoletti, I., Barra, A., Tucci, M., Parise, G., Vincenti, V., Grager, H., Viglietto,

G. and Persico, G. (1997). Placental growth factor-1 (PlGF-1). is chemotactic, mitogenic and angiogenic. Lab. Invest. *76*, 517–531.

Zigmond, S. H. and Hirsch, J. G. (1973). Leukocite locomotion and chemotaxis. J. Exp. Med. *137*, 387–410.

Zolotukhin, S., Potter, M., Hauswirth, W. W., Guy, J. and Muzyczka, N. (1996). A -humanized' green fluorescent protein cDNA adapted for high-level expression in mammalian cells. J. Virol. *70*, 4646–4654.

Zorzet, S., Perissin, L., Rapozzi, V and Giraldi, T. (1998). Restraint stress reduces the antitumor efficacy of cyclophosphamide in tumor-bearing mice. Brain Behav. Immun. *12*, 23–33.

Zoucas, E., Jorgensen, C. and Bengmark, S. (1986). Hemostasis following inoculation and during spreading of colon carcinoma in the rat. Cancer Res. *46*, 5662–5666.

Zvibel, I., Halpern, Z. and Papa, M. (1998). Extracellular matrix modulates expression of growth factors and growth factor receptors in liver- colonizing colon-cancer cell lines. Int. J. Cancer *77*, 295–301.

Subject index

Index for Cancer Metastasis

(Figures indicated by **bold type**)

Adhesion
    Blood-borne tumor cells 22–29
        Heterotypic aggregation with
            dissociated organ cell assay
            30–32
        Tumor-cell induced platelet
            aggregation 23–29
            Assay techniques 25–29
                In vitro platelet-tumor cell
                    adhesion assay 27–29
                Tumor-cell induced
                    aggregation assay 25–27
            Platelet activation 24
    Intercellular cohesion 10–22 (see also
        Intercellular cohesion)
    Target organ adhesion 29–64 (see also
        Target organ adhesion)
    Tumor cell surface glycoconjugates 9
    Tumor endogenous lectins 9
Angiogenesis 243–269
    Assay techniques
        Chick embryo chorioallantoic
            membrane (CAM) assay
            252–262
            Assessment of antiogenic or
                anti-angiogenic agents
                253–255
                Implants 255, **256**
                Gelatin implants 255, **258**
            Evaluation of the
                vasoproliferative response
                259–261, **260**

            Semiquantitative methods
                259–261, **260**
            Quantitative methods 261
        Histogenesis and structure 252
        Limitations 262
        Preparation of CAM
            Standard technique
                252–253, **254**
            Other techniques 255, **257**
    Corneal assay 244–252, **246**
        Histology 247, 249
        Quantitation 247–248, **248**
        Rabbit model as a preference
            249–251, **251**
        Sample preparation 245–247
        Surgery 245
    Subcutaneous implant assays
        263–269
        Reconstituted basement
            membrane use 263–264
        Chemoinvasion assay
            264–266
        In vivo sponge assay
            266–269
        Morphological studies 266
    Hepatocyte growth factor/scatter factor
        68
    Role in metastatic cascade 2

Cell deformation 90–96, 181–182
    Assay tecnhiques 92–96

Filtration methods for passive
deformability 92–96, **95**
Active deformability 96
Selection of metastatic variants
181–182
Active deformability selection
procedure 181–182

Endothelial cell interactions 34–64
Adhesion to endothelial cells 34–35
Assay techniques 36–64
Attachment assay 47–49
Characterization of endothelial
cells 41–43
Nonthrombogeneic cell
surface 42
Uptake of acetylated LDL
42
Angiotensin-converting
enzyme activity 43
Microvascular endothelial cell
preparation 38–41
From liver 39–41
From lung 39–41
Phenotypic modulation of
endothelial cells 43–47
Expression of adhesion,
growth and lytic
molecules 43–44
Cytokine-mediation 44–45
Organ-specific modulation
45–47
Bovine aortic
endothelial cell assay
45–47
Vascular endothelial cell
preparation 36–38
Adhesion to endothelial cell
monolayers 49–50
Extracellular matrix 45, 50–59, 67, 69,
85–86, 97–122, 127–128, 182–183,
252–262, 264–266
Degradation 97–112
Enzymatic degradation 97, 107–109
(Tables)
Markers of degradation 98
Assay techniques 99–113

Enzyme detection 99–104, 106
(Table)
Substrate-capture ELISA
99–101
Hydrolase assays 101–104
Inverse zymography 103
Solid-phase substrate
assay for glycosidase
104
Zymography 101–103,
103
Northern blotting and
RT-PCR 104, 106 (Table)
Viable cell degradation of ECM
105, 107–109 (Table),
110–112
Immobilized ECM
technique 105, 107–109
(Table), 110–111
Co-polymerized ECM
technique 111–113, **113**
Endothelial cell contact 45
Adhesion to endothelial cells
50–59, **55**
Assay techniques 53–54
ECM coating 53–59, **55**
Organ-specific endothelial cell
modulation 45
Tumor cell invasion 50–54
Invasion 98, 112–122, 182–183
Assay techniques 112–122
Amnion invasion assay
113–117, **116–117**
Cell monolayer assay 119–120
Chemoinvasion (Matrigel^{TM})
assay 117–119
Use in angiogenesis
quantitation 264–266
Use in selecting metastatic
variants 182–183
Chick chorioallantoic
membrane use 120–121
Use in angiogenesis
quantitation 252–262
In vivo monitoring of invasion
pathways 121–122
Organ culture assays 120

Motility of tumor cells 52, 67, 69,
  85–86
    Chemotaxis 52, 67
    ECM migration track assay 85–86
    Haplotaxis 52, 67
  Signalling 52
  Tumor proliferation 127–128
Extravasation 4, 34–64
  Assay techniques 36–64
    Adhesion to endothelial cells 34–35
      Attachment assay 47–49
      Characterization of endothelial
        cells 41–43
          Nonthrombogeneic cell
            surface 42
          Uptake of acetylated LDL
            42
          Angiotensin-converting
            enzyme activity 43
      Microvascular endothelial cell
        preparation 38–41
          From liver 39–41
          From lung 39–41
      Phenotypic modulation of
        endothelial cells 43–47
          Cytokine-mediation 44–45
          Expression of adhesion,
            growth and lytic
            molecules 43–44
          Organ-specific modulation
            45–47
              Bovine aortic
                endothelial cell assay
                45–47
      Vascular endothelial cell
        preparation 36–38
    Adhesion to endothelial cell
      monolayers 49–50

In vivo metastasis assays 207–242,
  216–217 (Tables)
  Animal use 214–219, 216–217 (Tables)
    Immunodeficient animals 215, 218
    Infection surveillance 218–219
    Stress effects 236–242
    Syngeneic animals 215–217, 216
      (Table)
    Transgenic animals 216 (Table)
  Assay techniques 222–228, **228**,
    229–231, **229**
    Experimental metastasis assay 227,
      229–231, **229**
      Inoculum 227, 229
      Intravenous inoculation
        229–231, **229**
    Spontaneous metastasis assay
      222–228, **227**
      Anesthesia 225–228
      Inoculum 224
      Intradermal inoculation
        226–228, **227**
      Mammary pad inoculation 224,
        226, **228**
      Primary tumor removal
        223–224
  Cell line use 210–214
    Inoculum preparation 212–214
      Detachment technique 213–214
      Single cell suspension 212–214
    Phenotypic drift 211–212
    Viability 211
  Criteria for a good model 209–210
  Enumeration of metastasis 231–235,
    233
    Euthanasia 231–232
    Genetic tagging 235
    Image analysis 234
    Lung metastasis 232–237, **233**
  Site of inoculation 219–222
    Multiple sites 222
    Orthotopic site injection 220–221
    Vasculature 219–220
  Statistical considerations 236
Intercellular cohesion 10–22
  Assay techniques 12–16
    Aggregation of cells in suspension
      13–16
    Monolayer adhesion assay 16
    Shearing of cell aggregates 12–13,
      14
  Detachment of cells from the primary
    tumor 4, 10–22
    Adheren junction perturbation 11
    Cadherin expression 11

Catenin expression 11
Cell surface changes 10
Gap junction communication
    disruption 11, 16–22
    Assay techniques 16–22
        Cell microinjection of
            fluorescent dyes 19
        Permeabilization with
            carboxylfluorescein-
            diacetate
            21
        Scrape loading 19–20
        Scratch loading 20
    Loss of cell adhesion genes 10
        Fat 10
        DCC 10
        Von Hippel-Lindau 10
Invasion 98, 112–122, 182–183
    Assay techniques 112–122
        Amnion invasion assay 113–117, **116–117**
        Cell monolayer assay 119–120
        Chemoinvasion (Matrigel^{TM}) assay
            117–119
            Use in angiogenesis quantitation
                264–266
            Use in selecting metastatic
                variants 182–183
        Chick chorioallantoic membrane
            use 120–121
            Use in angiogenesis quantitation
                252–262
        In vivo monitoring of invasion
            pathways 121–122
        Organ culture assays 120

Metastatic cascade 2–5 (review), **3**
    Angiogenesis 2
    Cadherins 4
    Extravasation 4, 34–64 (see also
        Extravasation)
    Metastatic variant selection 163–183
        (see also Metastatic variant
        selection)
    Organ-specific metastasis (see
        Organ-specific metastasis)
    Selectivity 2, 161–162

Metastatic variant selection 161–183
    Animal model systems 164–168, **165**
    Assay techniques
        Active deformability as selection
            criteria 181–182
        Cell surface properties as selection
            criteria 169–176
            Lectin resistance 170–175
                Adhesion to specific
                    substrates 173–175
                    Laminin 173–174
                    Fibronectin 174–175
                Panning 172–173
                Selection procedure
                    174–175
            Complement-mediated cell
                killing selection procedure
                179–181
        Immune resistance as selection
            criteria 176–178
            Loss of MHC antigen
                expression 176–177
            Natural/nonadaptive immunity
                176–178
                NK activity 176–178
                Macrophage-mediated lysis
                    177–178
        Invasion as selection criteria
            182–183
            Chemoinvasion assay 182–183
Motility of tumor cells 65–96
    Animal cell lines 65–66
    Assay techniques 70–90
        Agarose droplet assay 78–80, **79**
        Filter membrane assay 73–77, **76**
            Checkerboard analysis 75, **76**
        In vitro wound assay 86–87, **88–89**
        Migration track assays 83–85
            Phagokinetic track assay 83–85
            Extracellular matrix track assay
                85–86
        Scatter assay 87, 90, **91**
        Time-lapse videomicroscopy 70–73
        Under agarose assay 80–83, **81**
    Deformation (see Cell deformation)
    Human cancer cell lines 66
    Motility as a therapeutic target 66

Motility-related protein 66
Stimulation of motility 66–69
    Chemokinesis 67
        Assay techniques 73–76, 78–83, 86–87
    Chemotaxis 67
        Assay techniques 73–76, 78–83
    Cytokines 67–69, 129, 200
        Autocrine mobility factor 67–69
        Hepatocyte growth factor/scatter factor 68, 128, 200
        Migration stimulation factor 67
        Scatter factor-like factor 68
    Growth factors 69
    Haplotaxis 67
        Assay techniques 77, 86–87
    Traction force 69
        Adhesion molecule expression 69
        Extracellular matrix involvement 69

Organ-specific metastasis
    Evidence in humans (peritoneovenous shunts) 7, 164
    Ewing's theory 5–7, 163
    Multicellular metastatic units 22
    Paracrine interactions 7, 44, 123, 126–128
    Seed and soil theory 5–7, 163
    Target organ adhesion (see Adhesion, target organ adhesion)

Progression 185–206
    Assay techniques 186–199
        Genetic tagging to visualize metastasis 186
            Fluorescent tags 195–199
                Green fluorescent protein (GFP) 195–199, **196**
            Histochemical marker genes 190–195
                Double staining lacZ and alkaline phosphatase reporter gene 193–195
                E. coli lacZ reporter gene staining 191–192
                Placental alkaline phosphatase reporter gene staining 192–193
        Proviral tagging 203–206
        Restriction enzyme/southern blot analysis 186–187, **187**
        Retroviral vector use 188–189
    Clonal dominance 185–189, **187**
    Genetic control of metastasis 199–206
        CD44 splice variants 203
        EF-1α 201
        G-met 200
        mta1 201
        mts1 201
        nm23 202
        pGm21 201
        WDNM1 & 2 202
Proliferation 123–159
    Assay techniques 130–143
        Cell count 130–132
        Incorporation of labelled DNA precursors 132–136
        Enzymatic conversion of tetrazolium dyes 136–138
        Fluorescence cell viability assay 138–139
        Crystal violet staining of cell nuclei 135–141
        Colony formation ability in soft agar 141–143
    Growth factors 123–124, 143–159
        Assay techniques 143–159
            Autocrine effects 143–145
            Paracrine effects 145–147
                Organ extracts 146–147
                Organ perfusates 147
                Organ conditioned medium 148
            Tissue-specific cell co-culture 149–154
                Without cell-cell contact 149–150
                With cell-cell contact 150–154
        Autocrine stimulation 124–126, 143–145
            Extracellular release 125

Ligand-receptor interactions
125
Paracrine stimulation 126–128
ECM/basement membrane
growth factors 127–128
Soluble growth factors 126–127
Growth inhibitory factors 128–129
Juxtacrine growth interaction 150–154

Target organ adhesion 29–64
Assay techniques 36–64
Adhesion to endothelial cells 34–35
Attachment assay 47–49
Characterization of endothelial
cells 41–43
Nonthrombogeneic cell
surface 42
Uptake of acetylated LDL
42
Angiotensin-converting
enzyme activity 43
Microvascular endothelial cell
preparation 38–41
From liver 39–41
From lung 39–41
Phenotypic modulation of
endothelial cells 43–47

Cytokine-mediation 44–45
Expression of adhesion,
growth and lytic
molecules 43–44
Organ-specific modulation
45–47
Bovine aortic
endothelial cell assay
45–47
Vascular endothelial cell
preparation 36–38
Adhesion to cyrostat sections 32–34
Adhesion to endothelial cell
monolayers 49–50
Adhesion to extracellular matrix
assays 50–60
Heterotypic aggregation with
dissociated organ cell assay
30–32
Tumor-cell induced platelet aggregation
23–29
Assay techniques 25–29
In vitro platelet-tumor cell adhesion
assay 27–29
Tumor-cell induced aggregation
assay 25–27
Platelet activation 24

Assays According to their Use

(Figures indicated by **bold type**)

Adhesion
Adhesion to endothelial cells 34–35
Attachment assay 47–49
Characterization of endothelial cells
41–43
Microvascular endothelial cell
preparation 38–41
Phenotypic modulation of endothelial
cells 43–47
Bovine aortic endothelial cell assay
45–47
Vascular endothelial cell preparation
36–38
Adhesion to cyrostat sections 32–34

Adhesion to endothelial cell monolayers
49–50
Adhesion to extracellular matrix assays
50–59, **55**
Aggregation of cells in suspension 13–16
Gap junction communication disruption
assays 11, 16–22
Cell microinjection of fluorescent dyes
19
Permeabilization with
carboxylfluorescein-diacetate 21
Scrape loading 19–20
Scratch loading 20
Heterotypic aggregation with dissociated

organ cell assay 30–32
In vitro platelet-tumor cell adhesion assay 27–29
Monolayer adhesion assay 16
Shearing of cell aggregates 12–13, **14**
Tumor-cell induced aggregation assay 25–27

Angiogenesis
Chick embryo chorioallantoic membrane (CAM) assay 252–262
Corneal assay 244–252, **246**
Subcutaneous implant assays 263–269

Cell deformation
Active deformability 96
Filtration methods for passive deformability 92–96, **95**

Endothelial cell interactions
Adhesion to endothelial cells 34–35
    Attachment assay 47–49
Adhesion to endothelial cell monolayers 49–50
Characterization of endothelial cells 41–43
Microvascular endothelial cell preparation 38–41
Phenotypic modulation of endothelial cells 43–47
    Bovine aortic endothelial cell assay 45–47
Vascular endothelial cell preparation 36–38

Extracellular matrix degradation
Enzyme detection 99–104, 106 (Table)
    Substrate-capture ELISA 99–101
    Hydrolase assays 101–104
    Inverse zymography **103**
    Solid phase substrate assay for glycosidase 104
    Zymography 101–103, **103**
    Northern blotting and RT-PCR 104, 106 (Table)
    Viable cell degradation of ECM 105, 107–109 (Table), 110–112
    Immobilized ECM technique 105, 107–109 (Table), 110–111
    Co-polymerized ECM technique 111–113, 113

Extravasation
Attachment assay 47–49
Characterization of endothelial cells 41–43
Microvascular endothelial cell preparation 38–41
Phenotypic modulation of endothelial cells 43–47
    Bovine aortic endothelial cell assay 45–47
Vascular endothelial cell preparation 36–38

Gap junction communication disruption
Cell microinjection of fluorescent dyes 19
Permeabilization with carboxylfluorescein-diacetate 21
Scrape loading 19–20
Scratch loading 20

In vivo metastasis
Experimental metastasis assay 227, 229–231, **229**
Spontaneous metastasis assay 222–228, **227**

Intercellular cohesion 10–22
Aggregation of cells in suspension 13–16
Monolayer adhesion assay 16
Shearing of cell aggregates 12–13, **14**
Gap junction communication disruption 11, 16–22
    Cell microinjection of fluorescent dyes 19
    Permeabilization with carboxylfluorescein-diacetate 21
    Scrape loading 19–20
    Scratch loading 20

Invasion
Amnion invasion assay 113–117, **116–117**
Cell monolayer assay 119–120
Chemoinvasion (Matrigel^{TM}) assay 117–119
In vivo monitoring of invasion pathways 121–122

Organ culture assays 120

Metastatic variant selection
Active deformability as selection criteria 181–182
Cell surface properties as selection criteria 169–176
    Lectin resistance 170–175
    Complement-mediated cell killing selection procedure 179–181
Immune resistance as selection criteria 176–178
Invasion as selection criteria 182–183
    Chemoinvasion assay 182–183

Motility of tumor cells
Agarose droplet assay 78–80, **79**
Filter membrane assay 73–77, **76**
    Checkerboard analysis 75, **76**
In vitro wound assay 86–87, **88–89**
Migration track assays 83–85
Phagokinetic track assay 83–85
Extracellular matrix track assay 85–86
Scatter assay 87, 90, **91**
Time-lapse videomicroscopy 70–73
Under agarose assay 80–83, **81**
Chemokinesis assay techniques 73–76, 78–83, 86–87
Chemotaxis assay techniques 73–76, 78–83
Haplotaxis assay techniques 77, 86–87

Progression
Genetic tagging to visualize metastasis 186
Fluorescent tags 195–199
    Green fluorescent protein (GFP) 195–199, **196**

Histochemical marker genes 190–195
    Double staining lacZ and alkaline phosphatase reporter gene 193–195
    E. coli lacZ reporter gene staining 191–192
    Placental alkaline phosphatase reporter gene staining 192–193
Proviral tagging 203–206
Restriction enzyme/southern blot analysis 186–187, **187**
Retroviral vector use 188–189

Proliferation
Cell count 130–132
Incorporation of labelled DNA precursors 132–136
Enzymatic conversion of tetrazolium dyes 136–138
Fluorescence cell viability assay 138–139
Crystal violet staining of cell nuclei 135–141
Colony formation ability in soft agar 141–143
Growth factor assays 143–159
    Autocrine effects 143–145
    Paracrine effects 145–147
Tissue specific cell co-culture 149–154
    Without cell-cell contact 149–150
    With cell-cell contact 150–154

Tumor-cell induced platelet aggregation
In vitro platelet-tumor cell adhesion assay 27–29
Tumor-cell induced aggregation assay 25–27

Alphabetical List of Techniques

(Figures indicated by **bold type**)

Active deformability assays 96
Active deformability to select metastatic variants 181–182
Adhesion to cyrostat sections 32–34
Adhesion to endothelial cell monolayers 49–50

Adhesion to extracellular matrix assays 50–60
Agarose droplet motility assay 78–80, **79**
Aggregation of cells in suspension 13–16
Amnion invasion assay 113–117, **116–117**

SUBJECT INDEX

Attachment assay for endothelial cells 47–49
Autocrine effects assays 143–145

Cell count 130–132
Cell microinjection of fluorescent dyes 19
Cell monolayer invasion assay 119–120
Cell surface properties to select metastatic variants 169–176
Chemoinvasion (Matrigel^{TM}) assay 117–119
Chemokinesis assay techniques 73–76, 78–83, 86–87
Chemotaxis assay techniques 73–76, 78–83
Chick embryo chorioallantoic membrane (CAM) assay 252–262
Colony formation ability in soft agar 141–143
Complement-mediated cell killing to select metastatic variants 179–181
Co-polymerized ECM technique 111–113, **113**
Corneal assay for angiogenesis 244–252, **246**
Crystal violet staining of cell nuclei 135–141

Double staining lacZ and alkaline phosphatase reporter genes 193–195

E. coli lacZ reporter gene staining 191–192
Endothelial cell characterization assays for 41–43
Enzymatic conversion of tetrazolium dyes 136–138
Experimental metastasis assay 227, 229–231, **229**
Extracellular matrix coating in assays for adhesion to endothelial cells 50–59
Extracellular matrix degradation hydrolase assays 101–104
Extracellular matrix track assay 85–86

Filter membrane motility assay 73–77, **76**
Filtration methods for passive deformability 92–96, **95**
Fluorescence cell viability assay 138–139

Fluorescent tagging to visualize metastasis 195–199

Gap junction communication disruption assays 11, 16–22
Genetic tagging to visualize metastasis 186
Green fluorescent protein (GFP) tagging 195–199, **196**

Haplotaxis assay techniques 77, 86–87
Heterotypic aggregation with dissociated organ cell assay 30–32
Histochemical marker gene tagging 190–195
Hydrolase assays for extracellular matrix degradation 101–104

Immobilized extracellular matrix technique 105, 107–109 (Table), 110–111
Immune resistance to select metastatic variants 176–178
Incorporation of labelled DNA precursors 132–136
Invasion assays to select metastatic variants 182–183
In vitro platelet-tumor cell adhesion assay 27–29
In vitro wound motility assay 86–87, **88–89**
In vivo metastasis assays 207–242, 216–217 (Tables)

Lectin resistance assays to select metastatic variants 170–175

Microvascular endothelial cell preparation 38–41
Migration track assays 83–85
Monolayer adhesion assay 16

Northern blotting and RT-PCR for extracellular matrix degradation 104, 106 (Table)

Organ culture invasion assays 120

Paracrine effects assays 145–147

Passive deformability filtration methods
  92–96, **95**
Permeabilization with
  carboxylfluorescein-diacetate 21
Phagokinetic track assay 83–85
Phenotypic modulation assays for
  endothelial cells 43–47
Placental alkaline phosphatase reporter
  gene staining 192–193
Proviral tagging 203–206

Restriction enzyme/southern blot analysis
  186–187, **187**

Scatter assay 87, 90, **91**
Scrape loading assay 19–20
Shearing of cell aggregates 12–13, **14**
Spontaneous metastasis assay 222–228,
  227

Subcutaneous implant assays for
  angiogenesis 263–269
Substrate-capture ELISA for extracellular
  matrix degradation 99–101

Time-lapse videomicroscopy for measuring
  motility 70–73
Tissue specific cell co-culture 149–154
Tumor-cell induced aggregation assay
  25–27

Under agarose motility assay 80–83, **81**

Vascular endothelial cell preparation 36–38
Viable cell degradation of ECM 105,
  107–109 (Table), 110–112